Modern Applications of Quantum Dots

Modern Applications of Quantum Dots

Edited by **Eva Murphy**

New York

Published by NY Research Press,
23 West, 55th Street, Suite 816,
New York, NY 10019, USA
www.nyresearchpress.com

Modern Applications of Quantum Dots
Edited by Eva Murphy

International Standard Book Number: 978-1-63238-331-0 (Hardback)

Printed in the United States of America.

Contents

Preface   VII

Part 1 Introduction   1

Chapter 1 **Photothermal Spectroscopic Characterization
in CdSe/ZnS and CdSe/CdS Quantum Dots:
A Review and New Applications**   3
Viviane Pilla, Egberto Munin, Noelio O. Dantas,
Anielle C. A. Silva and Acácio A. Andrade

Part 2 Optics and LASER   25

Chapter 2 **Silicon Quantum Dots for Photovoltaics: A Review**   27
Georg Pucker, Enrico Serra and Yoann Jestin

Chapter 3 **Capping of InAs/GaAs Quantum
Dots for GaAs Based Lasers**   61
Alice Hospodková

Chapter 4 **Factors Affecting the Relative Intensity
Noise of GaN Quantum Dot Lasers**   81
Hussein B. AL-Husseini

Chapter 5 **Quantum Dots as a Light Indicator
for Emitting Diodes and Biological Coding**   93
Irati Ugarte, Ivan Castelló, Emilio Palomares and Roberto Pacios

Chapter 6 **Ultrafast Nonlinear Optical Response in GaN/AlN Quantum
Dots for Optical Switching Applications at 1.55 μm**   113
S. Valdueza-Felip, F. B. Naranjo, M. González-Herráez,
E. Monroy and J. Solís

Part 3   Electronics   **135**

Chapter 7   **Quantum Dots as Global Temperature Measurements**   **137**
            Hirotaka Sakaue, Akihisa Aikawa, Yoshimi Iijima,
            Takuma Kuriki and Takeshi Miyazaki

Chapter 8   **Silicon Oxide Films Containing Amorphous
            or Crystalline Silicon Nanodots for Device Applications**   **153**
            Diana Nesheva, Nikola Nedev, Mario Curiel, Irina Bineva,
            Benjamin Valdez and Emil Manolov

Chapter 9   **Magnetic Mn_xGe_{1-x} Dots for Spintronics Applications**   **177**
            Faxian Xiu, Yong Wang, Jin Zou and Kang L. Wang

Chapter 10 **Spin-Based Quantum Dot Qubits**   **207**
            V. N. Stavrou and G. P. Veropoulos

Part 4   Quantum Dots in Biology   **223**

Chapter 11 **Quantum Dots-Based Biological
            Fluorescent Probes for *In Vitro* and *In Vivo* Imaging**   **225**
            Yao He

Chapter 12 **Energy Transfer-Based Multiplex
            Analysis Using Quantum Dots**   **245**
            Young-Pil Kim

Chapter 13 **II-VI Quantum Dots as Fluorescent Probes
            for Studying Trypanosomatides**   **261**
            Adriana Fontes, Beate. S. Santos,
            Claudilene R. Chaves and Regina C. B. Q. Figueiredo

            Permissions

            List of Contributors

Preface

Researchers have been examining the properties of quantum dots for a long time. This book describes the modern applications of quantum dots with the help of valuable information. It provides a compilation of various practical applications of quantum dots. The book contains an overview of the thermo-optical characterization of CdSe/ZnS core-shell nanocrystal solutions and also discusses new optical and lasing applications along with a few examples of quantum dot systems for distinct applications in electronics. Examples regarding the usage of this system for biological applications have also been provided. It contains latest research work contributed by researchers and scientists related to this field from across the world, providing the readers with general research like the one conducted in basic sciences like chemistry, medicine, biology and physics with a valuable base text presenting latest research in the field of quantum-dot systems.

The information shared in this book is based on empirical researches made by veterans in this field of study. The elaborative information provided in this book will help the readers further their scope of knowledge leading to advancements in this field.

Finally, I would like to thank my fellow researchers who gave constructive feedback and my family members who supported me at every step of my research.

Editor

Part 1

Introduction

Photothermal Spectroscopic Characterization in CdSe/ZnS and CdSe/CdS Quantum Dots: A Review and New Applications

Viviane Pilla[1], Egberto Munin[2], Noelio O. Dantas[1],
Anielle C. A. Silva[1] and Acácio A. Andrade[1]
[1]Federal University of Uberlândia–UFU, Uberlândia, MG
[2]University Camilo Castelo Branco- UNICASTELO, São José dos Campos, SP
Brazil

1. Introduction

Nanostructured semiconductors or Quantum Dots (QDs) are materials in continuous development that hold potential for a variety of new applications, including uses in fluorescent labels for biomedical science, photonic devices and sensor materials (Bruchez et al., 1998; Prasad, 2004; Sounderya & Zhang, 2008). In biomedical applications, several nanodiagnostic assays have been developed that use QDs. They have been applied to diagnostics, the treatment of diseases, bioimaging, drug delivery, engineered tissues and biomarkers (Sounderya & Zhang, 2008). For example, CdSe/ZnS dendron nanocrystals have been used as biosensor systems for detection of pathogens such as *Escherichia Coli* and *Hepatitis B* (Liu, 2007). CdSe/ZnS core-shell nanocrystals have been shown to be useful for tailoring the fluorescence of dental resin composites (Alves et al., 2010). Core-shell quantum dots (CS) have been used as heteronanocrystals, structures that allow optical amplification because of their stimulated emission of single-exciton states (Klimov et al., 2007), and high-quality CdSe/ZnS doped titania and zirconia optical waveguides have been prepared (Jasieniak et al., 2007).

CdSe/ZnS and CdSe/CdS core-shell semiconductor nanocrystallites are II-VI semiconductor systems and Type I QDs; i.e., both electrons and holes are confined to the core of the core-shell QDs (Jasieniak et al., 2007; Kim et al., 2003). A CdSe nanocrystal core with band gap of (1.7-1.76) eV (Kortan et al., 1990; Lee et al., 2006) is often covered by another semiconductor shell with a high-energy band gap, such as ZnS or CdS, to improve the radiative quantum efficiency by passivating the nonradiative recombination sites at the surface (Dabbousi et al., 1997; Kortan et al., 1990). ZnS and CdS present band gap of (3.8-4.1) eV (Lippens & Lannoo, 1989; Rathore et al., 2008) and (2.5-3.9) eV (Banerjee et al., 2000; Martínez-Castañón et al., 2010), respectively. In such core-shell nanocrystals, the shell provides a physical barrier between the optically active core and the surrounding medium, making the nanocrystals less sensitive to environmental changes, surface chemistry, and photo-oxidation. Therefore the shell provides an efficient passivation of the surface trap states, giving rise to a strongly enhanced fluorescence quantum yield.

Recently, a new class of CdSe QDs called magic-sized nanocrystals (MSNs), with sizes from 1 to 2 nm and well-defined structures, has attracted considerable attention because of its novel physical properties. The most notable of these new properties are: a high stability during and after growth (Dagtepe et al., 2007), closed-shell structures (Chen et al., 2005), the presence of few unit cells, demonstrating strong quantum confinement similar to atoms (Soloviev et al., 2000), and different thermodynamically stable structures (Jose et al., 2006a; Nguyen et al., 2010). Moreover, these MSNs exhibit a high surface to volume ratio, which means that most of the atoms are located at the surface. For this reason, the passivating ligands contribute a significant portion of the total number of the atoms of the MSN, so that the dispersing medium can dramatically affect their properties (McBride et al., 2010). Therefore, the synthesis of stable nanocrystals via colloidal aqueous solutions is extremely important because it prevents changes in dispersion media, thereby preserving the surface properties of the MSNs. This approach yields a physiologically compatible medium that is useful in medical and biotechnological applications.

The ideal QDs for use in nanobiotechnology trials should be thermodynamically stable and have homogeneous dispersion, high radiative quantum efficiency, a very broad absorption spectrum, low levels of nonspecific links to biological compounds and, most importantly, stability in aqueous media. Obtaining all these characteristics simultaneously has been extremely difficult. Therefore, the fabrication of nanocrystals directly in an aqueous phase is of great interest because it would be highly reproducible, inexpensive, minimally toxic and capable of forming products that are easily dissolved in water. Close control of the optical and structural properties of MSNs (for example, CdSe) that are synthesized directly in a colloidal aqueous solution is possible because these properties are strongly affected by various parameters of the synthesis process, such as the Cd/Se molar ratio (Qu & Peng, 2002), the cadmium precursor type (Peng & Peng, 2002), the precursor concentration (Ouyang et al., 2008), reflux time (Jose et al., 2006b), the stabilizer type (Kilina et al., 2009; Park et al., 2010), and the reaction medium (Baker & Kamat, 2010; Dagtepe & Chikan, 2010; Yu et al., 2010). Furthermore, QD biomarkers have several advantages over organic dyes, such as high molar absorption, a broad absorption spectrum, a narrow emission spectrum and high photostability (Resch-Genger et al., 2008).

The solvent used to suspend the solute samples can exert important influence on such properties as the radiative quantum efficiency, absorption and emission spectra, stabilization and thermal parameters (including thermal diffusivity, thermal conductivity and the thermal coefficient of the refractive index) (Pilla et al., 2007; Lo et al., 2009) of the investigated materials. In this way, for a nanoparticle to be a candidate for practical applications, it is important to characterize their thermo-optical properties. The present work reports the thermo-optical properties of cadmium selenide/zinc sulfide (CdSe/ZnS) core-shell colloidal solutions measured with two techniques: the well-known Thermal Lens (TL) technique (Baesso et al., 1994; Shen et al., 1992; Snook & Lowe, 1995) and analysis of the laser-excited ring patterns caused by thermally induced self-phase modulation (SPM) effects, an alternative method that is referred to in this work as the thermal spatial self-phase modulation (TSPM) technique (Dabby et al., 1970; Du & Liu, 1993; Durbin et al., 1981; Hickmann et al., 1992; Khitrova et al., 1993; Ono & Kawatsuki, 1997; Ono & Saito, 1999; Torruellas et al., 1995; Valley et al., 1990; Whinnery et al., 1967; Yang et al., 2005). TSPM may be described by Kirchhoff's diffraction integral (KDI) applied to the propagation of a laser beam in a nonlinear medium (Catunda & Cury, 1990; Pilla et al., 2003). The theoretical calculation predicted by KDI corroborates the TSPM experimental results (Pilla et al., 2009). Thermo-optical characterization was performed on samples of CdSe/ZnS core-shell QDs incorporated into Poly(methyl methacrylate) (PMMA) suspended in tetrahydrofuran (THF), toluene or chloroform solvent and CdSe/ZnS core-shell QDs suspended in toluene or an aqueous colloidal solution. The thermo-optical properties, such as the refractive index temperature coefficient (dn/dT), fraction thermal load (φ) and radiative quantum efficiency (η) of the QD samples were determined. For CdSe/ZnS core-shell QDs suspended in toluene, the characterization was performed as a function of core size (~2-4 nm). The present study reports on the process of synthesis as well as the morphologic and η results of core–shell CdSe/CdS MSNs as a new study. The results obtained via Raman, absorbance and fluorescence spectroscopy, as well as the measurement of the morphology by atomic force microscopy, strongly confirm the formation of core-shell CdSe/CdS MSNs via aqueous colloidal solution.

2. Theoretical basis for photothermal spectroscopic characterization

2.1 Thermal Lens (TL) technique

In TL experiments (Baesso et al., 1994; Lima et al., 2000; Shen et al., 1992; Snook & Lowe, 1995) in a two-beam (pump and probe) configuration, the heat source profile, Q(r), is proportional to the Gaussian intensity profile of the excitation beam, which is expressed as $I_e(r) = (2P_e/\pi w_e^2) \exp(-2r^2/w_e^2)$, where P_e is the power of the excitation beam with radius w_e at the sample. The temporal evolution of the temperature profile, $\Delta T(r, t)$, of the sample can be obtained by the heat conduction equation. In experiments with short excitation pulses, heat diffusion can be neglected and $\Delta T(r, t)$ is proportional to the Gaussian intensity profile of the excitation beam, $I_e(r)$. For long pulse or continuous waves (cw) experiments, however, the effect of heat diffusion is important, and, consequently, $\Delta T(r, t)$ is wider than $I_e(r)$. For t >> τ_c (where τ_c is the characteristic heat diffusion time), the on-axis temperature rise is proportional to the absorbed excitation power ($P_{e,abs}$) and inversely proportional to the thermal conductivity K ($\Delta T(0, t) \propto P_{e,abs}/K$), but it does not depend on w_e.

The TL effect is created when the excitation laser beam passes through a sample of thickness L, and the absorbed energy is converted into heat. Heating changes the refractive index of the material and causes a thermally-induced phase change, $\Delta \phi_{TH}$, expressed as (Shen et al., 1992):

$$\Delta\phi_{TH} = \frac{\theta}{\tau_c} \int_0^t \frac{1}{1+2t'/\tau_c} \left[1 - \exp\left(-\frac{2r_2^2/w_e^2}{1+2t'/\tau_c}\right)\right] dt' , \qquad (1a)$$

and

$$\theta = -\varphi \left(\frac{P_{e,abs}}{K\lambda_p} \frac{dn}{dT}\right), \qquad (1b)$$

where $P_{e,abs} = P_e\alpha L_{eff}$, α (cm^{-1}) is the optical absorption coefficient at the excitation wavelength (λ_e), $L_{eff} = (1-e^{-\alpha L})/\alpha$ is the effective length, L (cm) is the sample thickness, λ_p is the wavelength of the probe beam, dn/dT is the refractive index temperature coefficient, and φ is the absolute nonradiative quantum efficiency, which represents the fraction of the absorbed energy converted into heat. $\tau_c = w_e^2/4D$, where w_e is the excitation beam radius at the sample, $D=K/\rho C$ is the thermal diffusivity (cm²/s), K is the thermal conductivity (W/cmK), ρ is the density (g/cm³), and C is the specific heat (J/gK).

The electric field of the probe beam as it leaves the sample can be expressed as $\varepsilon_s(\rho_1) = \varepsilon(\rho_1) \times exp(-i\Delta\phi_{NL})$, where $\Delta\phi_{NL}$ is the phase change caused by the nonlinearity of the sample (which may include Kerr and thermal components), $\varepsilon(\rho_1)$ is the field of the probe beam at the entrance face of the sample, $\rho_1 = [(x_1^2+y_1^2)/w_1^2]^{1/2}$, and w_1 is the beam waist. In this case, $\Delta\phi_{NL}$ is approximately equal to $\Delta\phi_{TH}$ and is expressed by Eq. 1 (a-b). The field $\varepsilon(r_2)$ at a point $(x_2, y_2, z+d)$ in the observation plane P, located at a distance d away from the sample (S plane), is given by the sum of the optical fields caused by all points in the S plane and is given by the following expression (Pilla et al., 2003):

$$\varepsilon(r_2) = \frac{i2\pi}{\lambda_p d} exp\left[-\frac{i2\pi}{\lambda_p}\left(d+\frac{r_2^2}{2d}\right)\right]\int_0^\infty exp\left(-\frac{i\pi r_1^2}{\lambda_p d}\right)\varepsilon_s(r_1) J_0\left(\frac{2\pi r_1 r_2}{\lambda_p d}\right)r_1\, dr_1 \qquad (2)$$

where $r_i^2 = x_i^2 + y_i^2$ (i= 1 or 2) and $J_0 (2\pi r_1 r_2/\lambda_p d)$ is the zero-order Bessel function. The variation of the probe beam on-axis intensity, $I(t) \approx |\varepsilon(r_2=0)|^2$, can be calculated using Eq. (2) at $r_2=0$ (central part of the probe laser beam) in the cw excitation regime (Lima et al., 2000; Shen et al., 1992; Snook & Lowe, 1995):

$$I(t) = I(0)\left[1 - \frac{\theta}{2}\tan^{-1}\left(\frac{2mV}{\left[(1+2m)^2 + V^2\right]\tau_c/2t + 1 + 2m + V^2}\right)\right]^2, \qquad (3)$$

where I(0) is the on-axis intensity when t is zero; m = $(w_1/w_e)^2$, V= z_1/z_{op}, z_1 is the distance between the sample and probe beam waist, $z_{op} = \pi w_{op}^2/\lambda_p$ is the probe beam Rayleigh range, $z_{op} \ll z_2$ (where z_2 (cm) is the distance between the sample and TL detector) and w_{op} is probe beam radius at the focus with wavelength λ_p.

The thermally induced distortion of the laser beam as it passes through the sample is described by the optical path-length (S) change $(ds/dT = L^{-1} dS/dT)$, which results in lensing at the sample. The propagation of a probe laser beam through the TL will result in either spreading (ds/dT<0) or focusing (ds/dT >0) of the beam, depending mainly on the temperature coefficients of the electronic polarizability of the sample, stress and thermal expansion (in the case of liquid samples, $ds/dT \approx dn/dT$).

In the dual beam mode-mismatched configuration with excitation and probe beams, the normalized transient signal amplitude is approximately the phase difference (θ) of the probe beam between $r = 0$ and $r = \sqrt{2}$ w_e induced by the pump beam, given by Eq. (1b). The normalized parameter, $\Theta = -\theta/P_e\alpha L_{eff}$, for liquid samples is defined as (Lima et al., 2000):

$$\Theta = \varphi\left(\frac{1}{K\lambda_p}\frac{dn}{dT}\right), \qquad (4a)$$

and
$$\varphi = 1 - \eta \frac{\lambda_e}{\langle \lambda_{em} \rangle},$$
(4b)

where $\langle \lambda_{em} \rangle$ is the average emission wavelength and η is the fluorescence quantum efficiency or quantum yield.

Sheldon et al. determined the optimization of the TL technique for samples located at position $Z_1 = 1.7\, z_{op}$ (Sheldon et al., 1982). However, Fang and Swofford noted that the TL technique becomes more sensitive when using a system with two beams, a probe and a pump, with waists that do not coincide (Fang & Swofford, 1979). Berthoud et al. derived a theoretical model for an arrangement with two beams in the mode-mismatched configuration (Berthoud et al., 1985). The results of these studies showed that the increased sensitivity of the technique is a result of placing the sample in the focus of the excitation beam ($w_e = w_{oe}$) and outside the focus of the probe beam. A quantitative model was developed (Shen et al., 1992) for this technique in mode-mismatched configuration so that it could be used for steady-state and time-resolved measurements, in both the single-beam and dual beam configurations.

2.2 Self-induced phase modulation technique

Thermally induced self-phase-modulation (SPM) effects can be understood as the ability of the excitation beam to induce spatial variations of the refractive index, leading to a phase shift that depends on the transverse distance from the beam axis. The transverse self-phase modulation (Dabby et al., 1970; Du & Liu, 1993; Durbin et al., 1981; Hickmann et al., 1992; Khitrova et al., 1993; Ono & Kawatsuki, 1997; Ono & Saito, 1999; Petrov et al., 1994; Torruellas et al., 1995; Valley et al., 1990; Whinnery et al., 1967; Yang et al., 2005) is also implicate in the emergence of rings in the pattern of transmitted light when $\Delta\phi_{NL} \gg 2\pi$. The number of rings $N \approx \Delta\phi_{NL}/2\pi$ (Durbin et al., 1981; Ono & Saito, 1999), in the case where $\Delta\phi_{NL} = \Delta\phi_{TH}$ (Eq. 1a-b) can be determined as a function of P_e by using the following expression (Andrade et al., 1998; Catunda et al., 1997; Pilla et al., 2009):

$$N = 1.3 \frac{\varphi \alpha L_{eff}}{2\pi K \lambda_p} \frac{dn}{dT} P_e.$$
(5)

Mathematically, the equation that describes the deflection path (β) of a beam that spreads in two components (r_1 and r_2) because of a change in the refraction index $\Delta n(r)$ caused by the laser beam at a distance r from its propagation axis, is defined as follows (Dabby et al., 1970):

$$\beta(r) = \left| \frac{\partial}{\partial r} \int_0^L \frac{\Delta n(r,z)dz}{n_0} \right|$$
(6)

where n_0 is the intensity-independent part of the refractive index. The interference rays that constitute the laser beam emerge parallel with different phases after crossing the nonlinear medium, i.e., $\beta(r_1) = \beta(r_2)$, with the same wavevector. The interference is constructive or destructive in the plane of the observation, P, if $\Delta\phi_{NL}(r_1)$- $\Delta\phi_{NL}(r_2)$= pπ, where p is an even or odd integer, respectively. This phenomenon is the origin of the appearance of diffraction rings (Dabby et al., 1970; Durbin et al., 1981).

3. Experimental

3.1 Nanostructured semiconductors

The CdSe/ZnS core-shell nanocrystal suspension was prepared by mixing poly(methyl methacrylate) PMMA-encapsulated CdSe/ZnS quantum dots with THF (C_4H_8O) solvent under constant magnetic stirring at a concentration of (12-60) mg/mL. The same procedure was applied for PMMA-encapsuled CdSe/ZnS core-shell particles using chloroform ($CHCl_3$) and toluene ($C_6H_5CH_3$). Evident Technologies (Evident, 2005) supplied the CdSe/ZnS encapsulated in PMMA with a molecular weight of 74 µg/nmol and an average crystal diameter of 3.7 nm (Pilla et al., 2007), as well as samples of CdSe/ZnS suspended in toluene solution with a core size between 2.4-4.1 nm (Pilla et al. 2008) and CdSe/ZnS suspended in water (2.6 µmol/L and core size 4.0 nm).

Magic-sized CdSe/CdS nanocrystals were synthesized by combining the following aqueous solutions in a nitrogen atmosphere: an aqueous solution containing selenium dioxide (SeO_2) and sodium hydroxide in ultra-pure water was magnetically stirred to form Na_2SeO_3, and 1-thioglycerol was immediately added to the solution; an aqueous solution containing cadmium acetate dihydrate in dimethyl formamide (DMF). The two solutions were then mixed. The function of thioglycerol was to limit nanocrystal growth by Ostwald ripening, and also to prevent aggregation and minimize surface dangling bonds, thereby decreasing the level of defects. This procedure was carried out at 80 ^{0}C and yielded a CdSe/CdS based colloidal aqueous solution. The solution of magic-sized CdSe/CdS nanocrystals had a molar concentration of 3.5 µ mol/L and a core size of 1.9 nm.

3.2 Thermal-optical measurements

The thermo-optical properties of core-shell nanoparticle solutions were investigated by the TL method (Cruz et al., 2010; Pilla et al., 2007, 2008) and by TSPM (Pilla et al., 2009). TL transient measurements were performed in the mode-mismatched dual-beam (excitation and probe) configuration (Fig. 1(a-b)). A He-Ne laser (λ_p= 632.8 nm) was used as the probe beam, and an Argon ion laser (λ_e= 514.5 nm) or He-Ne (λ_e= 594 nm) was used as the excitation beam. The excitation and probe beam radii at the sample were measured to be w_e= (53 ± 1) µm at 514.5 nm (or w_e= (81 ± 2) µm at 594 nm) and w_1= (172 ± 3) µm, respectively. Absorption of the excitation beam generated a TL heat profile and induced a phase shift proportional to θ. Modulation of the pump beam with a mechanical chopper allowed for time-resolved measurement. The transient curve was obtained from the weak probe beam, which counter-propagated in a nearly collinear with the excitation beam.

Figure 2 presents the experimental setup used for TSPM technique. In this case, an Argon ion laser (λ_e= 514.5 nm) was used as the excitation beam. All measurements, as a function of laser beam power, were carried out with solutions in 2-mm glass cuvette that was positioned in the focal region of a lens with 20 cm focal length.

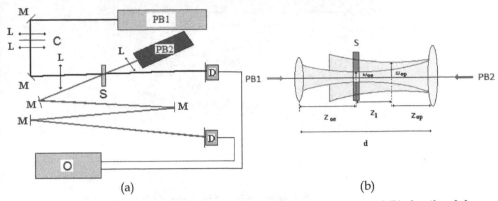

(a)   (b)

Fig. 1. (a) Schematic diagrams of the TL experimental configuration and (b) details of the
geometric configuration of the probe (PB2) and pump (PB1) beams. Here M, L and C are
mirror, convergent lens and chopper; S, D and O are sample, detector and oscilloscope. z_o
and w_o are the beam Rayleigh range focal lens and the radius for the pump (e) or probe (p)
beams, respectively. d is the separation between the lens of pump and probe beams.

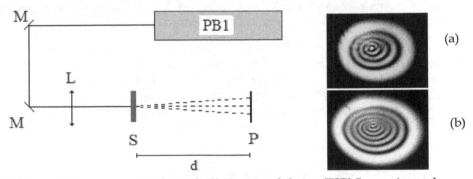

Fig. 2. Schematic diagrams of the thermal self-phase modulation (TSPM) experimental
apparatus. Here M and L are mirror and convergent lens; S is sample. P indicates the observation
plane, which shows the typical ring patterns for CdSe/ZnS QDs samples solutions at (a) α=
0.57 cm^{-1} and $P_e \approx$ 59 mW; (b) α= 0.35 cm^{-1} and $P_e \approx$ 88 mW, λ_e= 514.5 nm (Pilla et al., 2009).

4. Results and discussions

4.1 Spectroscopic characterization of core-shell QDs

The typical absorbance spectra of the core-shell CdSe/ZnS with different core sizes are
present in Fig. 3, and their corresponding transitions can be found elsewhere (Norris et
al., 1996; Valerini et al., 2005). The size of the core D_{CdSe} (in nm) can be calculated by
applying the value of the wavelength λ (in nm) for the first excitonic absorption peak of
the sample, using the experimental fitting function obtained for the CdSe nanocrystal (Yu
et al., 2003):

$$D_{CdSe} = (1.6122 \times 10^{-9})\, \lambda^4 - (2.6575 \times 10^{-6})\, \lambda^3 + (1.6242 \times 10^{-3})\, \lambda^2 - (0.4277)\, \lambda + (41.57) \qquad (7)$$

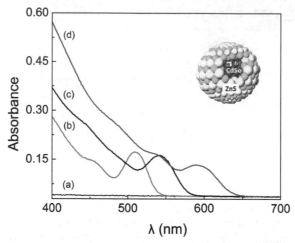

Fig. 3. Absorbance spectra for (a) toluene solvent and suspension of CdSe/ZnS in toluene for core sizes (b) (2.4 ± 0.2) nm, (c) (2.9 ± 0.5) nm and (d) (4.1 ± 0.9) nm at concentration ~0.046 mg/mL (Pilla et al., 2008). The inset shows the nanostructure of the QD Type I, with a CdSe core surrounded by a ZnS shell. In this case electrons (e-) and holes (h+) are confined to the core of the core-shell QDs (Jasieniak et al., 2007; Kim et al., 2003).

The fluorescence spectra of the CdSe/ZnS are red-shifted as the core size increases (Fig. 4); the values obtained for $<\lambda_{em}>$ are 525 nm, 560 nm and 619 nm for α (at 514.5 nm) of 0.31, 0.199 and 0.35 cm^{-1}, respectively. $<\lambda_{em}>$ values are dependent on QD concentrations (Pilla et al., 2008), solvents (Pilla et al., 2007) and aging effect (Cruz et al., 2010). For CdSe/CdS MSNs synthesized and studied here, the absorbance and fluorescence spectra are present in Fig. 5. The absorption band at ~447 nm obtained is typical for core-shell CdSe/CdS MSNs

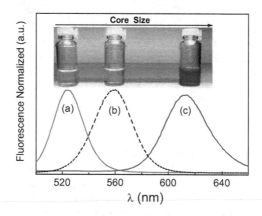

Fig. 4. Fluorescence spectra for CdSe/ZnS solutions as a function of the core sizes: (a) 2.4 nm, (b) 2.9 nm and (c) 4.1 nm. Fluorescence measurements were performed at λ_e= 457 nm (Pilla et al., 2008). The colors of the solutions are illustrative.

(Chen et al., 2008; Deng et al., 2010), and the value of $<\lambda_{em}>$= 588.9 nm for α= 1.25 cm^{-1} at 514.5 nm was obtained from fluorescence measurement. Peaks related to longitudinal optical phonons (LO) and (2LO) at 206.6 and 411 cm^{-1} of the synthesized CdSe MSNs are clearly observable in the Raman spectrum shown in Fig. 5 (Nien et al., 2008). Another peak at 280 cm^{-1}, related to the CdS vibrational mode, can also be seen (Leite & Porto, 1966). The frequency of the bulk CdS LO phonon at ~305 cm^{-1} decreases to 280-300 cm^{-1} in thin layers because of phonon confinement effects (Zou & Weaver, 1999). Therefore, the peak observed at approximately 280 cm^{-1} can be attributed to the LO phonon of the CdS shell (Chen et al., 2010; Dzhagan et al., 2007; Singha et al., 2005) which confirms the formation of core-shell CdSe/CdS MSNs.

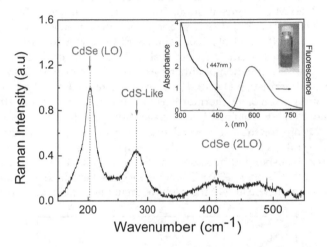

Fig. 5. Raman spectra for core–shell CdSe/CdS MSNs (λ_e= 514 nm and 300 K). The inset shows the absorbance and fluorescence of CdSe/CdS MSNs (λ_e= 457.9 nm for fluorescence).

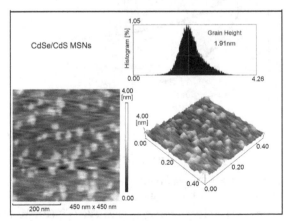

Fig. 6. Two-dimensional and three-dimensional AFM images and corresponding histograms revealing height and size dispersions of the core-shell CdSe/CdS MSNs.

The size and dispersions of CdSe/CdS MNSs were determined by examining AFM bi-dimensional and tri-dimensional images and their corresponding histograms. The average core sizes (D_{CdSe}) of the CdSe/CdS MSNs were determined from the peaks (Ong & Sokolov, 2007; Rao et al., 2007) of the histogram (Fig. 6). The average size of the CdSe/CdS MNSs (1.9 ± 0.4) nm, determined by AFM measurements, is in good agreement with the values obtained from Eq. (7) and the absorption spectrum of the Fig. 5 (Yu et al., 2003).

4.2 Thermal diffusivity and φdn/dT as a function of concentration, solvent and core size

TL transient signals for the PMMA-encapsulated CdSe/ZnS nanocrystals (Fig. 7) were fit to Eq. (3), which supplies τ_c and the amplitude parameter θ. From $D = w_e^2/4\tau_c$ and the measured value w_e, the average values of D for the core-shell nanocrystal solutions can be determined as a function of CdSe/ZnS PMMA concentration. These values are shown in Table 1 for the three different solvents. The thermal diffusivity results are in agreement with typical values obtained for pure solvents (Pilla et al., 2007) and for PMMA, which has a reported thermal diffusivity value of $D \sim 1.0 \times 10^{-3}$ cm^2/s (Agari et al., 1997; Goyanes et al., 2001). No significant variation in the D values as a function of either the concentration or the QD size (Table 2) was observed. The thermal parameters $\Theta = -\theta/P_e\alpha L_{eff}$ of the core-shell quantum dot suspensions were determined through transient thermal lens measurements. Using Eq. 1b with K= 0.12 W/mK (Nikogosyan, 1997), we obtained an average value of φ dn/dT= - 1.8 x 10^{-4} K^{-1} at λ_e= 594 nm and λ_p= 632.8 nm for the nanocrystal suspended in chloroform in the concentration range of (12-60) mg/mL. Table 1 shows the average values of φ dn/dT for PMMA-encapsulated CdSe/ZnS suspended in toluene and THF with K= 0.135 W/mK (Nikogosyan, 1997) and K= 0.14 W/mK (Ge et al., 2005). The transient curves obtained with the TL technique indicate that TL promotes defocusing of the probe beam in the far field; i.e., dn/dT is negative.

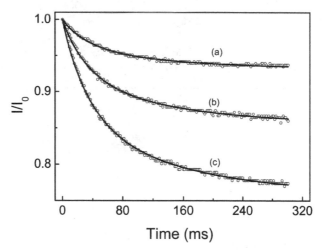

Fig. 7. Transient TL signal for PMMA-encapsulated CdSe/ZnS suspended in chloroform (CHCl$_3$) at (a) 12, (b) 20, (c) 40 mg/mL (λ_e= 594 nm, P_e = 1.03 mW and λ_p= 632.8 nm). Using θ and τ_c results by fitting with Eq. 3, D (10^{-3} cm^2/s) and -φdn/dT (10^{-4} K^{-1}) values obtained were the following: (a) (0.84 ± 0.02) and 1.59; (b) (0.84 ± 0.02) and 1.83; and (c) (0.83 ± 0.01) and 2.27, respectively.

In order to calculate $\varphi = \Theta/\Theta_s$, where Θ_s is the value for the solvent presented in Table 1, we used Eq. 4a and assumed a negligible fluorescence for the solvent (chloroform, THF or toluene) in the TL measurements, i.e., $\varphi_s = 1$. The calculated value of φ is (0.30 ± 0.04) for the core-shell suspended in chloroform. For the PMMA-encapsulated CdSe/ZnS suspended in THF or toluene, the φ values are shown in Table 1. Using the values of φ obtained for the core-shell suspended in THF, chloroform and toluene (Table 1), we obtained $dn/dT = -4.9$, -6.0 and -5.3×10^{-4} K^{-1}, respectively. These values are in agreement with those reported in the literature for THF, chloroform and toluene solvents (El-Kashef, 2002; Nikogosyan, 1997).

PMMA CdSe/ZnS	Θ_s (W^{-1})	D (10^{-3} cm^2/s)	$-\varphi dn/dT$ (10^{-4} K^{-1})	φ
THF	5644	(1.00 ± 0.03)	(1.7 ± 0.4)	(0.35 ± 0.07)
Chloroform	8165	(0.84 ± 0.02)	(1.8 ± 0.4)	(0.30 ± 0.04)
Toluene	6564	(1.01 ± 0.03)	(2.6 ± 0.3)	(0.49 ± 0.03)

Table 1. TL results for PMMA-encapsulated CdSe/ZnS suspended in different solvents (λ_e= 594 nm and λ_p= 632.8 nm). Average values obtained for different concentrations (12-60) mg/mL.

The temperature coefficient of the refractive index (dn/dT) is affected by two competing factors: thermal expansion, which leads to decreased density and causes the refractive index to decrease because of increased inter-molecular spacing, and increased electronic polarizability, which causes the refractive index to gradually increase. In liquid and film samples, dn/dT is usually negative, indicating that expansion is the dominant contributor. For example, for common solvents such as toluene and water, the values of dn/dT are -5.4 and -0.8×10^{-4} K^{-1}, respectively (Dovichi & Harris, 1979; El-Kashef, 2002; Kohanzadeh et al., 1973; Nikogosyan, 1997). For glasses and transparent crystals, dn/dT can be either positive or negative, depending on the glass (Andrade et al., 2003). In general, dn/dT is negative for highly expansive and loosely bound networks and positive for tightly bound networks. Polymers usually have high expansion coefficients and, therefore, negative values of dn/dT (Pilla et al., 2002).

CdSe/ZnS	core size (nm)	D (10^{-3} cm^2/s)	$-\varphi dn/dT^*$ (10^{-4} K^{-1})	φ^*
THF	3.5 ± 0.8	1.00 ± 0.03	(2.2 ± 0.2)	(0.44 ± 0.04)
Toluene	4.1 ± 0.9	0.99 ± 0.01	(3.4 ± 0.6)	(0.61 ± 0.06)
Toluene	2.9 ± 0.5	1.01 ± 0.06	(2.7 ± 0.3)	(0.52 ± 0.05)
Toluene	2.4 ± 0.2	0.94 ± 0.03	(2.1 ± 0.3)	(0.37 ± 0.06)

Table 2. TL and TSPM results for suspended CdSe/ZnS with different core sizes (λ_e= 514.5 nm and λ_p= 632.8 nm). *Average values obtained for TL and TSPM methods (Pilla et al., 2009)

For CdSe/ZnS nanocrystal solution suspended in THF with α= (0.57 ± 0.03) cm^{-1} at 514.5 nm, the value for $\varphi dn/dT$= -(2.1 ± 0.2) K^{-1} was obtained from the TL measurements. Alternatively, using a single laser beam for the excitation of the core-shell QD solution at λ_e = 514.5 nm, we observed the formation of ring patterns caused by thermal spatial self-phase modulation (TSPM) effects ($\Delta\phi_{TH}$ described by Eq. 1). Typical ring patterns were observed when the sample was positioned at the focus of the pump beam (Pilla et al., 2009). The numbers of rings (N) as a function of the beam power (P_e) are shown in Fig. 8 for the QDs suspended in different solvents at different concentrations. The experimental N versus P_e results were fit using a linear equation, and the values of the angular coefficients, A, of the fitting line were determined. Using Eq. (5) and the K values of the solvents, the value of $|\varphi dn/dT| = (2.2 \pm 0.2) \times 10^{-4}$ K^{-1} was determined for PMMA-encapsulated CdSe/ZnS suspended in THF. For QDs suspended in toluene with different core sizes, Table 2 presents the average values for $\varphi dn/dT$ determined by the TL and TSPM methods, which are in good agreement. The dynamic response time of ring formation for the CdSe/ZnS soltuion (at $\lambda_e = 514.5$ nm, $P_e \approx 92$ mW) is shown in Fig. 8. This result was obtained with the detector in the far field (~ 116 cm after the sample), and an iris placed in front of the detector, in the position of the formation of the first ring.

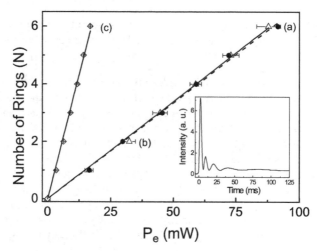

Fig. 8. Number of rings (N) as a function of excitation beam power (P_e) for encapsulated-PMMA CdSe/ZnS (a) and CdSe/ZnS (b-c) at $\lambda_e = 514.5$ nm. The linear trend was obtained by fitting the equation N = A*P_e, where obtained A values are: (a) (66.9 ± 0.7) W^{-1} and $\alpha = 0.57$ cm^{-1}; (b) (67.7 ± 0.7) W^{-1} and $\alpha = 0.35$ cm^{-1}; and (c) (347 ± 5) W^{-1} and $\alpha = 2.86$ cm^{-1} (Pilla et al., 2009). The inset shows the time evolution of ring formation for CdSe/ZnS QDs.

Several transverse effects can be observed when an intense laser beam propagates through a nonlinear medium (Dabby et al., 1970; Durbin et al., 1981; Ono & Saito, 1999; Valley et al., 1990). The laser-matter interaction modifies the spatial profile of the incident beam so strongly that rings surrounding the central spot of the beam may be observed in the far-field region. The ring structure exhibits circular symmetry when distortion caused by thermal convection is absent or negligible (Dabby et al., 1970; Whinnery et al., 1967). Spatial self-phase modulation (SPM) may be described by Kirchhoff´s diffraction integral, applied to the propagation of a laser beam inside a nonlinear medium (Catunda & Cury, 1990; Pilla et al., 2003). The effects of the nonlinearity include induced phase variation caused by the transverse variation of the refractive index and intensity-dependent absorption (Petrov et al., 1994). The literature describes studies of SPM effects in several materials, including liquid crystals, crystals, glasses, transparent media, and polymers (Du & Liu, 1993; Durbin et al., 1981; Torruellas et al., 1995; Yang et al., 2005). SPM effects have been observed in semiconductor-doped glasses in studies on fast-electronic nonlinearity (Hickmann et al., 1992) and thermal nonlinearity (Petrov et al., 1994). Other phenomena related to SPM include self-focusing and self-defocusing, the generation of spatial solitons, simultaneous spatial-temporal focusing, and self-guided propagation (Catunda & Cury, 1990; Khitrova et al., 1993).

Thermal spatial self-phase modulation can be predicted numerically using the Kirchhoff Diffraction Integral (Pilla et al., 2003). We consider the diffraction of a linearly polarized monochromatic plane wave in a finite region on the output sample face (S plane), where the geometrical points are described by the coordinates (x_1, y_1, z). The intensity of the excitation beam is proportional to $|\varepsilon(r_2)|^2$, and their spatial profiles as a function of the pump beam intensity are presented in Fig. 9 and 10 for two different concentrations. Here, $\rho_2 = r_2/w_2$, where w_2 is the probe beam waist in the observation plane. All curves in Figs. 9-10 are normalized to the maximum intensity, which occurs at $\rho_2 = 0$ for $P_e = 0$ mW.

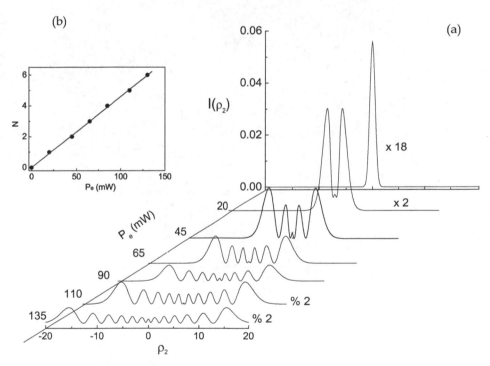

Fig. 9. (a) Numerical predictions of the excitation beam intensity profile, $I(\rho_2)$, as a function of $\rho_2 = r_2/w_2$, using Eq. (2) and (1a-b), with $\alpha = 0.35$ cm^{-1}, dn/dT= -5.4 x 10^{-4} K^{-1}, and λ_p= 514.5 nm. The curves are normalized to the value corresponding to $\rho_2 = 0$ at P_e= 0 mW. (b) Number of dark Rings (N) versus P_e (mW) obtained from numerical calculations. The linear trend was obtained by fitting the equation N=0.05 *P_e.

Figures 9 (b) and 10 (b) shows N versus P_e obtained from numerical calculations; the linear fits obtained are in good agreement with the experimental measurements (Fig. 8) on CdSe/ZnS core-shell nanocrystals suspended in toluene.

In Table 2, we present the $\varphi dn/dT$ values obtained for CdSe/ZnS solutions suspended in toluene as a function of core sizes (~2.4–4.1 nm). The experimental measurements were performed by applying the TL and TSPM techniques at λ_e= 514.5 nm. Supposing a constant dn/dT parameter, we obtained φ values that are dependent on the nanocrystal core size. φ increases with the increasing size of the nanoparticle cores (~39 %), and the best result was obtained for the smallest analyzed core (~2.4 nm). In fact, the obtained dependence of the φ values could be attributed to quantum confinement effects, which are dependent on the size of the semiconductor nanocrystal relative to their exciton Bohr radius (~5.3 nm for CdSe) (Katz et al., 2002; Nirmal et al., 1994; Prasad, 2004). The CdSe/ZnS nanocrystals studied (Table 2) are in a strong quantum confinement regime. In this form, thermally induced depopulation of the lowest electronic states is inhibited when the quantum dot size is reduced, because the energy spacing of the atom-like states is greater than the available thermal energy (Klimov, 2003).

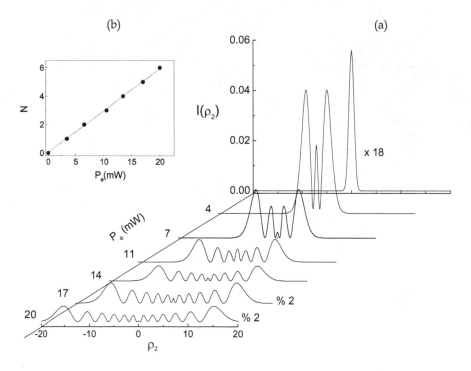

Fig. 10. (a) Numerical predictions of the excitation beam intensity profile, $I(\rho_2)$, as a function of $\rho_2=r_2/w_2$, using Eq. (2) and (1a-b), with α= 2.86 cm^{-1}, dn/dT= -5.4 x 10^{-4} K^{-1}, and λ_p= 514.5 nm. (b) Number of dark Rings (N) versus P_e (mW) obtained from numerical calculations. The linear trend was obtained by fitting the equation N=0.30 *P_e.

4.3 Fluorescence quantum efficiency: Measurements as a function of concentration, solvent and core size

To determine the radiative quantum efficiency, η, solvents of negligible fluorescence were used (S). Thus, for all solvents, the absorbed energy is converted into heat in the material, i.e., $\varphi_S = 1$ ($\eta_S = 0$). Applying Eq. 4 (b), it is possible to obtain the following expression:

$$\eta = \left(1 - \varphi\right)\left(\frac{< \lambda_{em} >}{\lambda_e}\right) \tag{8}$$

The radiative quantum efficiency values for core-shell colloidal nanocrystal solutions are presented in Table 3. The η results are presented for core-shell PMMA-encapsulated CdSe/ZnS suspended in toluene, chloroform and THF, for CdSe/ZnS suspended in toluene, and for the aqueous colloidal suspension of CdSe/ZnS and CdSe/CdS, respectively. For PMMA-encapsulated CdSe/ZnS, η is dependent on concentration (Pilla et al., 2007), and the aging effect (Cruz et al., 2010). The value of η decreased by ~ 60% over a 200-hour interval of UV-vis TL measurements (Cruz et al., 2010). For CdSe/ZnS QDs, the η values were dependent on the core size and concentration (Pilla et al. 2008); for CdSe/ZnS suspended in toluene, the η results were nearly independent of the pump wavelength (335-543) nm (Cruz et al., 2010). The values of η increased for solutions without PMMA, and it can be concluded that PMMA plays a significant role in the fluorescence quenching of the complex CdSe/ZnS PMMA solution (Pilla et al., 2008).

Core/shell	Core Size (nm)	Concentration	η
PMMA-CdSe/ZnS chloroform[1]	(3.7 ± 0.9)	(12-60) g/L	(0.71 ± 0.03)
PMMA- CdSe/ZnS THF[1]	(3.5 ± 0.8)	(12-60) g/L	(0.65 ± 0.07)
PMMA-CdSe/ZnS toluene[1]	(3.8 ± 0.9)	(12-60) g/L	(0.50 ± 0.03)
CdSe/ZnS toluene[2]	(2.4 ± 0.2)	3.14 µmol/L	(0.76 ± 0.02)
CdSe/ZnS water	(4.0 ± 0.8)	2.6 µmol/L	(0.61 ± 0.04)
CdSe/CdS water	(1.9 ± 0.4)	3.5 µmol/L	(0.43 ± 0.07)

[1]Average values obtained for different concentrations (12-60) mg/mL (Pilla et al., 2007).
[2]Average values at UV-vis excitation range (335-528) nm (Cruz et al., 2010).

Table 3. Core-shell QDs suspended in toluene, chloroform, THF or aqueous solutions.

In addition, radiative quantum efficiency values are presented for CdSe/ZnS and CdSe/CdS colloidal aqueous solutions, and high quantum efficiencies were obtained. As a comparison, for Rhodamine B dye dissolved in methanol, the η values decreased from 0.8 to ~0.1 with increasing concentration in the range of ~10^{-5} -10^{-3} molar (Bindhu et al., 1996). For CdSe/ZnSe core/shell nanocrystals suspended in organic solvents, the values of η= (0.60-0.85) were obtained, as a result of the thicker deposit shell concentration (Reiss et al., 2002); quantum yield values were reported as 0.4-0.6 for CdSe QDs (core-sizes between 2.8 and 5.0 nm), 0.55 for CdSe/ZnSe/ZnS (Donegá et al., 2006) and up to 0.70 for CdTe/CdS MSNs aqueous colloidal solution (Deng et al., 2010).

5. Conclusions

A review of the thermo-optical characterization of CdSe/ZnS core-shell nanocrystal solutions was performed. The Thermal Lens (TL) technique was used, and the thermal self-phase Modulation (TSPM) technique was adopted as the simplest alternative method. The main parameter determined was $\varphi dn/dT$; this parameter permitted the calculation of the fraction of thermal load (φ) for different core-sizes of QDs suspended in different solvents. The thermo-optical results obtained using TSPM and predicted numerically using the Kirchhoff Diffraction Integral are in good agreement with those obtained from the TL technique. The thermal diffusivity parameters were characteristic of the solvents used to suspend the core-shell QDs. The radiative quantum efficiency, η, for CdSe/ZnS core-shells and its dependence on the concentration, solvent, core size and aging effects were reported. In addition, the thermo-optical characterization of CdSe/ZnS and CdSe/CdS core-shell nanocrystals suspended in aqueous solution is presented as a result of interest for applications in nanobiotechnology. As future research trends we could highlight the application of the Thermal Lens technique for the characterization of fluorescent nanostructures functionalized for biomarker applications.

6. Acknowledgement

This research was supported by CNPq (Proc. 473951/2010-0), FAPESP (Proc. 2006/01277-2), FAPEMIG (Proc. APQ-02878-09 and Proc. CEX-APQ-00576-11) and Nanobiotec-Brazil Network (Proc. 04/CII-2008).

7. References

Agari, Y.; Ueda, A.; Omura, Y. & Nagai, S. (1997). Thermal Diffusivity and Conductivity of PMMA/PC Blends. *Polymer* Vol. 38, N. 4, pp. 801-807, ISSN 0032-3861.

Alves, L. P.; Pilla, V.; Murgo, D. O. A. & Munin, E. (2010). Core–shell Quantum Dots Tailor the Fluorescence of Dental Resin Composites. *Journal of Dentistry* Vol. 38, N. 2, pp. 149-152, ISSN 0300-5712.

Andrade, A. A.; Catunda, T.; Lebullenger, R.; Hernandes, A. C. & Baesso, M. L. (1998). Time-Resolved Study of Thermal and Electronic Nonlinearities in Nd^{+3} doped Fluoride Glasses. *Electronics Letters* Vol. 34, N. 1, pp. 117-119, ISSN 0013-5194.

Andrade, A. A.; Lima, S. M.; Pilla, V.; Sampaio, J. A.; Catunda, T. & Baesso, M. L. (2003). Fluorescence Quantum Efficiency Measurements using the Thermal Lens Technique. *Review of Scientific Instruments* Vol. 74, N. 1, pp. 857-859, ISSN 0034-6748.

Baesso, M. L.; Shen, J. & Snook, R. D. (1994). Mode-mismatched Thermal Lens Determination of Temperature Coefficient of Optical Path Length in Soda Lime Glass at different Wavelengths. *Journal of Applied Physics* Vol. 75, N. 8, pp. 3732-3737, ISSN 0021-8979.

Baker, D.R. & Kamat, P.V. (2010). Tuning the Emission of CdSe Quantum Dots by Controlled Trap Enhancement. *Langmuir* Vol. 26 N. 13, pp. 11272–11276, ISSN 0743-7463.

Banerjee, R.; Jayakrishnan, R. & Ayyub, P. (2000). Effect of the Size-Induced Structural Transformation on the Band Gap in CdS Nanoparticles. Journal of Physics: Condensed Matter Vol.12, N. 50, pp. 10647–10654, ISSN 0953-8984.

Berthoud, T.; Delorme, N. & Mauchien, P. (1985). Beam Geometry Optimization in Dual-Beam Thermal Lensing Spectrometry. Analytical Chemistry Vol. 57, N. 7, pp. 1216-1219, ISSN 0003-2700.

Bindhu, C.V.; Harilal, S. S.; Varier, G. K.; Issac, R. C.; Nampoori, V. P. N. & Vallabhan, C. P. G. (1996). Measurement of the Absolute Fluorescence Quantum Yield of Rhodamine B Solution using a Dual-Beam Thermal Lens Technique. Journal of Physics D Applied Physics Vol 29, N. 4, pp. 1074-1079, ISSN 0022-3727.

Bruchez Jr., M.; Moronne, M.; Gin, P.; Weiss, S. & Alivisatos, A. P. (1998). Semiconductor Nanocrystals as Fluorescent Biological Labels. Science. Vol. 281, N. 5385, pp. 2013-2016, ISSN 0036-8075.

Catunda, T. & Cury, L. A. (1990). Transverse Self-Phase Modulation in Ruby and GdAlO$_3$: Cr^{+3} Crystals. Journal of the Optical Society of America B Vol. 7, N. 8, pp. 1445-1455, ISSN 0740-3224.

Catunda, T.; Baesso, M. L.; Messaddeq, Y. & Aegerter, M. A. (1997). Time-resolved Z-scan and Thermal Lens Measurements in Er^{+3} and Nd^{+3} doped Fluoroindate Glasses. Journal of Non-Crystalline Solids Vol. 213-214, pp. 225-230, ISSN 0022-3093.

Chen, X.; Samia, A. C. S.; Lou, Y. & Burda, C. (2005). Investigation of the Crystallization Process in 2 nm CdSe Quantum Dots. Journal of the American Chemical Society V. 127, N. 12, pp. 4372–4375, ISSN 0002-7863.

Chen, Y.; Vela, J.; Htoon, H.; Casson, J. L.; Werder, D. J.; Bussian, D. A.; Klimov, V. I. & Hollingsworth, J. A. (2008). Giant Multishell CdSe Nanocrystal Quantum Dots with Suppressed Blinking. Journal of the American Chemical Society, Vol. 130, N. 15, pp. 5026-5027, ISSN 0002-7863.

Chen, I. C.; Weng, C. L.; Lin, C. H. & Tsai, Y. C. (2010). Low-frequency Raman Scattering from Acoustic Vibrations of Spherical CdSe/CdS Nanoparticles. Journal of Applied Physics Vol. 108, N. 8, pp. 083530 (1-9), ISSN 0021-8979.

Cruz, R. A.; Pilla, V. & Catunda, T. (2010). Quantum Yield Excitation Spectrum (UV-visible) of CdSe/ZnS Core-Shell Quantum Dots by Thermal Lens Spectrometry. Journal of Applied Physics Vol. 107, N. 8, pp. 083504 (1-6), ISSN 0021-8979.

Dabbousi, B. O.; Rodrigues-Viejo, J.; Mikulec, F. V., Heine, J. R., Mattoussi, H.; Ober, R.; Jensen, K. F. & Bawendi, M. G. (1997). (CdSe)ZnS Core-Shell Quantum Dots: Synthesis and Characterization of a Size Series of Highly Luminescent Nanocrystallites. The Journal of Physical Chemistry B Vol. 101, N. 46, pp. 9463-9475, ISSN 1520-6106.

Dabby, F. W.; Gustafson, T. K.; Whinnery, J. R.; Kohanzadeh, Y. & Kelley, P.L. (1970). Thermally Self-Induced Phase Modulation of Laser Beams. Applied Physics Letters Vol. 16, N. 9, pp. 362-365, ISSN 0003-6951.

Dagtepe, P.; Chikan, V.; Jasinski, J. & Leppert, V. J. (2007). Quantized Growth of CdTe Quantum Dots; Observation of Magic-Sized CdTe Quantum Dots. The Journal of Physical Chemistry C Vol. 111, N. 41, pp. 14977–14983, ISSN 1932-7447.

Dagtepe, P. & Chikan, V. (2010). Quantized Ostwald ripening of Colloidal Nanoparticles. The Journal of Physical Chemistry C Vol. 114 , N. 39 , pp. 16263-16269, ISSN 1932-7447.

Deng, Z.; Schulz, O.; Lin, S.; Ding, B.; Liu, X., Wei, X., Ros, R., Yan, H. & Liu, Y. (2010). Aqueous Synthesis of Zinc Blende CdTe/CdS Magic-Core/Thick-Shell Tetrahedral-Shaped Nanocrystals with Emission Tunable to Near-Infrared. *Journal of the American Chemical Society*, Vol. 132, N. 16, pp. 5592-5593, ISSN 0002-7863.

Donegá, C. M.; Bode, M. & Meijerink, A. (2006). Size- and Temperature-dependence of Exciton Lifetimes in CdSe Quantum Dots. *Physical Review B* Vol.74, N.8 , pp. 085320 (1-9), ISSN 0163-1829.

Dovichi, N. J. & Harris, J. M. (1979). Laser Induced Thermal Lens Effect for Calorimetric Trace Analysis. Analytical Chemistry Vol. 51, N. 6, pp. 728-731, ISSN 0003-2700.

Du, W. & Liu, S. (1993). Single Beam Self-Interaction in Langmuir-Blodgett Films. *Optics Communications* Vol. 98, N. 1-3, pp. 117-119, ISSN 0030-4018.

Durbin, S. D.; Arakelian, S. M. & Shen, Y. R. (1981). Laser-induced Diffraction Rings from a Nematic-Liquid-Crystal film *Optics Letters* Vol. 6, N. 9, pp. 411-413, ISSN 0146-9592.

Dzhagan, V. M.; Valakh, M. Y.; Raevskaya, A. E.; Stroyuk, A. L.; Kuchmiy, S. Y.; Zahn, D. R. T. (2007). Resonant Raman Scattering Study of CdSe Nanocrystals Passivated with CdS and ZnS. *Nanotechnology* Vol.18, N. 28, pp. 285701(1-7), ISSN 0957-4484.

El-Kashef, H. (2002). Study of the Refractive Properties of Laser Dye Solvents: Toluene, Carbon Disulphide, Chloroform, and Benzene. *Optical Materials* Vol. 20, N. 2, pp. 81-86, ISSN 0925-3467.

Evident Technologies Inc. (September 2005). Nanomaterials Catalog V7.

Fang, H. L. & Swofford, R. L. (1979). Analysis of the Thermal Lensing Effect for an Optically Thick Sample-A Revised Model. *Journal of Applied Physics* Vol. 50, N. 11, pp. 6609-6615, ISSN 0021-8979.

Ge, Z.; Kang, Y.; Taton, T. A.; Braun, P. V. & Cahill, D. G. (2005). Thermal Transport in Au-core Polymer-shell Nanoparticles. *Nano Letters* Vol. 5, N. 3, pp. 531- 535, ISSN 1530-6984.

Goyanes, S. N.; Marconi, J. D.; Konig, P. G; Rubiolo, G. H.; Matteo, C. L. & Marzocca, A. J. (2001). Analysis of Thermal Diffusivity in Aluminum (particle) - filled PMMA Compounds. *Polymer* Vol. 42, N. 12, pp. 5267-5274. ISSN 0032-3861.

Hickmann, J. M.; Gomes, A. S. L. & Araújo, C. B. (1992). Observation of Spatial Cross-Phase Modulation Effects in a Self-Defocusing Nonlinear Medium. *Physical Review Letters* Vol. 68, N. 24 , pp. 3547- 3550, ISSN 0031-9007.

Jasieniak, J.; Pacifico, J.; Signorini, R.; Chiasera, A.; Ferrari, M.; Martucci, A. & Mulvaney, P. (2007). Luminescence and Amplified Stimulated Emission in CdSe–ZnS-Nanocrystal-Doped TiO2 and ZrO2 Waveguides. *Advanced Functional Materials* Vol. 17, N. 10, pp. 1654-1662, ISSN 1616-301X.

Jose, R.; Zhanpeisov, N. U.; Fukumura, H.; Baba, Y. & Ishikawa, M. (2006a). Structure Property Correlation of CdSe Clusters Using Experimental Results and First-Principles DFT Calculations. *Journal of the American Chemical Society*, Vol. 128, N. 2, pp. 629-636, ISSN 0002-7863.

Jose, R.; Zhelev, Z.; Bakalova, R.; Baba, Y. & Ishikawa, M. (2006b). White-light-emitting CdSe Quantum Dots Synthesized at Room Temperature. *Applied Physics Letters* V. 89, N. 1, pp. 013115(1-3), ISSN 0003-6951.

Katz, D.; Wizansky, T.; Millo, O.; Rothenberg, E.; Mokari, T. & Banin, U. (2002). Size-Dependent Tunneling and Optical Spectroscopy of CdSe Quantum Rods. Physical Review Letters Vol. 89, N. 8, pp. 86801 (1-4), ISSN. 0031-9007.

Khitrova, G.; Gibbs, H. M.; Kawamura, Y.; Iwamura, H.; Ikegami, T.; Sipe, J. E. & Ming, L.
(1993) Spatial Solitons in a Self-focusing Semiconductor Gain Medium. *Physical
Review Letters* Vol. 70, N. 7, pp. 920-923, ISSN 0031-9007.

Kilina, S.; Ivanov, S. & Tretiak, S. (2009). Effect of Surface Ligands on Optical and Electronic
Spectra of Semiconductor Nanoclusters, *Journal of the American Chemical Society* Vol.
131, N. 22, pp. 7717–7726, ISSN 0002-7863.

Kim, S.; Fisher, B.; Eisler, H.J. & Bawendi, M. (2003). Type-II Quantum Dots: CdTe/CdSe
(Core/Shell) and CdSe/ZnTe (Core/Shell) Heterostructures. *Journal of the American
Chemical Society* Vol. 125, N. 38, pp. 11466-11467, ISSN 0002-7863.

Klimov, V. I. (2003). Nanocrystal Quantum Dots from Fundamental Photophysics to
Multicolor Lasing. Los Alamos Science Vol. 28, N. 28, pp. 214-220, ISSN 0273-7116.

Klimov, V. I.; Ivanov, S. A.; Nanda, J.; Achermann, M.; Bezel, I.; McGuire, J. A. & Piryatinski,
A. (2007). Single-Exciton Optical Gain in Semiconductor Nanocrystals. *Nature* Vol.
447, N. 7143, pp. 441-446, ISSN 0028-0836.

Kohanzadeh, Y.; Ma, K. W. & Whinnery, J. R. (1973). Measurement of Refractive Index
Change with Temperature Using Thermal Self-Phase Modulation. *Applied Optics*
Vol. 12, N. 7, pp. 1584-1587, ISSN 1559-128X.

Kortan, A. R.; Hull, R.; Opila, R. L.; Bawendi, M. G.; Steigerwald, M. L.; Carroll, P. J. & Brus,
L. E. (1990). Nucleation and Growth of CdSe on ZnS Quantum Crystallite Seeds,
and Vice Versa, in Inverse Micelle Media. *Journal of the American Chemical Society*
Vol. 112, N. 4, pp. 1327-1332, ISSN 0002-7863.

Lee,Y. J.; Kim, T. G. & Sung Y. M. (2006). Lattice Distortion and Luminescence of
CdSe/ZnSe Nanocrystals. *Nanotechnology* Vol. 17, N. 14, pp. 3539-3542, ISSN 0957-
4484.

Leite, R.C.C. & Porto, S.P.S. (1966). Enhancement of Raman Cross Section in CdS due to
Resonant Absorption. *Physical Review Letters* Vol. 17, N. 1, pp. 10-12, ISSN 0031-
9007.

Lima, S. M.; Sampaio, J. A.; Catunda, T.; Bento, A. C.; Miranda, L. C. M. & Baesso, M. L.
(2000). Mode-mismatched Thermal Lens Spectrometry for Thermo-optical
Properties Measurement in Optical Glasses: a Review, *Journal of Non-Crystalline
Solids* Vol. 273, N. 1-3, pp. 215-227, ISSN 0022-3093.

Lippens, P. E. & Lannoo, M. (1989). Calculation of the Band Gap for Small CdS and ZnS
Crystallites. *Physical Review B* Vol. 39, N. 15, pp. 10935-10942, ISSN 0163-1829.

Liu, Y.; Brandon,R.; Cate, M.; Peng, X.; Stony, R & Johnson, M. (2007). Detection of
Pathogens Using Luminescent CdSe/ZnS Dendron Nanocrystals and a Porous
Membrane Immunofilter. *Analytical Chemistry* Vol. 79, N. 22, pp. 8796-8802, ISSN
0003-2700.

Lo, S. S.; Khan, Y.; Jones, M. & Scholes, G. D. (2009). Temperature and Solvent Dependence
of CdSe/CdTe Heterostructure Nanorod Spectra. *The Journal of Chemical Physics*
Vol. 131, N. 8, pp. 084714 (1-10), ISSN 0021-9606.

Martínez-Castañón, G. A.; Loyola-Rodríguez, J. P.; Reyes-Macías, J. F. & Ruiz, N.N.M.F.
(2010). Synthesis and Optical Properties of Functionalized CdS Nanoparticles with
Different Sizes. *Superficies y Vacío* Vol. 23, N. 4, pp. 1-4, ISSN 1665-3521.

McBride, J. R.; Dukes, A. D.; Schreuder, M. A. & Rosenthal, S. J. (2010). On Ultrasmall
Nanocrystals. *Chemical Physics Letters* Vol. 498, N. 1-3, pp. 1–9, ISSN 0009-2614.

Nguyen, K. A.; Day, P. N. & Pachter, R. (2010). Understanding Structural and Optical Properties of Nanoscale CdSe Magic-Size Quantum Dots: Insight from Computational Prediction. *The Journal of Physical Chemistry C* Vol. 114, N. 39, pp. 16197-16209, ISSN 1932-7447.

Nien, Y. T.; Zaman, B.; Ouyang, J.; Chen, I. G; Hwang, C. S. & Yu, K. (2008). Raman Scattering for the Size of CdSe and CdS Nanocrystals and Comparison with Other Techniques. *Materials Letters* Vol. 62, N. 30, pp. 4522-4524, ISNN 0167-577X.

Nikogosyan, D. N. (1997). Properties of Optical and Laser-Related Materials: A Handbook, Wiley, ISBN-13: 978-0-471-97384-3, Chichester, UK, pp. 469-470.

Nirmal, M.; Murray, C. B. & Bawendi, M. G. (1994) Fluorescence-line Narrowing in CdSe Quantum Dots: Surface Localization of the Photogenerated Exciton, *Physical Review B* Vol. 50, N. 4, pp. 2293-2300, ISSN 0163-1829.

Norris, D. J.; Efros, A. L.; Rosen, M. & Bawendi, M. G. (1996). Size Dependence of Exciton Fine Strucutre in CdSe Quantum Dots. *Physical Review B* Vol. 53, N. 24, pp. 16347-16354, INSS 0163-1829.

Ong, Q. K. & Sokolov, I. (2007). Attachment of Nanoparticles to the AFM Tips for Direct Measurements of Interaction Between a Single Nanoparticle and Surfaces. *Journal of Colloid and Interface Science* Vol. 310, N. 2, pp. 385-390, ISSN 0021-9797.

Ono, H. & Kawatsuki, N. (1997). Self-Phase Modulation Induced by a He-Ne Laser in Host-Guest Liquid Crystals with Different Nematic-Isotropic Transition Temperatures. *Japanese Journal of Applied Physics* Vol. 36, N. 3B, pp. L353-L356, ISSN 0021-4922.

Ono, H. & Saito, I. (1999). Characterization of Self-Phase Modulation in Liquid Crystals on Dye-Doped Polymer Films. *Japanese Journal of Applied Physics* Vol. 38, N. 10 A, pp. 5971-5976, ISSN 0021-4922.

Ouyang, J.; Zaman, M. B.; Yan, F. J.; Johnston, D.; Li, G.; Wu, X.; Leek, D.; Ratcliffe, C. I.; Ripmeester, J. A. & Yu, K. (2008). Multiple Families of Magig-Sized CdSe Nanocrystals with Strong Bandgap Photoluminescence via Noninjection One-Pot Syntheses. *The Journal of Physical Chemistry C* Vol. 112, N. 36, pp. 13805-13811, ISSN 1932-7447.

Park, Y.S.; Dmytruk, A.; Dmitruk, I.; Kasuya, A.; Takeda, M.; Ohuchi, N.; Okamoto, Y.; Kaji, N.; Tokeshi, M. & Baba, Y. (2010). Size-Selective Growth and Stabilization of Small CdSe Nanoparticles in Aqueous Solution. *ACS Nano*, Vol. 4, N. 1, pp. 121-128, ISSN 1936-0851.

Peng, Z. A. & Peng, X. (2002). Nearly Monodiperse and Shape-Controlled CdSe Nanocrystals via Alternative Routes: Nucleation and Growth. *Journal of the American Chemical Society* Vol. 124, N. 13, pp. 3343-3353, ISSN 0002-7863.

Petrov, D. V.; Gomes, A. S. L. & Araújo, C. B. (1994). Spatial Phase Modulation due to Thermal Nonlinearity in Semiconductor-doped Glasses. *Physical Review B* Vol. 50, N. 13, pp. 9092-9097, ISSN 0163-1829.

Pilla, V.; Catunda, T.; Balogh, D. T.; Faria, R. M. & Zilio, S. C. (2002). Thermal Lensing in Poly(vinyl alcohol)/polyaniline Blends. *Journal of Polymer Science. Part B: Polymer Physics* Vol. 40, N. 17, pp. 1949-1956, ISSN 0887-6266.

Pilla, V.; Menezes, L. S.; Alencar, M. A. R. C. & Araújo, C. B. (2003). Laser-Induced Conical Diffraction due to Cross-Phase Modulation in a Transparent Medium. *Journal of the Optical Society of America B* Vol. 20, N. 6, pp. 1269-1272, ISSN 0740-3224.

Pilla, V.; Alves, L. P.; Munin, E. & Pacheco, M. T. T. (2007). Radiative Quantum Efficiency of CdSe/ZnS Quantum Dots Suspended in Different Solvents. *Optics Communications* Vol. 280, N. 1, pp. 225-229, ISSN 0030-4018.

Pilla, V.; Alves, L. P.; Pacheco, M. T. T. & Munin, E. (2008). Radiative Quantum Efficiency of CdSe/ZnS Core-shell Colloidal Solutions: Size-dependence. *Optics Communications* Vol. 281, N. 23, pp. 5925-5928, ISSN 0030-4018.

Pilla, V.; Munin, E. & Gesualdi, M. R. R. (2009). Measurement of the Thermo-optic Coefficient in Liquids by Laser-induced Conical Diffraction and Thermal Lens Techniques. *Journal of Optics A: Pure and Applied Optics* Vol. 11, N. 10, pp. 105201 (1-7), ISSN 1464-4258.

Prasad, P. N. (2004). Nanophotonics, John Wiley & Sons, ISBN-13: 9780471649885, New York.

Qu, L. & Peng, X. (2002). Control of Photoluminescence Properties of CdSe Nanocrystals in Growth. *Journal of the American Chemical Society*, Vol. 124, N. 9, pp. 2049-2055, ISSN 0002-7863.

Rao, A.; Schoenenberger, M.; Gnecco, E.; Glatzel, T.; Meyer, E.; Brändlin, D. & Scandella, L. (2007). Characterization of Nanoparticles using Atomic Force Microscopy. *Journal of Physics: Conference Series* Vol. 61, N. 1, pp. 971–976, ISSN 1742-6588.

Rathore, K. S.; Patidar D.; Janu, Y.; Saxena, N. S.; Sharma, K. & Sharma, T. P. (2008). Structural and Optical Characterization of Chemically Synthesized ZnS Nanoparticles. *Chalcogenide Letters* Vol.5, N. 6, pp. 105-110, ISSN 1584-8663.

Reiss, P.; Bleuse, J. & Pron, A. (2002) Highly Luminescent CdSe/ZnSe Core/Shell Nanocrystals of Low Size Dispersion. *Nano Letters* Vol. 2, N. 7, pp. 781-784, ISSN 1530-6984.

Resch-Genger, U.; Grabolle, M.; Cavaliere-Jaricot, S.; Nitschke, R.; Nann, T. (2008). Quantum Dots versus Organic Dyes as Fluorescent Labels. *Nature Methods* Vol. 5, N. 9, pp. 763-775, ISSN 1548-7091.

Sheldon, S. J.; Knight, L. V. & Thorne, J. M. (1982). Laser-induced Thermal Lens Effect: a New Theoretical Model. *Applied Optics* Vol. 21, N. 9, pp. 1663-1669, ISSN 1559-128X.

Shen, J.; Lowe, R. D. & Snook, R. D. (1992). A Model for cw Laser Induced Mode-mismatched Dual-beam Thermal Lens Spectrometry. *Chemical Physics* Vol. 165, N. 2-3, pp. 385-396, ISSN 0301-0104.

Soloviev, V. N.; Eichhofer, A.; Fenske, D. & Banin, U. (2000). Molecular Limit of a Bulk Semiconductor: Size Dependence of the "Band Gap" in CdSe Cluster Molecules. *Journal of the American Chemical Society* Vol.122, N. 11, pp. 2673–2674, ISSN 0002-7863.

Singha, A.; Satpati, B.; Satyam, P.V. & Roy, A. (2005). Electron and Phonon Confinement and Surface Phonon Modes in CdSe-CdS Core–Shell Nanocrystals. *Journal of Physics: Condensed Matter* Vol. 17, N. 37, pp. 5697–5708, ISSN 0953-8984.

Snook, R. D. & Lowe, R. D. (1995). Thermal Lens Spectrometry. A Review. *Analyst* Vol. 120, N. 8, pp. 2051-2068, ISSN 0003-2654.

Sounderya, N. & Zhang, Y. (2008) Use of Core/Shell Structured Nanoparticles for Biomedical Applications. *Recent Patents on Biomedical Engineering*. Vol. 1, N. 1, pp. 34-42, ISSN 1874-7647.

Torruellas, W. E.; Wang, Z.; Hagan, D. J.; Vanstryland, E. W.; Stegeman, G. I.; Torner, L. & Menyuk, C. R. (1995). Observation of Two-Dimensional Spatial Solitary Waves in a

Quadratic Medium. *Physical Review Letters* Vol. 74, N. 25, pp. 5036-5039, ISSN 0031-9007.

Valerini, D.; Cretí, A.; Lomascolo, M.; Manna L.; Cingolani, R. & Anni, M. (2005). Temperature Dependence of the Photoluminescence Properties of Colloidal CdSe/ZnS core/shell Quantum Dots Embedded in a Polystyrene Matrix. *Physical Review B* Vol. 71, N. 23, pp. 235409 (1-6), ISSN 0163-1829.

Valley, J. F.; Khitrova, G.; Gibbs, H. M.; Granthan, J. W. & Jiajin, X. (1990). Cw Conical Emission: First Comparison and Agreement between Theory and Experiment. *Physical Review Letters* Vol. 64, N. 20, pp. 2362-2365, ISSN 0031-9007.

Yang, X.; Qi, S.; Zhang, C.; Chen, K.; Liang, X.; Yang, G.; Xu, T.; Han, Y. & Tian, J. (2005). The study of self-diffraction of mercury dithizonate in polymer film *Optics Communications* Vol. 256, N. 4-6, pp. 414-421, ISSN 0030-4018.

Yu, K.; Hu, M. Z.; Wang, R.; Piolet, M. L.; Frotey, M.; Zaman, M. B.; Wu, X.; Leek, D. M.; Tao, Y.; Wilkinson, D. & Li, C. (2010). Thermodynamic Equilibrium-Driven Formation of Single-Sized Nanocrystals : Reaction Media Tuning CdSe Magic-Sized versus Regular Quantum Dots, *The Journal of Physical Chemistry C* Vol. 114, n. 8, pp. 3329-3339, ISSN 1932-7447.

Yu, W. W.; Qu, L.; Guo, W. & Peng, X. (2003). Experimental Determination of the Extinction Coefficient of CdTe, CdSe and CdS Nanocrystals. *Chemistry of Material* Vol. 15, N. 14, pp. 2854-2860, ISSN 0897-4756.

Whinnery, J.; Miller, D. & Dabby, F. (1967). Thermal Convection and Spherical Aberration Distortion of Laser Beams in Low-Loss Liquids. *IEEE Journal of Quantum Electronics* Vol. 3, N. 9, pp. 382-383, ISSN 0018-9197.

Zou, S. & Weaver, M. J. (1999). Surface-Enhanced Raman Scattering of Ultrathin Cadmium Chalcogenide Films on Gold Formed by Electrochemical Atomic-Layer Epitaxy: Thickness-Dependent Phonon Characteristics. *The Journal of Physical Chemistry B* Vol. 103, N. 13, pp. 2323-2326, ISSN 1932-7447.

Part 2

Optics and LASER

2

Silicon Quantum Dots for Photovoltaics: A Review

Georg Pucker, Enrico Serra and Yoann Jestin
Bruno Kessler Foundation, Center for Materials and Microsystems,
Italy

1. Introduction

Controlling size and shape of materials on the nanoscale allows to engineer their physical properties. Indeed nanomaterials offer unique opportunities in engineering, micro-electronics, life-science, and renewable energies. Conversion of sunlight to electricity is assumed to be of major importance for the supply of electrical energy for future generations. The actual photovoltaic market is dominated by silicon solar cells. However, to address future needs in electricity in the order of ~30 Terawatts, the cost of solar energy will have to be significantly lower and the efficiency will have to be increased significantly with respect to actual values. With these preconditions it is understandable that the interest in exploring the properties of silicon nanostructures in solar cells is enormously increased. Especially, since for silicon, a similar revolution already happened regarding the shrinkage of the size of transistors accompanied by an exponential increase in speed. The review summarizes the most important properties of silicon quantum dots (Si-QDs), gives an overview of solar cell concepts, in which Si-QDs can play a role and indicates major achievements to be envisioned for another technological revolution based on silicon technology.

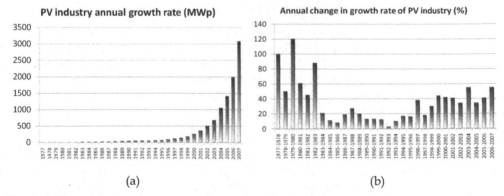

Fig. 1. a) Photovoltaic industry annual growth (1977-2007), b) annual change in growth rate (data from (Marketbuzz, 2010)).

1.1 Current trends in photovoltaics

Since over 2 decades the photovoltaic market shows an exponential growth with annual growth rates in the range of 40-50% (Hirshman et al., 2007). It took, however, more than hundred years, since the discovery of the photovoltaic effect by Edmond Becquerel (Becquerel, 1867) in 1839, until the 1st viable commercial demonstration of a solar cell. These c-Si based solar cell invented in 1954 by researchers at Bell laboratories had an efficiency of ~4% (Chapin et al., 1954). Only after the oil crisis of the 1970s photovoltaic energy was considered as an alternative source of electricity. The rapid exponential growth of cumulative installed capacity, which occurred in the last 2 decades is illustrated in Fig.1. Indeed world solar photovoltaic (PV) market installations reached a record high of 7.3 gigawatt (GW) in 2009, representing growth of 20% over the previous year (Marketbuzz, 2010). Although, PV electricity can't yet compete directly with electricity from conventional sources on the wholesale or retail prices, it is already competitive during peak power times (Wolfsegger & Stierstorfer, 2007), and it is expected to reach grid parity with retail prices in Southern Europe before 2015 and in most of Europe by 2020 (European Photovoltaic Technology Platform, 2007). Current photovoltaic technology is dominated by silicon.

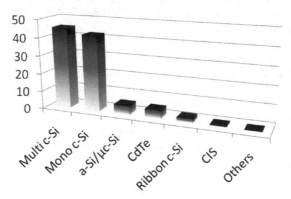

Fig. 2. Market share of photovoltaic technology (Teske & Bitter, 2008).

Figure 2 shows the market share of PV technology (Teske & Bitter, 2008). The overwhelming part of the market is based on crystalline silicon ~85% (including both large grain multi-crystalline [multi cSi] and mono-crystalline [mono cSi] materials). These solar modules based on silicon wafers, either multi-crystalline or mono-crystalline, have been termed "first generation" photovoltaic technologies. The devices are based on a fairly simple p-n junction and have a thermodynamic efficiency limit of ~30% (Shockley & Queisser, 1961) (taking into account the shape of the solar cell, the refractive index of the cell, realistic solar spectrum, concentration, and radiative recombination) – this limit is often called Shockley-Queisser limit. Best laboratory c-Si solar cells arrive at an efficiency of about ~25% (Green et al., 2010) with highest efficiencies for modules of ~21.4% (Sunpower Corp.), while efficiencies for most of the commercial c-Si modules on the market are in the range 16-20%. A second class of solar cells "2nd generation" attempts to reduce costs by using thin film semiconductors like amorphous silicon (a-Si), CdTe, Cu(In,Ga)Se$_2$ with about ~14% of market share and cell efficiencies in the range 10 - 20% depending on the system. In addition, in the same class of cells fall dye-sensitized solar cells (DSSC) and organic photovoltaic cells (OPV) based on conducting polymers and fullerenes. DSSCs introduced by M. Grätzel and co-workers in 1991 (O'Regan, & Grätzel, 1991), are based on a concept of light harvesting and charge separation similar to the one known from nature in photosynthesis. Best DSSCs have an efficiency of ~10-12% (Green et al., 2010). Current most efficient dye-sensitized cells make use of ruthenium organic complexes and liquid electrolytes. The use of ruthenium-organic complexes is critical for large scale production due to the high cost and low abundance of ruthenium, while the use of liquid-electrolytes for carrier transport reduces lifetime of the modules because of corrosion and evaporation of electrolyte. Therefore, intense research is performed on the use of alternative dyes and solid-state electrolytes (Wei et al., 2010). OPV cells consist of a mixture of small organic molecules and polymers structured on the nanometer scale. Light absorption results in the formation of excitons, which are subsequently separated at the interfaces of the distinct organic materials working as electron acceptors or donors, respectively. The created charges diffuse then eventually to the electrodes and are collected in the external circuit. The organic molecules and polymers used are inherently inexpensive and cells are rapidly increasing in efficiency. OPV cells can be produced on flexible substrates with large throughput by low-cost roll to roll processing (Brabec et al., 2008). Efficiencies of OPV cells rapidly increased and highest reported efficiencies are ~8% (Solamer Inc., Konarka Technologies). One key issue for the success of OPV is device degradation caused by oxygen or water. Development of encapsulations that are stable for 10-25 years will be a significant challenge.

1.2 Scenario of future photovoltaics

The market for photovoltaic technology - mainly based on silicon - showed exponential growth rates in the last twenty years, and manufacturing costs of silicon based solar models in the range of ~1\$/W_p should be achieved in the near future (Del Cañizo et al., 2009). With an estimate of about ~40 GW of electricity produced by solar photovoltaic power for the year 2010 the contribution to worldwide electricity generation is still negligible, if compared with the average total worldwide power consumption of 15 Terawatts (TW). Capturing and transforming in electrical power even a small fraction of the 122,000 TW of energy from the sun could give a significant contribution to present and future power generation. According to (Feltrin, & Freundlich, 2008) it is expected that between 2050 and 2070 fossil fuels (gas, oil and coal) will play a minor role for energy production and, even projecting a robust growth of nuclear and wind energy, solar power generation should play an important role. Scaling current photovoltaic technologies from present gigawatt energy production up to the terawatt range, which means that the energy production by photovoltaics has to increase by nearly 3 orders of magnitude. For this reason abundancy of materials and materials shortage will play a more important role than in current photovoltaics. Nearly all of the current photovoltaic technologies are based on materials, which will be severely affected by material shortage. In the following we give some examples: Indium, which is currently widely used in touch screens and in photovoltaics i.e. in a-Si, DSSC, and organic solar cells as electrode material and in multi-junction cells as active material; silver and gold are used as electrode materials, germanium and gallium find application in multi-junction cells, and tellurium is used as cell material in CdTe solar cells. Interestingly, in another study, which considered the materials availability their material extraction costs, and the annual electricity potential, only a limited number of poorly investigated materials like FeS_2, CuO_2, and Zn_2P_3 were considered favorable with respect to a-Si solar technology for TW photovoltaics (Wadia et al., 2009). The problem of materials availability together with the still too large costs of power generation of most photovoltaics technologies has led to search for a new generation of cells (3rd generation photovoltaics), aiming at the same time to keep the costs/Watts as low as possible and to push the cell efficiency significantly over the Shockley-Queisser limit of 30% for single junction solar cells. A series of different concepts and physical phenomena are currently investigated for this purpose: tandem and multi-junction cells, hot-carrier cells, up-and down conversion processes, etc.

1.3 Why silicon quantum dots for solar cells?

Si-QDs differ in a series of properties considerably from bulk silicon: i) Si-QDs show an increase in the energy gap with shrinking dimensions of the dot. Therefore by controlling the size it is in principle possible to shift the energy bandgap of the material to the optimum energy gap (i.e. 1.6 eV) for maximum efficiency of single junction solar cells, or to stack cells with different bandgap and hence different spectral response together to form tandem cells. ii) Si-QDs show strong photoluminescence in the visible-red region of the solar spectrum and external quantum efficiencies as high as 60% were reported in literature. This strong photoluminescence can be used in photoluminescence down-shifters to increase the efficiency. iii) in quantum dots due to the widening of the distance between the energy states the relaxation of excited carriers is slowed down with respect to bulk silicon. If the carriers can be extracted from the dot before relaxing to the edge of the band gap, one could increase the open circuit voltage of the solar cells. This idea is explored in the concepts for hot carrier solar cells. iv) another phenomena, which is investigated for use in solar cells is the generation of multiple excitons in silicon quantum dots. An exciton with an energy much larger than the bandgap (at least twice), has an increased probability to split into 'n' excitons instead of losing the excess energy in form of heat and relaxing to the edge of the bandgap. These multiple excitons, if extracted successfully from the dot, would lead to an increased current of the solar cell, or if both multiple excitons recombine radiatively the material would show an increased photoluminescence quantum efficiency. The review is organized in the following way: in the 1st section we give a brief summary on the properties of Si-QDs focusing especially on the properties relevant for use in solar cells such as band-gap opening, photoluminescence properties, electrical transport etc. (an enormous amount of research was done in this field since the observation of strong photoluminescence from porous silicon in 1990 (Canham, 1990) and a series of excellent books and reviews are available for the interested reader (Amato et al., 1997; Cullis et al., 1997; Canham, 2007), while in the 2nd part we present a series of ideas and concepts, which aim on the realization of solar cells based on silicon quantum dots. Some of these concepts were already successfully demonstrated with other type of semiconductors. For example tandem cells consisting of stacked solar cells with each cell harvesting a different portion of the solar spectrum, but they rely on relatively rare elements like Ga, In, or As and will therefore not play a significant role for the generation of TW of electrical energy from sun light. Other concepts like hot-carrier cells are very promising in terms of their high thermodynamic efficiency limit (~85%), but no hot carrier-solar cell has been fabricated so far. Experimental research in this field therefore concentrates on the study of hot carriers in different bulk and quantum confined semiconductors and on the development of energy selective contacts for carrier extraction.

 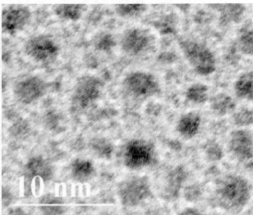

Fig. 3. (left) Upper part of an ingot of single crystalline silicon. The diameter of the ingot shown in the photograph is 100 mm. Largest silicon ingots grown by the Czrochalski method have a diameter of 30 mm and a length of ~2m. (right) HRTEM image of tiny silicon quantum dots in a SiO$_2$ matrix. Si-dots appear as dark structures and the (111) lattice plans are clearly seen. The dimension of the Si-quantum dots shown in the image is between 2 and 5 nm.

2. Properties of silicon quantum dots

2.1 Optical properies

In Figure 3 a piece of a silicon single crystal ingot with a diameter of 10 cm and a high resolution TEM image of tiny silicon nanocrystals embedded in SiO$_2$ with dimensions of about 2-4 nm are shown. While the properties of crystalline silicon can be considered uniform for dimensions ranging from some tenths of nanometers up to centimeters, the properties of Si-QDs as we will see in the following are (i) strongly dependent on size, (ii) influenced by the properties of the interface between QDs and matrix and (iii) the electrical transport in the material depends strongly on the distance between the QDs. For a better understanding of the properties of Si-QDs, it is useful to remember some of the fundamental properties of silicon. In the crystalline phase silicon forms a diamond structure. In this structure every silicon atom is surrounded by four nearest neighbour atoms forming a tetrahedron. The unit cell of silicon consists of 8 atoms and the lattice constant "a" is 0.543 nm. Due to the small distance of the atoms in the solid state the energy levels of the atoms overlaps and causes the formation of bands. The outermost electrons form two bands referred as the valence and the conduction band. At zero degrees Kelvin the valence band is completely filled with electrons and the conduction band is completely empty of electrons. A forbidden gap (band gap) separates the conduction from the valence band. For silicon this bandgap is of 1.12 eV (or ~1107 nm) at room temperature. Silicon is an indirect band gap semiconductor, which means that the bottom of the valence band and the top of the conductance band are not aligned in-k space (see the electronic band structure of silicon based on k · p calculations in Figure 4).

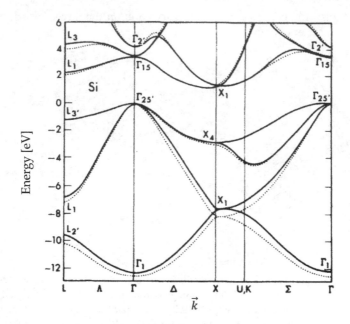

Fig. 4. Theoretical band structures of Si. In the case of Si, results are shown for nonlocal (solid line) and local (dashed line) pseudo-potential calculations (Chelikowsky & Cohen, 1976). Reprinted figure with permission from [Chelikowsky, J.R.; Cohen, M.L. Physical Review B, Vol.14, pp. 556-582, 1976. Copyright (1976) by the American Physical Society].

The maximum of the valence band is at the Brillouin zone center or Γ point (at k=0), while the minimum of the conduction band is close to the X point (k $\neq$ 0) at the Brillouin zone boundary. Since optical transitions conserve both energy and momentum the excitation of an electron from the top of the valence band to the bottom of the conduction band can occur only via the assistance of another process to conserve the momentum. In pure silicon this occurs via the transfer of the electron momentum to a phonon (the transition is therefore called indirect and the absorption close the band edge is very low, which can be nicely seen from the absorption coefficient k in Figure 5 (left). For the same reason the probability of relaxation of an excited electron from the conduction band to the valence accompanied by the spontaneous emission of a photon is low and radiative lifetimes in silicon as long as milliseconds can be observed. The absorption increases at higher energies due to the onset of the lowest energy direct transitions. The absorption reaches a first maximum at 359 nm assigned to the E_1 transition, a strong maximum at 296 nm due to the E_0 transition and another weaker maximum at 225 nm assigned to the E_1' transition. Figure 5 (right) shows the absorption edge of silicon, where structures can be identified as exitonic transitions. An exciton is formed due to the interaction of the excited electron in the conduction band with the remaining electrons (hole in the valence band). Due to the indirect nature of the edge also excitonic transitions become partially allowed through absorption or emission of phonons and the related transitions are seen as sharp structures in the absorption spectrum (Absorption edge of c-Si measured at 1.8 K from (Shaklee & Nahory, 1970). The exciton binding energy in bulk silicon is relatively small (~15 meV). If we reduce the dimensions of a semiconductor like silicon in one, two or three dimensions on nanometric scale the properties of the material will be influenced by the spatial confinement i.e. the energy states of the electrons and holes lose their continuity and become discrete. Therefore, the absorption and emission of light when electrons or holes move from one state to another are affected. Nanostructures are often classified according the number of dimensions in which carriers (electrons) are free to move in 0-D (nanocrystals, quantum dots, porous silicon), 1-D (quantum wires and pillars), and 2-D (quantum wells).

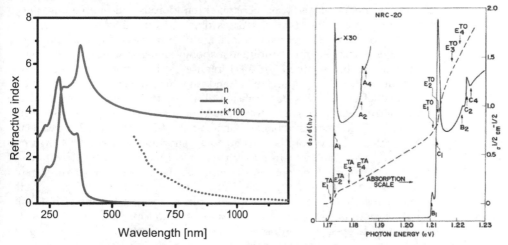

Fig. 5. Experimental dielectric function of Si (blue – real part of the refractive index, red imaginary part (or absorption coefficient) (left). Absorption edge of c-Si measured at 1.8 K from (Shaklee & Nahory, 1970). Reprinted figure with permission from [Shaklee, K.L. & Nahory, R.E. Physical Review Letters, Vol.24, pp. 942-945, 1970]. Copyright (1970) by the American Physical Society.

In the following, we will concentrate mainly on results obtained on 0-D silicon quantum structures, although silicon quantum-wires and quantum wells are very similar in their properties and a series of interesting and encouraging results for use in solar cells were especially obtained on silicon quantum wires (Tsakalakos et al., 2007; Kelzenberg et al., 2010). One of the fundamental effects of the confinement is its influence on the electronic properties of the system especially on the energy of the band gap. A simple estimation of the effect of the confinement on the energy of the electron gap E_{gap} can be obtained by using the effective-mass approximation,

$$E_{gap} = E_{Si\_bulk} + \frac{\hbar^2 \pi^2}{2} \left[\frac{1}{L_x^2} + \frac{1}{L_y^2} + \frac{1}{L_z^2} \right] \left[\frac{1}{m_e^*} + \frac{1}{m_h^*} \right] \tag{1}$$

Where E_{Si_bulk} stands for the energy gap in bulk silicon (1.12 eV), L_x, L_y, L_z are the dimensions of the nanoparticle and m_e^* and m_h^* are the effective electron and hole masses in bulk silicon. In a first approximation the hole mass is essentially half of the electron mass, the simple theory predicts that the valence band should shift twice the shift of the conductance band. In Figure 6 the increase of the energy gap for shrinking size of silicon quantum structures is shown (Lockwood et al. 1992). From the figure we can see that an appreciable effect on the energy gap is observed for Si-QDs smaller than 5nm and for dots of 2nm of size the gap is predicted to be as large as ~2eV. The shifts of the conduction and valence bands can be experimentally observed in x-ray absorption techniques and confirmed with good approximation the shifts of conduction and valence band foreseen from the effective mass approximation (Lockwood et al., 1996). In addition to changing bandgap energy the quantum confinement influences also the transition probabilities. Due to the localisation of the carriers (excitons) in a very small volume the k selection rules are relaxed and the transition probabilities are increased (quasi direct transitions). A lot of the early work on silicon nanocrystals was performed on porous silicon. Porous silicon discovered more than 50 years ago (Uhlir, 1956) is obtained typically by electrochemical etching of bulk silicon with HF. The porosity can range in dimensions from nanometers to micrometers and under correct etching conditions a spongy material with wall dimensions of a few nanometers can be obtained. In 1990 intense visible photoluminescence was observed from porous silicon (Canham, 1990) and the luminescence was ascribed to quantum confinement effects (Cullis & Canham, 1991). In the following years intense research on porous silicon was performed, for application in silicon photonics and sensing. These research is reviewed in a series of books (Amato et al., 1997; Bisi et al., 2000; Canham, 2007; Feng & Tsu, 1994). Highly porous silicon has an enormous surface area, which is available for interacting with the ambient either by chemical reactions or physisorption of molecules. For example oxidation of porous silicon tends to continue for weeks at room temperature, if the surface is not accurately passivated. In addition, due to the porosity the material is also mechanically instable and fragile. For this reason the scientific community looked for alternative methods to prepare better passivated and mechanically more stable Si-QDs. Nowadays a wide spectrum of fabrication techniques exists which allows to form Si-QDs in a gas phase, in solution or in solid state. Common to most of the fabrication methods is that the silicon quantum dots show a certain variation in size and shape. As an example we show in Figure 7 the absorption coefficient and the photoluminescence spectra of silicon quantum-dots obtained by thermal annealing of a silicon rich silicon oxide film (silicon rich or excess of silicon means that the film contains an excess of silicon with respect to the ratio of 1: 2 between Si and oxygen in stoichiometric silicon dioxide). In this type of measurements due to the dimensions of the light spot and the large density of silicon quantum dots a huge number of quantum dots is measured at the same time (~10^{11} quantum-dots in the photoluminescence and about ~10^{12} Quantum-dots in the ellipsometric measurements from which the absorption coefficients were obtained). The critical points of energy transitions, which are well defined in the absorption spectrum of c-Si (for example the E1 transition at 3.395 eV), are broadened and poorly defined due to this averaging over an enormous number of slightly different silicon quantum dots. The absorption coefficient increases with increasing silicon content (Γ-value, which refers to the ratio of the gases SiH_4 and N_2O used during the deposition of the films). The photoluminescence spectra measured at room temperature of these samples show a broad photoluminescence band in the VIS-NIR region, which shifts from 800 to 850 nm for decreasing silicon content. This type of behaviour is reported by numerous studies and only some of them are cited here (Rinnert & Vergnat, 2003; Umezu et al., 2000; Yang et al.,1997; Pucker et al., 2000).

Further inside in the structure of Si quantum dots comes from the analysis of photoluminescence (PL) decay curves. Similar to what is found in porous silicon the PL decay shows a dispersion in decay lifetime not only as function of emission wavelength, but also for a fixed wavelength. The decay curves are influenced by the dispersion in size and shape of the nanocrystals resulting in different radiative transition probabilities and by migration and trapping of excitons. Decay times at room temperature are typically in the range of 10 to 300 μs (Linnros et al., 1999; Wilson et al., 1993) and the decay time increases up to milliseconds for low temperatures (<20 K) (Dovrat et al., 2004; Brongersma et al., 2000; Vinciguerra et al., 2000). The temperature dependence of the photoluminescence decay curves can be explained (Calcott et al., 1993) by ascribing the luminescence to the recombination of strongly localised excitons in the Si quantum dot. Due to the exchange interaction between the electron and the hole, the excitonic levels are split by an energy Δ. The lower level corresponds to a triplet state (threefold degenerate), while the upper state is a singlet state. The ground state of the exciton is a singlet state and therefore the transition to the triplet state is forbidden (the non-zero transition probability is due to spin-orbit interaction mixing slightly the singlet and triplet states) therefore the radiative decay rate is small at low temperatures.

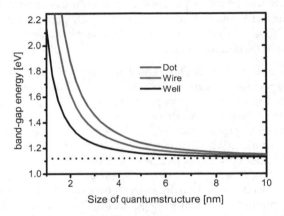

Fig. 6. Optical gap in Si quantum wells, wires, and dots, versus systems diameter. The transition energy is calculated for the lowest electron and heavy hole energies for infinite confining potentials after Ref (Lockwood et al., 1992).

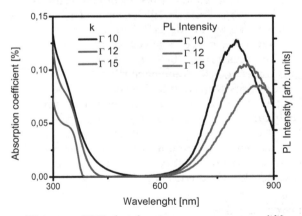

Fig. 7. Absorption coefficient and RT photoluminescence spectra of films containing Si-QDs obtained by thermal annealing of silicon rich oxide films (larger Γ less silicon excess => smaller).

At higher temperatures the singlet state becomes populated and the decay rate increases. The exciton splitting energy Δ increased for lower wavelength of PL (smaller Si-dots) and is in the range of 8 to 15 meV and much larger than the 0.15 meV of exchange splitting for the exciton observed in bulk silicon. The use of energy selective-optical spectroscopy allows to gain further inside in the behaviour of Si-QDs (Kovalev et al., 1998): Resonant PL-spectra obtained by exciting a subset of Si-QDs (selected by excitation at the low energy side of the broad PL emission band) show emission bands, which can be assigned to transitions involving either no momentum conserving phonons, or transitions involving the absorption or emission of one or two transversal acoustical or transversal optical phonons. Observation of the ratio of the intensity of zero-phonon to phonon assisted transitions indicate that although the transition probability is enhanced due to the quantum confinement effects, the bandgap of the Si-QDs remains indirect. Finally by measuring the PL spectra of single Si-QDs at low temperature (down to 35K) emission line widths as low as 2meV were found. In addition dots with and without a Transverse Optical (TO) phonon replica were observed and the presence or absence of the phonon replica seems to depend on the degree of localisation of the exciton (Sychugov et al. 2005).

The description of properties of Si-QDs based on recombination of confined excitons explains a lot of the experiments performed, however not all the experimental findings can be exclusively explained within the model presented before: i) depending on the process of fabrication, the Si-QDs or quantum wells can be either crystalline or amorphous and the presence of both species makes the interpretation of experimental findings more complicated (Nesheva et al., 2002). ii) the photoluminescence properties observed can be influenced by defect states or photoluminescence active centres in the matrix or at the interface between the Si-QD and the matrix, iii) properties of Si-QDs are found to be strongly influenced by strain. Already from early studies on porous silicon it is well known that the intensity of photoluminescence depends on the quality of the surface of the nanocrystals and treatment with hydrogen at elevated temperatures was found to be very efficient in reducing the number of defects, which favour non-radiative recombination. In silicon nanocrystals embedded in SiO_2 a series of defect centres, such as O vacancy, O thermal donors and Si dangling bonds, were identified using electron-spin resonance spectroscopy (Prokes et al., 1998). While dangling bonds act as non radiative recombination centers, which can be passivated with hydrogen (Garrido Fernandez et al., 2002; Godefroo et al., 2008), oxygen related defects are believed to play also a role in the radiative recombination process. The presence of these defects results in a photoluminescence band typically at higher emission energy than the band related to the exciton recombination and differs from the later one in decay time and temperature dependence (Tsybeskov et al., 1994; Kanemitsu 1994; Min et al., 1996; Kanemitsu et al. 1997). The oxygen related surface states are also believed to play an important role in the observation of stimulated emission from Si-QDs (Pavesi et al., 2000, Ruan et al., 2003, Dal Negro et al., 2003). Finally in a series of studies it is observed that the photoluminescence properties of silicon containing films depend also on film stress and on the stress at the interface between the Si-QD and the matrix (Khriachtchev et al., 2001; Daldosso et al., 2003; Zatryb et al., 2011).

2.2 Electrical transport properties

Interestingly early work on conduction processes and charging in Si-QDs embedded in a oxide matrix was performed back in 1984 by DiMaria (DiMaria et al. 1984a; DiMaria et al.

1984b) some years before the observation of light emission from porous silicon initiated the boom of research in silicon photonics. DiMaria and coworkers studied the electrical and electro-optical properties of annealed silicon-rich-oxide containing Si-quantum dots, observed electroluminescence – for which they proposed carrier relaxation in quantum confined silicon as origin, and discussed the transport of carriers in the material. The conduction mechanism in dielectric matrices containing Si-QDs depends strongly on the type of matrix (silicon dioxide, silicon-nitride and silicon carbide are the most commons one) and on the distance between the Si-QDs. When Si-QDs are prepared by phase separation from silicon rich oxide two distinct regions can be found in conductivity measurements: a region in which the content of excess silicon is so large that the Si-QDs are not isolated anymore but form a percolation backbone and a lower Si-excess region in which the Si-QDs are isolated and conductivity is dominated by tunneling processes (Balberg et al., 2007). In the region of large Si-excess the conductivity is in the order of 10^{-7} (Ohm cm)$^{-1}$, which is about 3 orders of magnitude lower than the conductivity of high resistivity floating zone silicon ($\sim 10^{-4}$ (Ohm cm)$^{-1}$). For heavily boron doped films, conductivity values of about 0.1 (Ohm cm)$^{-1}$ were reported by Mirabella et al. (Mirabella et al., 2010). However as shown in (Balberg et al., 2007) in this high Si-excess regime (above the percolation threshold) the photoluminescence observed for isolated Si-QDs is absent due to the growth of the cluster size and hence relaxation of the quantum confinement effect. For this reason this clustered structures seem less favorable for many of the illustrated application of Si-QDs reviewed here. Conductivity in SiO_2 films with isolated Si-QDs is in the order of 10^{-10} (Ohm cm)$^{-1}$ (Balberg et al. 2007). In recent years films representing this type of structure were intensively investigated especially for realization of silicon base light emitting devices (Perálvarez et al., 2006; Franzò et al., 2002; Prezioso et al., 2008) and references therein. For the realization of light emitting diodes the oxide in a metal-oxide-semiconductor (MOS) structure is replaced by the Si-QDs containing material and the top electrode thickness reduced to obtain a semitransparent electrode. This diodes show under a forward bias, which strongly depends on the composition of the active material, electroluminescence in the spectral region around 730-830 nm (1.5-1.7eV), characteristic for the emission from the Si-QDs. In LED's the Si-QD has to be excited by the carriers injected from the external circuit. Silicon dioxide has an energy gap of ~9 eV and the barrier heights are about 3.2 eV and 4.7 eV for electrons and holes respectively. Due to the difference in the barrier height it is in case of a SiO_2 matrix much easier to inject electrons than holes. Direct tunneling of carriers through an oxide is the dominating conduction process only if the SiO_2 thickness is smaller than ~1.5 nm. For larger distances transport is assumed to take place by Fowler Nordheim tunneling, in which the height of the barrier is reduced by the applied electric field and the electrons are accelerated. These hot electrons can then result in the formation of exciton-hole pairs by impact ionization of the nanocrystals, which then relax radiatively. The important information from these studies for use of Si-QDs in solar cells is twofold: 1st conductivity is rather low, 2nd it is difficult to have a balanced situation regarding electron and hole current (at least if the embedding material is silicon dioxide). Several strategies are currently investigated to improve carrier transport in and from quantum confined Si-QDs: Very accurate placing of the Si-QDs controlling the distance between the dots within a range of 1.5 nm to 2 nm allows to obtain carrier injection by direct tunneling of both holes and electrons and in such structures electroluminescence was observed applying very low voltages (Anopchenko et al. 2009). A common route for fabrication of these multi-layer stacks with controlled Si-quantum dot distances is explained in section 3.3.1. An alternative route for improving the conductivity is to substitute silicon dioxide with other materials with smaller band-gap than silicon. Widely investigated matrices are Si_3N_4 and SiC with bandgaps of 5.3 and 2.85 eV, respectively. The conduction band offset is 2.4 eV for Si_3N_4 and 2.95 eV for SiC. Indeed, recently for Si quantum dots in silicon nitride matrix conductivities between 10^{-4} (Ohm cm)$^{-1}$ and 10^{-2} (Ohm cm)$^{-1}$ are reported (Wan et al., 2011), which is a huge increase with respect to the conductivity values found for Si-QDs in SiO_2. Finally, the conductivity of Si-QD containing materials can be increased by doping. Si-QDs in silicon dioxide matrix are typically doped with either boron or phosphor to obtain p-type and n-type conducting films, which would allow to engineer p-n junctions similar to the ones for standard silicon solar cells. For doping of Si-QDs large concentrations of dopants are needed. The typical dopant concentrations are around 1 atom%. An increase of the conductivity of several orders of magnitude is reported for both boron and phosphor doping. Si-QDs can be doped with boron or phosophor, which seem to act as electron acceptor or donor similar as in bulk-silicon. Interestingly boron and phosphor doping has opposite effects on the photoluminescence efficiency (Hao et al., 2009). While in phosphor doped films the photoluminescence intensity increases in doped films, it decreases for boron doped films. Current studies indicate that phosphor doping reduces the number of dangling bonds – acting as as non-radiative recombination centers - at the quantum-dot/SiO_2 interface (Stegner et al., 2008), while boron doping results in contrary in an increase of non-radiative recombination centers (Zatryb et al., 2010).

3. Si-quantum dots in solar cells

Nowadays, 3rd-generation solar cells as quantum-dots based solar cells, are targeting significant increases in energy conversion efficiency. With the perspective of using the energy confinement of silicon based quantum dot nanostructures, different concepts of solar cells can be found in the literature. Through these concepts, four promising approaches can be retained: (i) Down-conversion and down-shifting solar cells (Klampaftis et al., 2009; Richards, 2006a): In this case, the aim is to make a more efficient utilization of the short wavelength part of the solar spectrum, mainly by producing more than one low-energy photons from the absorption of one high-energy photon (for which the cell quantum efficiency (QE) is low), or by shifting high energy photons to a lower energy photon for which the cell QE is high. (ii) Hot carrier solar cells (Luque & Marti, 2010, Nozik et al., 2010): in this scheme, solar-to-electricity energy conversion is enhanced by reducing energy losses related to the absorption of solar photons with energy larger than the bandgap of the active photovoltaic material. (iii) All-silicon tandem or multi-junction solar cells (Conibeer et al., 2008; Di et al., 2010): here, the expected increase in efficiency is due to a multi-band gap approach. The cell with the highest band gap is placed on the very top, while the cell with the lowest band gap is positioned at the bottom of the tandem stack. Each cell absorbs the light it can most effectively convert, with the rest passing through to the underlying cells. (iv) Inorganic/organic solar cells (Buhbut et al., 2010): the principle scopes of the use of QDs in this approach are twofold: 1st the light harvesting can be improved by broadening the region in which light is absorbed through the use of QDs of different size, 2nd the lifetime of the cell is improved by replacing organic molecules with more stable inorganic ones.

3.1 Si-QDs as spectral converters for solar cells

Aim of spectral converters is the more efficient utilization of the short-wavelength (λ) part of the solar spectrum. Under a spectral converter we understand a material or film, which by absorption and re-emission changes the spectral distribution and numbers of photons of the solar spectrum impinging on the converting layer. Two distinct physical mechanisms are explored in this context: down-conversion and down-shifting. In down conversion one photon with energy $hv \geq 2 E_g$. yields 2 photons (in the ideal case) with energies $hv \geq E_g$. Down-conversion improves the performance of solar cells by minimizing the thermalisation losses in solar cells. The maximum efficiency improvement for a solar cell with a bandgap of 1.1 eV (crystalline silicon) was estimated to be 38.6% placing the spectral converter on top of the solar cell (Trupke et al., 2002). Down-conversion is mainly observed for rare-earth ion doped materials and occurs either on single ions or involves energy transfer processes among different rare earth ions. A description of rare-earth based down converting phosphors is found in (Richards B.S., 2006b). For Si-QDs up to now no down-conversion process is reported. However for Si-QDs the generation of multiple excitons (MEG) was reported, which is somehow similar to down-conversion since it results in the formation of more than 1 exciton of low energy from a high energy exciton. MEG in the context of Si-QDs will be discussed in section 3.2. In down-shifting a high energy photon is absorbed by the spectral converter and re-emitted at lower energy. The process of down-shifting in silicon QDs accompanied with a radiative energy transfer from the QDs to the solar cell is depicted in Figure 8. The absorption of incident high energy photons and re-emission at lower energy can be achieved with every three level systems. A first and fast non radiative relaxation takes place between the highest excited level and the intermediate level. In this case the emission of only one lower energy photon is accompanied by the radiative

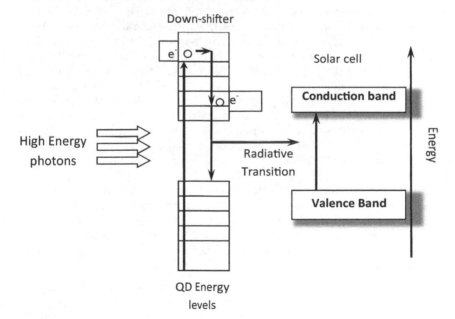

Fig. 8. Schematic energy diagram of a solar cell in combination with a spectral down-shifter.

recombination of the electron from the intermediate level to the lowest level. As a result of the luminescent process, a part of the incident photons are shifted to longer wavelength before reaching the active photovoltaic material of the device. Aim of the down-shifting process is to maximize the number of incoming photons in the spectral region in which the internal quantum efficiency (IQE) of the solar cell is at its maximum. In solar cells based on crystalline silicon the IQE is lower for photons with wavelengths shorter than ~450 nm (due to recombination of carriers generated close to the surface –surface recombination) and very high between ~500 and 1000 nm. For this reason a downshifting of UV-blue photons to the red-NIR region can result in a larger short circuit current and hence an improvement of the External Quantum Efficiency of the device. The two others electrical characteristics of the device i.e. the open-circuit voltage and the fill factor will not change significantly, since the electronic properties of the semiconducting material or the resistance of the device remain unchanged.

A typical schematic representation of a photovoltaic spectral converter is illustrated in Figure 9. It is constituted of four separate layers. In this case Silicon QDs i.e. the active material are located on the front surface of the device and the layer is electronically isolated from the solar cell by an insulator layer, i.e. the coupling between the active medium and the solar cell is purely radiative. A perfect mirror is located on the rear surface of the device to provide high internal surface reflectance for all angles of incidence of light. In order to be used as down-converter for all type of existing solar cells, Silicon QDs has to respect the following trend: (i) the absorption band has to be large enough to cover the region where the External Quantum Efficiency of the cell is low (UV-blue region in Silicon based solar cells); (ii) the absorption coefficient has to be as high as possible; (iii) the emission band has to cover a spectral band where the external quantum efficiency of the cell is the best (i.e. red region in Silicon based solar cells); (iv) the energy difference between the absorption band and emission band i.e. the difference between the conduction and valence band has to be large enough to avoid re-absorption phenomenon of the emitted photons. One of the main advantage of using quantum dots in general, and it is also true for Silicon QDs, as down-shifters in solar cells is that both light absorption and emission can be tuned by choice of the dot diameter. Furthermore Si-QDs have a large quantum efficiency (over 50%, Walters, R.J. 2006), which is another important requisite to obtain an enhancement in efficiency of the solar cell (see below). Preliminary simulation results made on ideal QDs by van Sark have demonstrated the great capacity of QDs as down-shifting species (van Sark et al., 2004). Indeed calculation have shown that quantum dots with an emission at ~600 nm increase the multicrystalline solar cell short-circuit current by nearly 10%. The simulation shows that the most beneficial effects will be accomplished if the spectral response of solar cells has a specific form such that the spectral response is low at low wavelengths and high at high wavelengths. When considering the critical parameters that influence the performance of down-conversion solar cells, we may pin down three essential factors: one is the energy distribution of the incident spectra the second is the presence or not of an anti reflecting coating on the front face of the solar cell and the third is the surface recombination. Jestin et al (Jestin et al., 2010) have shown on a c-Si solar cell, that the effect of a down-conversion layer containing Silicon QDs on the Power Conversion Efficiency (PCE) is rather low.

Simulation and experimental results have confirmed that an increase in the cell performance can be observed only in case of poor performance of the cell in the UV-blue region (e.g. very high surface recombination velocity). In this case solar cells with a luminescent down-conversion layer can outperform normal c-Si based solar cells. The first experimental tests with silicon QDs as a down-converter were performed by Svreck et al (Svreck et al., 2004) and described the embedding of Silicon QDs into spin-on-glass antireflecting SiO_2 based solution spun onto standard silicon solar cells. In this case the interest is to take advantage of the emission properties of the Silicon QDs that can be tuned by their size, as a result of quantum confinement (Alivisatos, 1996). The interest of quantum dots with respect to other down-converting species like organic dyes molecules (McIntosh & Richards, 2006; Maruyama et al., 2000; Richards & McIntosh, 2007) or rare-earth ion complexes (Ledonne et al., 2009, Ledonne et al., 2010) are their high brightness, stability and quantum efficiency (Bruchez et al., 1998). One main advantage of the application of a luminescent down-conversion layer on solar cells lies in the fact that the luminescence down-converter is only optically coupled to the solar cell and in principle spectral down converter and solar cell can be optimized independently. This down-conversion concept mainly shifts the photons from the blue to the red.

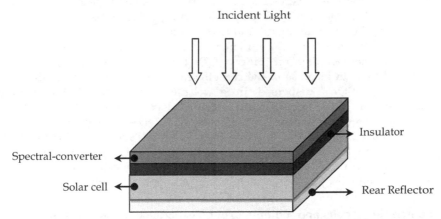

Fig. 9. Schematic representation of a photovoltaic down-converter based device.

Recent years saw several studies on the Si QD based downs-shifters: (most important results are summarized in Table 1):

Solar cell	Quantum dots	Host material	Performance difference [%]	Illuminating spectrum	Reference
a-Si	Ideal QDs	Ideal plastic	No improvement	AM1.5G	[Sark et al., 2004]
c-Si	Silicon QDs	SiO_2	No improvement	AM1.5G	[Jestin et al., 2010]
c-Si	Silicon QDs	Spin-on Glass	+0.4	AM1.5	[Svrcek et al., 2004]
c-Si	Silicon QDs	Organic solvent	+0.6	-	[Pi et al., 2011]
Poly-Si	Silicon QDs	Directly on cell	+10	Visible light	[Stupca et al. 2007]

Table 1. Performance results of Silicon QDs based luminescent down-shifters

Svrcek et al observe an efficiency increase of about 0.4%. In this case the Silicon QDs were prepared ex-situ (pulverizing of electrochemical etched porous silicon) embedded into a spin-on-glass anti-reflecting SiO_2 based solution and then onto a standard silicon solar cell. They claim that an improvement of the Internal Quantum Efficiency (IQE) has been detected in the region 300-500 nm. The difference observed in the solar cell performance points out the difficulty to compare the efficiency of silicon QDs in different conditions. Indeed Pi et al. who report an increase in efficiency of 0.6% claim that the efficiency improvement observed by Svreck et al may have been due to the improved optical coupling of the light into the cell by Silicon QDs embedded in the spin-on glass at the solar cell surface (Pi et al., 2011). The authors claim that it is the porous nature of the deposited film which could be at the origin of the observed efficiency increase (porous silicon layers are well known as anti-reflecting coating (Chaoui et al., 2008)) rather than a down-conversion effect of silicon QDs. In their paper, Pi et al propose to deposit a silicon QDs based ink at the solar cell surface by spin-coating. Silicon QDs are produced using a new synthesis approach (Mangolini et al., 2005) and dispersed in an organic solvent. In this case, an enhancement of 2.3% in short circuit current (Isc) is observed, and thus consistent with the fact that the Silicon QDs film induce an enhancement of light absorption and then an increase in the cell efficiency. Finally Stupca et al. (Stupca et al., 2007), deposited Si-QDs dispersed in a volatile solvent directly on a industrial polycrystalline solar cell. The Si-QDs used showed depending on size either strong blue (1nm diameter) or orange/red (2.85 nm) photoluminescence. A huge improvement of up to 70% was reported shining blue light on the cells (in this spectral region the IQE of the cell is poor) and an still notable efficiency improvement of 10% was obtained for visible light. Current results on Si-QDs as down shifters allow to envision that improvements of some percents in efficiency can be expected from this approach within the next years.

3.2 Hot carrier solar cells and multiple exciton generation in Silicon QDs

Here we address two concept which try to avoid the losses due to thermalisation of hot carriers. One is the so called hot carrier solar cells (HCSC), which aims on the extraction of hot carrier before they relax to the bandedges, the other is based on the generation of multiple excitons (MEG) from high energy excitons observed in a series of QDs.

Hot carrier solar cells (HCSC) presented in Figure 10 have been proposed in early 80′ as a promising approach to increase the efficiency of conventional solar cells (Ross & Nozik, 1982 ; Takeda et al., 2009; Würfel et al., 2005). The thermalisation of photoexcited involving lattice phonons reduces significantly the number of carriers which can be effectively separated through the p-n junction, thus, preventing to achieve high voltages in a photovoltaic cell. A possibility to reduce the thermalisation time, which is known to be the main loss mechanism in photovoltaic cells, would allow collecting more carriers when these are still excited to higher energetic states (hot carriers) and therefore increase the cell voltage. Generally speaking, in bulk semiconductor-based solar cells, the damping of the thermalisation rate is achieved at very high intensities (order of 1000 suns, see for example Ref. (Würfel, 1997)). On the other hand, in low dimensional nanostructures, such as 1D superlatticies, the larger (artificial) translation period with respect to the lattice period of an ordinary crystal results in a much smaller Brillouin zone and, as a consequence, in very small number of acoustic phonon modes, preventing fast thermalisation of excited carriers (an enhancement of the phonon bottleneck effect) (Nozik et al., 1990; Westland et al., 1988). In the last years 0D systems, quantum dots and semiconductor nanocrystals, have attracted much interest. As already discussed above, these material systems have the advantage to improve the efficiency of the absorbing layer of the photovoltaic cell. In case of HCSC, a quantum dot system can have an important role also in realizing energy selective contacts (Conibeer et al., 2006; Conibeer et al., 2008). The basic concept standing behind this idea consists in increasing the efficiency of an HCSC device by collecting the photogenerated in the absorbing layer hot carriers over a very small energy range with selective energy contacts (SEC). In Ref. (Conibeer et al., 2006) Conibeer and co-workers introduce the quantum dot approach for SEC where a double-barrier resonant tunnelling structure is described (here a QD-based layer with a discrete energy level is embedded between two insulating layers). An electrode with such a configuration should result in strongly peaked conduction at a discrete energy level (Figure 10).

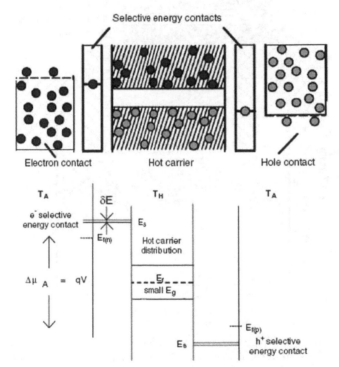

Fig. 10. The schematic of a hot carrier solar cell with energy selective electrodes. Reprinted from Thin Solid Films, Vol.511-512, Conibeer, G. et al., Films Silicon nanostructures for third generation photovoltaic solar cells, pp. 654-662, Copyright (2006), with permission from Elsevier. (Conibeer et al., 2006).

Proof-of-concept experiments have been realized by using silicon nanocrystals (Si-ncs) embedded in a silica matrix as double-barrier electrodes (Conibeer et al., 2008). The double barrier resonant tunneling structure used in this work comprises in a 4 nm thick Si-nc layer sandwiched between two 5 nm SiO$_2$ barrier layers. Si-ncs have been formed through a high temperature treatment of the multilayer structure, which is known to result in a phase separation of Si and SiO₂ inside the silicon rich silica layer. The layer thickness controls the size of the Si-ncs which in its turn is providing the specific discrete energy levels inside the double tunnelling barrier. The significant result in this study is the observation of room temperature negative differential resistance (NDR) feature in the I-V curve of the electrode (Figure 11). Despite the weak NDR resonance observed in these experiments, the results, as an evidence for 1D energy selection, are very encouraging. Recently, HCSC devices with Si-nc-based energy selective electrodes have been studied in much detail addressing the growth, compositional, structural and spectroscopic properties of the nanocrystalline system (Aliberti et al., 2010; Shrestha et al., 2010).

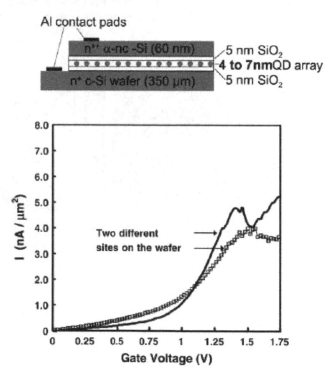

Fig. 11. (top) The sketch of SEC experiments and I-V curves at room temperature showing the NDR features. Reprinted from Thin Solid Films, Vol. 516, No.20, Conibeer, G. et al., Selective energy contacts for hot carrier solar cells, pp. 6968-6973, Copyright (2008), with permission from elsevier. (Conibeer et al., 2008).

The concept of using Si-nc-based energy selective electrodes proved to offer an important potential to improve the efficiency of HCSC photovoltaic cells, however, further work on significantly improved the NDR resonance quality and characteristics is expected in the next years.

Multiple exciton generation has been investigated and demonstrated by Beard et al in different types of semiconductor QDs (PbS, PbSe, PbTe, CdS, CdSe, InAs, and Si) (Beard et al., 2007). In a conventional photovoltaic semiconductor device, the absorption of a photon usually results in the production of a single electron-hole pair; energy of a photon in excess of the semiconductor's bandgap is efficiently converted to heat through electron and hole interactions with the crystal lattice. Recently, colloidal semiconductor QDs and QDs-based films have been shown to exhibit efficient multiple electron-hole pair generation from a single photon with energy greater than twice the effective band gap. This multiple carrier pair process, referred to as multiple exciton generation (MEG), represents one route to reducing the thermal loss in semiconductor solar cells. In the case of Silicon QDs, Beard et al. investigated carrier multiplication using femtosecond transient absorption sectroscopy to study three different sizes of colloidal Si-QDs suspended in a nonpolar solvent (Beard et al., 2007). The authors found that MEG occurs with considerably higher efficiency in Si-QDs as compared to bulk Si, showing a threshold in 9.5 nm diameter QDs (effective band gap Eg = 1.20 eV) to be 2.4 ± 0.1Eg, and this same size QD showed an exciton quantum yield of 2.6 ± 0.2 excitons per absorbed photon at 3.4Eg. The measurements were the first reported occurrence of MEG in an indirect-gap semiconductor. The authors also investigated larger QDs and found that even for relatively small confinement energy of 80 meV, high photon energies produced efficient MEG. To this point for usefull photovoltaics applications involving Si-QDs, the QDs must be coupled electronically to their environment. Solar energy applications attempting to harness the MEG-created electron-hole pairs there are several requirements that any approach must justify; (i) The QDs must be the active component, (ii) efficient transport of electrons and holes, or excitons, must occur over macroscopic distances and (iii) the coupling can not be so strong as to reduce the MEG efficiency yet must be strong enough to separate the electron-hole pairs either by undergoing charge transfer or diffusing one exciton to a nearby QD that does not have an exciton prior to non-radiative Auger recombination. Interestingly, the indirect nature of the silicon bandgap, which is preserved in Si-QDs, and which is detrimental to photonic applications, turns out to be beneficial here, as the relatively long exciton lifetime (in comparison with direct-badgap QDs) simplifies energy extraction for photovoltaic applications.

3.3 Multilayer tandem solar cells

Tandem solar cells are meant to absorb a major part of the solar spectrum and hence to increase the efficiency of the device by increasing the electric current. The approach is to fabricate p-n, p-i-n or any other diode structure with different band gap semiconductors and connect them to enhance the photovoltaic conversion. The fundamental limit of the performance of tandem structures has been studied by Meillaud et al (Meillaud et al., 2006). The radiative efficiency limit for a single junction silicon cell is 30%. This increases to 42.5% and 47.5% for 2-cell and 3-cell tandem stacks respectively. For an AM1.5G solar spectrum the optimal band gap of the top cell required to maximize conversion efficiency is ~ 1.7 to 1.8 eV for a 2-cell tandem with a Si bottom cell and 1.5 eV and 2.0 eV for the middle and upper cells for a 3- cell tandem. As shown in Figure 12, the first cell located at the front is fabricated with a wide band-gap material to convert high-energy photons from the blue-end whereas the third cell located at the back is fabricated with a narrow band-gap material to convert low-energy photons from the infra-red end. With this approach, the next step is to connect the cells together to form one device. Two types of connection can be envisioned: in series or in parallel. In series, devices are connected using a tunnel junction, and the n-type material of one device is connected to the p-type material of the adjacent device i.e. conduction band and valence band are connected to each others. In parallel, the material changes its conduction gradually from n to p.

With the aim of fabricating an "all-Si" tandem solar cell for third generation photovoltaics, silicon QDs in combination of an inexpensive silicon thin-film technology may allow the fabrication of higher band gap solar cells. Taking the advantage of quantum confinement effects in silicon, a band gap engineering can be achieved by optimizing the silicon QDs size. Indeed when silicon is made very thin (in the order of a few nanometers) in one or more dimensions, quantum confinement causes its effective band gap to increase. The strongest effect is obtained when silicon is confined in 3D (i.e., quantum dots). If the quantum dots are close to each other, carriers can tunnel between them to form quantum dot superlattices, which can be used as the higher band gap cells in a tandem stack.

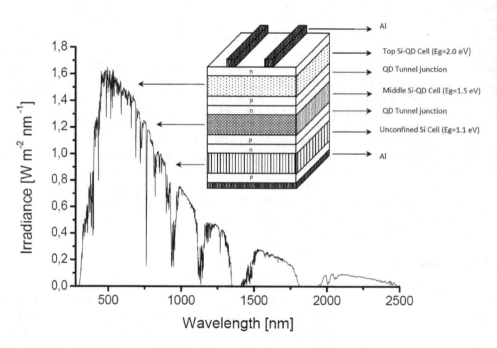

Fig. 12. Solar AM1.5G spectrum and arrangement of three different p-n junctions to utilize photons in the three regions. The cell materials have larger band gap in the cell located in the top than the cell located on the bottom.

3.3.1 Engineering of the silicon QDs stack

In order to obtain crystalline QDs of controlled size and distance, which allows also electrical carrier transport, layer thickness has to be controlled on nanometer scale. Quantum confined nanostructures of silicon with barriers of SiO_2, Si_3N_4 or SiC can potentially fulfil these criteria. Hence common thin film techniques deposition like PECVD are used to fabricate silicon QDs stacks. This multilayer deposition typically consist of the repetition of a tens of bi-layers of silicon rich dielectric and stochiometric dielectric. The deposition is followed by annealing in N_2 from 1050 to 1150°C inducing the precipitation of silicon nanocrystals or QDs. The process is depicted in Figure 13 together with TEM images of superlattices and silicon QDs. Similarly to a traditional p-n junction, a QD p-n junction is required to separate the photon-generated electron-hole pairs. This junction may be achieved by doping the silicon QDs with impurity atoms to form a nano-scale p-n junction. From a fabrication point of view, the introduction of a few impurity atoms into a QD that contains only a few hundred atoms may lead to their expulsion to the surface or compromise the crystal structure. This will inherently create a heavily doped QD under strong quantum confinement.

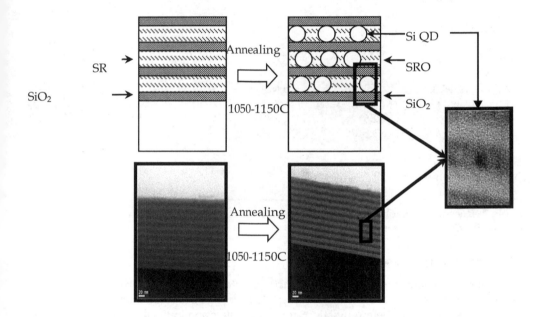

Fig. 13. Multilayer structure illustrating precipitation of silicon QDs in a Si-rich layer.

The electronic and optical properties in such circumstances are still unresolved. Thus, QDs doping still remains a critical step for tailoring their properties for specific applications in solar cells. To this date, no realization of an all silicon tandem cell has been done, nevertheless, important research in this field has lead to the development of a method to dope semiconductor nanocrystals with metal impurities, enabling control of the band gap and Fermi energy. In this case, a combination of optical measurements, scanning tunneling spectroscopy, and theory revealed the emergence of a confined impurity band and band-tailing (Mocata et al., 2011). This method yields n- and p-doped semiconductor nanocrystals, which have potential applications in all silicon tandem solar cells. In Table 2 recent results obtained on silicon solar cells with QDs as active material are summarized.

Solar cell configuration	Voc [mV]	Jsc [mA/cm²]	FF [%]	Efficiency [%]	
Doped Si-QD p-i-n homojunction diodes on quartz	493 (Si-QD:SiO₂) 83 (Si-QD:SiC)	0.003 mA (SiQd:SiO₂)*	Not determined	Not determined	[Lima, 2011]
Doped Si-QD:SiO₂ (n- or p-type) heterojunction	556 (n) 613 (p)	29.8 (n) 32.0 (p)	67.1 (n) 73.1 (p)	11.1 (n) 14.3 (p)	[Lima, 2011]
Doped Si-QD:SiC (p-type) heterojunction on c-Si	463	19.0	53.0	4.66	[Song et al., 2008]
Doped Si-QD (p-type) heterojunction on c-Si	540	28	66	9.5	[Hong et al., 2010]
Undoped Si-QD on quartz with poly-Si electrodes	220 (Si-QW:SiO₂) 286 (Si-QD:Si₃N₄)	0.6 (Si-QW:SiO₂) 0.0012 (Si-QD:Si₃N₄)	30 (Si-QW:SiO₂)	0.041 (Si-QW:SiO₂)	[Lima, 2011]
Undoped Si-QD:SiON heterojunction diodes on c-Si	485	12.5	11.4	0.69	[Prezioso et al., 2009]

Table 2. Recent results obtained on silicon solar cells with Si-QDs as active material *I not J as area is too small to define

3.4 Inorganic/organic solar cells

First reports on silicon/organic hetero-junctions date back to 1982 (Forrest et al., 1982). At this time, devices were based on small organic molecules evaporated on crystalline silicon wafers. The use of such hetero-junctions for photovoltaic applications has been first reported with pyrene as the organic compound (Mccaffrey & Prasad, 1984), thus opening the road to silicon/organic based solar cells. One can distinguish two type of organic solar cells: (i) the organic bulk hetero-junction (OBHJ) solar cell (Park et al., 2009) based on a hole conductor and an electron acceptor material. The mixture poly(3-hexylthiophene): 1-(3-methoxycarbonyl)propyl-1-phenyl[6,6]C61 (P3HT:PCBM) is currently the most prominent material system in organic photovoltaic, (ii) the dye sensitized solar cell (DSSC) (ORegan & Grätzel, 1991), based on a semiconductor formed between a photo-sensitized anode and an electrolyte. The introduction of silicon QDs in such devices finds its interest in increasing the electro-optical properties of the cell and thus increasing the external quantum efficiency of the device. In order to understand the behavior of silicon QDs with conducting polymer, Dietmueller et al (Dietmueller et al., 2009) have reported the charge transfer between silicon QDs and the component of a bulk hetero-junction solar cells (P3HT and PCBM). In figure 14 the conduction band (CB) and the valence band (VB) of silicon as well as the lowest

unoccupied molecular orbital (LUMO) and the highest occupied molecular orbital HOMO
of P3HT are displayed based on literature values.

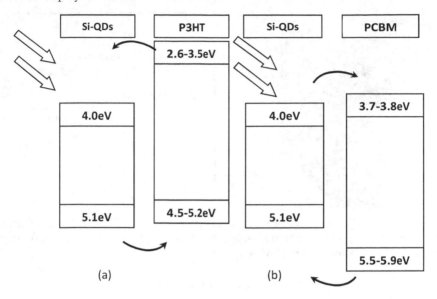

Fig. 14. Schematic view of the energy levels of Silicon-QDs and P3HT (a) and of Si-nc and
PCBM (b). The curved arrows indicate the transfer direction of electrons and holes.

The LUMO of P3HT lies clearly above the CB of Si (Figure 14 a), enabling dissociation of
excitons followed by an electron transfer to the Si CB. However, taking into account the
different literature values of the HOMO of P3HT (4.7–5.2 eV) (Al-Ibrahim et al., 2005) and
possible influences on the band alignment, such as Fermi level pinning or interface
dipoles, it is a priori not clear if a hole transfer from the Si VB to the P3HT HOMO can
take place. For the photogeneration of the positive polarons of P3HT (P+) in the Si-
nc/P3HT system, one can consider different scenarios: After the light induced generation
of excitons in P3HT, the excitons dissociate by electron transfer to Silicon QDs, while the
holes remain in the P3HT. Alternatively, an energy transfer might occur, whereby
excitons are transferred from P3HT to the Silicon QDs, followed by a back transfer of the
holes (Gowrishankar et al., 2006). A third possibility is that, after light absorption in the
Silicon, the holes are transferred from the Silicon QDs to the P3HT. In any case, for all the
processes it is required that the HOMO of P3HT lies energetically above the Si-QDs VB.
As we indeed do detect light induced P+ in the P3HT, we can conclude that this
requirement is fulfilled, which is a prerequisite for solar cells based on P3HT and Si-QDs.
The analogous argument holds for the alignment of the LUMO (3.7–3.8 eV) (Al-Ibrahim et
al., 2005) of PCBM and the CB of Silicon QDs (Figure 14 b) where we detect PCBM⁻ anions
in the Silicon QDs/PCBM composite during illumination. Thus the LUMO of PCBM must
be energetically lower than the Silicon QDs CB. Despite the fact that PCBM, in contrast to
P3HT, is not a good absorber for the solar spectrum, the Silicon QDs/PCBM material
combination could in principle be used for solar cell applications as well.

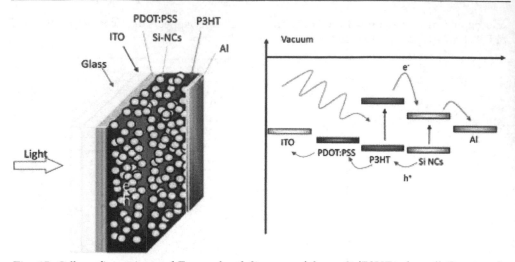

Fig. 15. Cell configuration and Energy level diagram of the nc-Si/P3HT solar cell. Reprinted with permission from (Liu et al., Nano Letters, Vol.9, No.1, (2009), pp. 449–452. Copyright (2009) American Chemical Society (Liu et al., 2009).

3.4.1 Performance of Silicon QDs in organic solar cells

Experimental data on the system silicon QDs/P3HT have been reported by Liu et al (Liu et al., 2009). They reported new hybrid solar cells based on blends of Silicon QDs and P3HT polymer in which a percolating network of the nanocrystals acts as the electron-conducting phase. A schematic representation of the organic bulk hetero-junction solar cell with all the energy level is depicted in Figure 15. The authors found that the open-circuit voltage and short-circuit current are observed to depend on the Silicon QDs size due to changes in the bandgap and surface-area-to-volume ratio. Under simulated one-sun A.M. 1.5D direct illumination (100 mW/cm²), devices made with 35 wt % with silicon QDs of 3−5 nm in size showed 1.15% power conversion efficiency.

Organic phase	Performance [%]	Illuminating spectrum	Reference
P3HT	1.15	AM1.5D	[Liu et al., 2009]
MDMO-PPV	0.49	AM1.5D	[Liu, 2009]
C$_{60}$	1.4 (under 400nm)	-	[Svreck et al., 2010]

Table 3. performances of Silicon QDs Hybrid cells.

Using the same solar cell configuration, Liu reported also some experiments on a cell consisting of 3-5 nm silicon QDs and poly [2-methoxy-5-(3',7'-dimethyloctyloxy)-1,4-phenylenevinylene] (MDMO-PPV) and fabricated from solution processes. The hybrid cells of 58wt% 3-5 nm Si NCs/MDMO-PPV showed a power conversion efficiency of 0.49% under simulated one sun A.M. 1.5D direct illumination (100 mW/cm2) (Liu, 2009). In comparison with the previous results based on P3HT cells, the low efficiency of the Silicon

QDs/MDMO-PPV may result from the lower hole mobility and narrow absorption spectrum of MDMO-PPV. The energy band alignment of Silicon QDs and MDMO-PPV may also be one of the factors limiting device efficiency. Using a different approach, Svreck et al (Svrcek et al., 2010) reported experimental data on Silicon QD/Fullerene (C_{60}) solar cells. They first demonstrate that Silicon QDs can be electronically coupled with fullerenes (C_{60}) without any additional surfactant or catalyst. The fabrication of an Hetero-junction solar cells made out of Silicon QDs and C_{60} show photovoltaic action with increased quantum efficiency in the region where the absorption of Silicon QDs appears. The cell shows and External Quantum Efficiency (EQE) over 1.4% at high energies (≥ 400 nm) and a power conversion efficiency up to 0.06% in the spectral region 550-700 nm. In Table 3 are resumed the performances of Si-QDs Hybrid cells reported in the literature.

4. Conclusions

In this review we presented recent results on the application and research on Si-QDs related to photovoltaics. This research is strongly related to future demands in electricity generation. Fascinating scientific results were shown in the field of hot-carrier cells and multi-exciton generation. It is far too early to understand if these phenomena and concepts will play an important role from a technological point of view, but deep scientific investigation of the underlaying processes will be necessary. However simpler concepts like down-shifting are on the way of being applyied on existing solar technology. In addition a series of studies on QDs in heterojunction solar cells showed cell efficiencies in the range of 5-10%. It is important to note that Si-QDs can also play a role in organic based solar cells development. Even if this research is still in an initial phase, encouraging results have been demonstrated. Although all these cells are still far away from outperforming best c-Si cells they are in an efficiency range which justifies deep investigation. A critical issue will be to achieve similar and even larger effiencies on QD solar cells which are not based on a crystalline silicon substrate.

5. Acknowledgment

Authors acknowledge financial support throught the European Commison by the project LIMA (Improve Photovoltaic efficiency by applying novel effects at the limis of light to matter interaction (FP7-ICT-2009.3.8-248909)).

6. References

Aliberti, P.; Shrestha, S.K.; Teuscher, R.; Zhang, B.; Green, M.A. & Conibeer, G.J. Study of silicon quantum dots in a SiO2 matrix for energy selective contacts applications. *Solar Energy Materials and Solar Cells*, Vol.94, No.11, (November 2010), pp. 1936-1941, ISSN 0927-0248.

Alivisatos, A.P. Perspectives on the Physical Chemistry of Semiconductor Nanocrystals. *Journal of Physical Chemistry*, Vol.100, No.31, (August 1996), pp. 13226–13239, ISSN 0047-2689.

Al-Ibrahim, M.; Roth, H.-K.; Schroedner, M.; Konkin, A.; Zhokhavets, A.; Gobsch, G.; Scharff, P. & Sensfuss, S. The influence of the optoelectronic properties of poly(3-

alkylthiophenes) on the device parameters in flexible polymer solar cells. *Organic Electronics*, Vol.6, No.2, (April 2005), pp. 65-77, ISSN 1566-1199.

Amato, G.; Delerue, C. & VonBardeleben, H.J. (1997). *Structural and optical properties of porous silicon nanostructures*, (Optoelectronic properties of semiconductors and superlattices) Gordon and Breach Science Publishers, ISBN 90-5699-604-5, Amsterdam.

Anopchenko, A.; Marconi, A.; Moser, E.; Prezioso, S.; Wang, M.; Pavesi, L.; Pucker, G. & Bellutti, P. Low-voltage onset of electroluminescence in nanocrystalline-Si/SiO2 multilayers. *Journal of applied physics*, Vol.106, No.3, (August 2009), pp. 033104-033112, ISSN 1089-7550.

I. Balberg, E. Savir, J. Jedrzejewski, A. G. Nassiopoulou and S. Gardelis. Fundamental transport processes in ensembles of silicon quantum dots. Physical review B, Vol.75, No.23, (June 2007), pp. 235329-235337, ISSN 1550-235X.

Becquerel, E. (1867) *La lumière, ses causes, ses effets, tome II*, Firmin diderot frères, ISBN 88-11-63036-3, Paris.

Bisi, O.; Osscini, S. & Pavesi, L. Porous silicon: a quantum sponge structure for silicon based optoelectronics. *Surface Science Reports*, Vol.38, No.1-3, (April 2000), pp. 1-126, ISSN 0167-5729.

Brabec, C.; Dyakonov, V. & Scherf, U. (2008). *Organic Photovoltaics – Materials, Devices, and Manufacturing Technologies*, Wiley-VCH Verlag Gmbh & Co. KGaA, ISBN 978-3-527-31675-5, Weinheim, Germany.

Brongersma, M. L.; Kik, P.G.; Polman, A.; Min, K.S. & Atwater, H.A. Size-dependent electron hole exchange interaction in Si nanocrystals. *Applied Physics Letters*, Vol. 76, No.3, (January 2000), pp. 351-1-351-3, ISSN 1077-3118.

Bruchez, M.; Moronne, M.; Gin, P.; Weiss, S. & Alivisatos, A.P. Semiconductor Nanocrystals as Fluorescent Biological Labels. *Science*, Vol.281, No. 5385, (September 1998), pp. 2013-2016, ISSN 1095-9203.

Buhbut, S.; Itzhakov, S.; Tauber, E.; Shalom, M.; Hod, I.; Geiger, T.; Garini, Y.; Oron, D. & Zaban, A. Built-in Quantum Dot Antennas in Dye-Sensitized Solar Cells. *ACS Nano*, Vol.4, No.3, (March 2010), pp. 1293-1298, ISSN 1936-086X.

Calcott, P.D.J.; Nash, K.J.; Canham, L.T.; Kane, M.J. & Brumhead, D. Identification of radiative transitions in highly porous silicon. *Journal of physics: Condensed Matter*, Vol.5, No.7, (February 1993), pp.91-98, ISSN 1361-648X.

Canham, L.T. Silicon quantum wire array fabrication by electrochemical and chemical dissolution of wafers. *Applied Physics Letters*, Vol.57, No.10, (September 1990), pp. 1046-1-1046-3, ISSN 0003-6951.

Canham, L.T (2007). *Properties of porous silicon*, (Leigh Canham), Institution of Engineering and Technology, ISBN 0863415555, 9780863415555), London.

Chaoui, R.; Mahmoudi, B. & Si Ahmed, Y. Porous silicon antireflection layer for solar cells using metal-assisted chemical etching. *Physica Status Solidi A*, Vol.205, No.7, pp. 1724–1728, (July 2008), ISSN 1862-6319.

Chapin, D.M.; Fuller, C.S. & Pearson, G.S. A new silicon p-n junction photocell for converting solar radiation into electrical power. *Journal of Applied Physics*, Vol.25, No.5, (May 1954), pp. 676- ISSN 1089-7550.

Chelikowsky, J.R. & Cohen, M.L. Nonlocal pseudopotential calculations for the electronic structure of eleven diamond and zinc-blende semiconductors. *Physical Review B*, Vol.14, No.2, (July 1976) pp. 556-582, ISSN 1550-235X.

Conibeer, G.; Jiang, C.W.; König, D.; Shrestha, S.; Walsh, T. & Green, M.A. Selective energy contacts for hot carrier solar cells. *Thin Solid Films*, Vol.516, No.20, (August 2008), pp. 6968-6973, ISSN 0040-6090.

Conibeer, G.; Green, M.A.; Corkish, R.; Cho, Y.; Cho, E.C.; Jiang, C.W.; Fangsuwannarak, T.; Pink, E.; Huang, Y.; Puzzer, T.; Trupke, T.; Richards, B.S.; Shalav, A. & Lin, K.L. Silicon nanostructures for third generation photovoltaic solar cells. *Thin Solid Films*, Vol.511-512, (July 2006), pp. 654-662, ISSN 0040-6090.

Conibeer, G.; Green, M.A.; Cho, E.C.; König, D.; Cho, Y.H.; Fangsuwannarak, T.; Scardera, G.; Pink, E.; Huang, Y.; Puzzer, T.; Huang, S.; Song, D.; Flynn, C.; Park, S.; Hao, X. & Mansfield, D. Silicon quantum dot nanostructures for tandem photovoltaic cells. *Thin Solid Films*, Vol.516, No.20, pp. 6748-6756, (August 2008), ISSN 0040-6090.

Cullis, A.C. & Canham, L.T. Visible light emission due to quantum size effects in highly porous crystalline silicon. *Nature*, Vol.353, No.6342, (September 1991), pp. 335-338, ISSN 1476-4687.

Cullis, A.C., Canham, L.T. & Calcott, J.P.D. The structural and luminescence properties of porous silicon. *Applied Physics Reviews*, Vol.82, No.3, (August 1997), pp. 909-965, ISSN 0021-8979.

Daldosso, N.; Luppi, M.; Ossicini, S.; Degoli, E.; Magri, R.; Dalba, G.; Fornasini, P.; Grisenti, R.; Rocca, F.; Pavesi, L.; Boninelli, S.; Priolo, F.; Spinella, C. & Iacona, F. Role of the interface region on the optoelectronic properties of silicon nanocrystals embedded in SiO_2. *Physical Review B*, Vol.68, No.8, (August 2003), pp. 085327-085335, ISSN 1550-235X.

Dal Negro, L.; Cazzanelli, M.; Pavesi, L.; Ossicini, S.; Pacifici, D.; Franzò, G.; Priolo, F. & Iacona, F. Dynamics of stimulated emission in silicon nanocrystals. *Applied Physics Letters*, Vol.82, No.26, (June 2003), pp. 4636-4639, ISSN 1077-3118.

Del Cañizo, C.; Del Coso, G. & Sinke, W.C. Crystalline silicon solar module technology: Towards the 1€ per watt-peak goal. *Progress in Photovoltaics: Research and Applications*, Vol.17, No.3, (May 2009), pp. 199-209, ISSN 1099-159X.

Di, D.; Perez-Wurfl, I.; Conibeer, G. & Green, M.A. Formation and photoluminescence of Si quantum dots in $SiO2/Si3N4$ hybrid matrix for all-Si tandem solar cells. *Solar Energy Materials and Solar Cells*, Vol.94, No.12, (December 2010), pp. 2238-2243, ISSN 0927-0248.

Dietmueller, R.; Stegner, A.R.; Lechner, R.; Niesar, S.; Pereira, R.N.; Brandt, M.S.; Ebbers, A.; Trocha, M.; Wiggers, H. & Stutzmann, M. Light-induced charge transfer in hybrid composites of organic semiconductors and silicon nanocrystals. *Applied Physics. Letters*. Vol.94, No.11, (March 2009), pp. 113301, ISSN 1077-3118.

DiMaria, D.J.; Dong, D.W.; Pesavento, F.L.; Lam, C. & Brorson, S. D. Enhanced conduction andminimized charge trapping in electrically alterable read-only memories using off stoichiometric silicon dioxide films. *Journal of Applied Physics*, Vol.55, No.8, (April 1984), pp. 3000-3020, ISSN 1089-7550.

DiMaria, D.J.; Kirtley, J.R.; Pakulis, E.J.; Dong, D.W.; Kuan, T.S.; Pesavento, F.L.; Theis, T.N.; Cutro, J.A. & Brorson, S.D. Electroluminescence studies in silicon dioxide films containing tiny silicon islands. *Journal of Applied Physics*, Vol.56, No.2, (July 1984), pp. 401-417, ISSN 1089-7550.

Dovrat, M.; Goshen, Y.; Jedrzejewski, J.; Balberg, I. & Sa'ar, A. Radiative versus nonradiative decay processes in silicon nanocrystals probed by time resolved photoluminescence spectroscopy. *Physical Review B*, Vol.69, No.15, (April 2004), pp. 155311-155319, 1550-235X.

European Photovoltaic Technology Platform. (2007). A strategic research agenda for photovoltaic solar energy technology, ISBN 978-92-79-05523-2, date of access: 10/08/11, Available from: www.eupvplatform.org.

Feltrin, A. & Freundlich, A. Material considerations for terawatt level deployment of photovoltaics. *Renewable Energy*, Vol.33, No.2, (February 2008), pp. 180-185, ISSN 0960-1481.

Feng, J.C. & Tsu, R. (1994). *Porous silicon* (Tsu, R.), World Scientific, ISBN 981-02-1634-3, Singapore.

Forrest, S. R.; Kaplan, M. L.; Schmidt, P. H.; Feldmann, W. L. & Yanowski, E. Organic-on-inorganic semiconductor contact barrier devices. *Applied Physics Letters*, Vol.41, No.1, (July 1982), pp. 90–93, ISSN 0003-6951.

Franzò, G.; Irrera, A.; Moreira, E.C.; Miritello, M.; Iacona, F.; Sanfilippo, D.; Di Stefano, G.; Fallica, P.G. & Priolo, F. Electroluminescence of silicon nanocrystals in MOS structures. Applied Physics A, Vol.74, No.1, (January 2002) pp. 1-5, ISSN 1432-0630.

Garrido Fernandez, B.; Lopez, M.; García, C.; Pérez-Rodríguez, A.; Morante, J.R.; Bonafos, C.; Carrada, M. & Claverie, A. Influence of average size and interface passivation on the spectral emission of Si nanocrystals embedded in SiO_2. *Journal of Applied Physics*, Vol.91, No.2, (January 2002), pp. 798-808, ISSN 1089-7550.

Godefroo, S.; Hayne, M.; Jivanescu, M.; Stesmans, A.; Zacharias, M.; Lebedev, O.I.; Van Tendeloo, G. & Moshchalkov, V.V. Classification and control of the origin of photoluminescence from Si nanocrystals. *Nature Nanotechnology*, Vol.3, No.3 (March 2008), pp. 174-178, ISSN 1748-3387.

Gowrishankar, V; Scully, S. R.; McGehee, M.D.; Wang, Q. & Branz, H. M. Exciton Splitting and Carrier Transport across the Amorphous-Silicon/Polymer Solar Cell Interface. *Applied Physics Letters*, Vol.89, No.25, (December 2006), pp. 252102, ISSN 1077-3118.

Green, M.A.; Emery, K.; Hishikawa, Y. & Warta, W. Solar cell efficiency tables (version 36). *Progress in Photovoltaics: Research and Applications*, Vol.18, No.5, (August 2010), pp. 346-352, ISSN 1099-159X.

Hao, X.J.; Cho, E.C.; Scardera, G.; Shen, Y.S.; Bellet-Amalric, E.; Bellet, D.; Conibeer, G. & Green, M.A. Phosphorus-doped silicon quantum dots for all-silicon quantum dot tandem solar cells. *Solar Energy Materials and Solar Cells*, Vol.93, No.9, (September 2009), pp. 1524-1530, ISSN 0927-0248.

Hirshman, W.P.; Hering, G. & Schmela, M. Gigawatts–the measure of things to come. *Photon International*, Vol.3, No.12, (March 2007), pp. 136-166.

Hong, S.H.; Park, J.H.; Shin, D.H.; Kim, C.O.; Choi, S.H. Doping and size dependent photovoltaic properties of p-type Si-Quantum dot heterojunction solar cell: correlation with photoluminescence. *Applied Physics Letters*, Vol.97, No.7, (August 2010), pp. 072108-1-072108-3, ISSN

Jestin, Y.; Pucker, G.; Ghulinyan, M.; Ferrario, L.; Bellutti, P.; Picciotto, A.; Collini, A.; Marconi, A.; Anopchenko, A.; Yuan, Z. & Pavesi, L. Silicon solar cells with nano-crystalline silicon down shifter: experiment and modeling. *Proceeding SPIE optics and photonics*, pp. 77721-77727, ISBN 9780819483119, San Diego, USA, August 2-5, 2010.

Kanemitsu, Y. Luminescence properties of nanometer-sized Si crystallites: Core and surface states. *Physical Review B*, Vol.49, No.23, (June 1994), pp. 16845-16848, ISSN 1550-235X.

Kanemitsu, Y.; Okamoto, S.; Otobe M. & Oda, S. Photoluminescence mechanism in surface-oxidized silicon nanocrystals. *Physical review B*, Vol.55, No.12, (March 1997), pp. R7375-R7378, ISSN 1550-235X.

Kelzenberg, M.D.; Boettcher, S.W.; Petykiewicz, J.A.; Turner-Evans, D.B.; Putnam, M.C.; Warren, E.L.; Spurgeon, J.M.; Briggs, R.M.; Lewis, N.S.; Atwater, H.A. Enhanced absorption and carrier collection in Si wire arrays for photovoltaic applications. Nature Materials, Vol.9. No.3, (March 2010), pp. 239-244, ISSN 1476-1122.

Khriachtchev, L.; Kilpelä, O.; Karirinne, S.; Keränen, J. & Lepistö, T. substrate-dependent crystallization and enhancement of visible photoluminescence in thermal annealing of Si/SiO₂ superlattices. *Applied Physics Letters*, Vol.78, No.3, (January 2001), pp. 323-326, ISSN 1077-3118.

Klampaftis, E.; Ross, D.; McIntosh, K.R. & Richards, B.S. Enhancing the performance of solar cells via luminescent down-shifting of the incident spectrum: A review. *Solar Energy Materials and Solar Cells*, Vol.93, No.8, (August 2009), pp. 345-351, ISSN 0927-0248.

Kovalev, D.; Heckler, H.; Ben-Chorin, M.; Polisski, G.; Schwartzkopff, M. & Koch, F. Breakdown of the k-conservation rule in Si nanocrystals. *Physical Review letters*, Vol.81, No.13, (September 1998), pp. 2803-2806, ISSN 1079-7114.

Le Donne, A.; Acciarri, M.; Narducci, D.; Marchionna, S. & Binetti, S. Encapsulating Eu3+ complex doped layers to improve Si-based solar cell efficiency. *Progress in Photovoltaics*, Vol.17, No.8, (December 2009), pp. 519–525, ISSN 1099-159X.

Le Donne, A.; Dilda, M.; Crippa, M.; Acciarri, M. & Binetti, S. Rare earth organic complexes as down-shifters to improve Si-based solar cell efficiency. *Optical Materials*, Vol.33, No.7, (May 2011), pp. 1012-1014, ISSN 0925-3467.

Lehmann, V. & Gosele, U. Porous silicon formation: A quantum wire effect. *Applied Physics Letters*, Vol. 58, No.8 (February 1991), pp. 856-1-856-3, ISSN 0003-6951.

Lima, European project Improve Photovoltaic efficiency by applying novel effects at the limis of light to matter interaction (FP7-ICT-2009.3.8-248909) internal report 2011.

Linnros, J.; Lalic, N.; Galeckas, A. & Grivckas, V. Analysis of the stretched exponential photoluminescence decay from nanometer-sized silicon nanocrystals. *Journal of Applied Physics*, Vol.86, No.11, (December 1999), pp. 6128-6136, ISSN 1089-7550.

Liu, C.Y; Holman, Z.C. & Kortshagen, U.R. Hybrid Solar Cells from P3HT and Silicon Nanocrystals. *Nano Letters*, Vol.9, No.1, (January 2009), pp. 449–452, ISSN 1530-6992.

Liu, C.Y. Hybrid Solar Cells from Polymers and Silicon Nanocrystals. PhD Tesis, University of Minnesota, (December 2009).

Lockwood, D.J.; Aers, G. C.; Allard, L. B.; Bryskiewicz, B.; Charbonneau, S.; Houghton, D.C.; McCaffrey, J. P. & Wang, A. Optical properties of porous silicon. *Canadian Journal of Physics*, Vol.70, No.10-11, (November 1992), pp. 1184-1193, ISSN 1208-6045.

Lockwood, D.J.; Lu, Z.H. & Baribeau, J.-M. Quantum Confined Luminescence in Si/SiO2 Superlattices. *Physical Review letters*, Vol.76, No.3, (March 1996), pp. 539-541, ISSN 1079-7114.

Luque, A. & Marti, A. Electron–phonon Energy transfer in hot-carrier solar cells. *Solar Energy Materials and Solar Cells*, Vol.94, No.2, (February 2010), pp. 287-296, ISSN 0927-0248.

Mangolini, L.; Thimsen, E. & Kortshagen, U. High-yield plasma synthesis of luminescent silicon nanocrystals. *Nano Letters*, Vol.5, No.4, (April 2005), pp. 655–659, ISSN ISSN 1530-6992.

Marketbuzz®. (March 2011). Shine a light on the global PV supply/demand market balance, In: The leading annual world solar PV Industry Report, date of access: 07/09/11, Available from: http://www.solarbuzz.com.

Maruyama, T.; Enomoto, A. & Shirasawa, K. Solar cell module colored with fluorescent plate, *Solar Energy Materials and Solar Cells*, Vol.64, No.3 , (October 2000), pp. 269-278, ISSN 0927-0248.

McCaffrey, R.R. & Prasad, P.N. Organic-thin-film-coated solar cells: Energy transfer between surface pyrene molecules and the silicon semiconductor substrate. *Solar Cells*, Vol.11, No.4, (May 1984), pp. 401–409, ISSN 0927-0248.

McIntosh, K.R. & Richards, B.S. Increased mc-Si module efficiency using fluorescent organic dyes: a ray-tracing study, *Proceedings of IEEE 4th World Conference on Photovoltaic Energy Conversion*, pp. 2108-2111, ISBN 1-4244-0016-3, Waikoloa, Hawaii, USA, May 7-12, 2006.

Meillaud, F.; Shah, A.; Droz, C.; Vallat-Sauvarin, E. & Miazza, C. Efficiency limits for single-junction and tandem solar cells. *Solar Energy Materials and Solar Cells*. Vol.90, No.18-19, (November 2006), pp. 2952-2959, ISSN 0927-0248

Min, K.S.; Shcheglov, K.V.; Yang, C.M.; Atwater, H.A.; Brongersma, M.L. & Polman, A. Defect-related versus excitonic visible light emission from ion beam synthesized Si nanocrystals in SiO_2. *Applied Physics Letters*, Vol.69, No.14, (September 1996), pp. 2033-1-2033-3, ISSN 1077-3118.

Mirabella, S.; Di Martino, G.; Crupi, I.; Gibilisco, S.; Miritello, M.; Lo Savio, R.; Di Stefano, M. A.; Di Marco, S.; Simone, F. & Priolo, F. Light absorption and electrical transport in Si:O alloys for photovoltaics. Journal of Applied Physics, Vol.108, No.9, (November 2010), pp. 093507-093514, ISSN 1089-7550.

Mocatta, D.; Cohen, G.; Schattner, J.; Millo, O.; Rabani, E. & Banin, U. Heavily Doped Semiconductor Nanocrystal Quantum Dots. *Science*, Vol. 323, No.4, (April 2011), pp 77-81, ISSN 1095-9203.

Nesheva, D.; Raptis, C.; Perakis, A.; Bineva, I.; Aneva, Z.; Levi, Z.; Alexandrova, S.; Hofmeister, H. Raman scattering and photoluminescence from Si nanoparticles in annealed SiO_x thin films. *Journal of Applied Physics*, Vol.92, No.8, (October 2002), pp. 4678-4683, ISSN 1089-7550.

Nozik, A.J.; Beard, M.C.; Luther, J.M.; Law, M.; Ellingson, R.J. & Johnson, J.C. Semiconductor Quantum Dots and Quantum Dot Arrays and Applications of

Multiple Exciton Generation to Third-Generation Photovoltaic Solar Cells. *Chemical Review*, Vol.110, No.11, (November 2010), pp. 6873-6890, ISSN 1520-6890.

O'Regan, B. & Grätzel, M. A low-cost, high-efficiency solar cell based on dye-sensitized colloidal TiO2 films. *Nature*, Vol.353, No.6346, (October 1991), pp. 737-740, ISSN 1476-4687.

Park, S.H.; Roy, A.; Beaupre, S.; Cho, S.; Coates, N.; Moon, J.S.; Moses, D.; Leclerc, M.; Lee, K. & Heeger, A.J. Bulk heterojunction solar cells with internal quantum efficiency approaching 100%. *Nature Photonics*, Vol.3, No.5, (May 2009), pp. 297-303, ISSN 1749-4885.

Pavesi, L.; Dal Negro, L.; Mazzoleni, C.; Franzò, G. & Priolo, F. Optical Gain in Silicon Nanocrystals. *Nature*, Vol.408, No.6811, (November 2000), pp. 440-444, ISSN 0028-0836.

Perálvarez, M.; García, C.; López, M.; Garrido, B.; Barreto, J.; Domínguez, C. & Rodríguez, J.A. Field effect luminescence from Si nanocrystals obtained by plasma-enhanced chemical vapor deposition. *Applied physics Letters*, Vol.89, No.5, (July 2006), pp. 051112-051115, ISSN 1077-3118.

Pi, X.; Li, Q.; Li, D. & Yang, D. Spin-coating silicon-quantum-dot ink to improve solar cell efficiency. *Solar Energy Materials and Solar Cells*, Vol.95, No.10, (October 2011), pp. 2941-2945, ISSN 0927-0248.

Prezioso, S.; Anopchenko, A.; Gaburro, Z.; Pavesi, L.; Pucker, G.; Vanzetti, L. & Bellutti, P. Electrical conduction and electroluminescence in nanocrystalline silicon based light emitting devices. *Journal of applied physics*, Vol.104, No.6, (September 2008), pp. 063103-063111, ISSN 1089-7550.

Prezioso, S.; Hossain, S. M.; Anopchenko, A.; Pavesi, L.; Wang, M.; Pucker, G. & Bellutti, P. Superlinear photovoltaic effect in Si nanocrystals based metal-insulator-semiconductor devices. *Applied Physics Letters*, Vol.94, No.6, (February 2009), pp. 062108-1-062108-3, ISSN 1077-3118.

Prokes, S.M.; Carlos, W.E.; Veprek, S. & Ossadnik, Ch. Defect studies in as deposited and processed nanocrystalline Si/SiO_2 structures. *Physical Review B*, Vol.58, No.23, (December 1998), pp. 15632-15635, ISSN 1550-235X.

Pucker, G.; Bellutti, P.; Spinella, C.; Gatterer, K.; Cazzanelli, M. & Pavesi, L. Room temperature luminescence from $(Si/SiO_2)_n$ n=1,2,3 multilayers grown in an industrial low-pressure chemical vapour deposition reactor. *Journal of Applied Physics*, Vol.88, No.10, (November 2000) pp. 6044-6052, ISSN 1089-7550.

Richards, B.S. Luminescent layers for enhanced silicon solar cell performance: Down-conversion. *Solar Energy Materials and Solar Cells*. Vol.90, No.9, (May 2006), pp.1189-1207, ISSN 0927-0248.

Richards, B.S. Enhancing the performance of silicon solar cells via the application of passive luminescence conversion layers. *Solar Energy Materials and Solar Cells*, Vol.90, No.15, (September 2006), pp. 2329-2337, ISSN 0927-0248.

Richards, B.S. & McIntosh, K.R. Overcoming the poor short wavelength spectral response of CdS/CdTe photovoltaic modules via luminescence down-shifting: ray-tracing simulations, *Progress in Photovoltaics*, Vol.15, No.1, (January 2007), pp. 27-34, ISSN 1099-159X.

Rinnert, H. & Vergnat, M. Influence of the temperature on the photoluminescence of silicon clusters embedded in a silicon oxide matrix. *Physica E*, Vol.16, No.3-4, (March 2003), pp. 382-387, ISSN 1386-9477.

Ross, R.T. & Nozik, A.J. Efficiency of hot-carrier solar energy converters. *Journal of Applied Physics*, Vol.53, No.5, (May 1982), pp. 3813-3819, ISSN 1089-7550.

Ruan, J.; Fauchet, P.M.; Dal Negro, L.; Cazzanelli, M. & Pavesi, L. Stimulated emission in nanocrystalline silicon superlattices. *Applied Physics Letters*, Vol.83, No.26, (December 2003), pp. 5479-1-5479-3, ISSN 1077-3118.

van Sark, W.G.J.H.M.; Meijerink, A.; Schropp, R.E.I.; van Roosmalen, J.A.M. & Lysen, E.H. Modeling improvement of spectral response of solar cells by deployment of spectral converters containing semiconductor nanocrystals. *Semiconductors*, Vol.38, No.8, (August 2004), pp. 962–969, ISSN 1090-6479.

Shaklee, K.L. & Nahory, R.E. Valley-Orbit Splitting of Free Excitons? The Absorption Edge of Si. *Physical Review Letters*, Vol.24, No.17, (April 1970) pp. 942-945 ISSN 1079-7114.

Shockley, W. & Queisser, H.J. Detailed Balance Limit of Efficiency of p-n Junction Solar Cells. *Journal of Applied Physics*, Vol.32, No.3, (March 1961), pp. 510-520, ISSN 1089-7550.

Shrestha, S.K.; Aliberti, P. & Conibeer, G. Energy selective contacts for hot carrier solar cells. *Solar Energy Materials and Solar Cells*, Vol.94, No.9, (September 2010), pp. 1546-1550, ISSN 0927-0248.

Song, D.; Cho, E.C.; Conibeer, G.; Flynn, C.; Huang, Y. & Green, M.A. Structural, electrical and photovoltaic characterization of Si nanocrystals embedded SiC matrix and Si nanocrystals/c-Si heterojunction devices. *Solar Energy Materials and Solar Cells*, Vol.92, No.4, (April 2008), pp. 474-481, ISSN 0927-0248.

Stegner, A.R.; Pereira, R.N.; Klein, K.; Lechner, R.; Dietmueller, R.; Brandt, M.S.; Stutzmann, M. & Wiggers, H. Electronic Transport in Phosphorus-Doped Silicon Nanocrystal Networks. *Physical review letters*, Vol.100, No.2, (January 2008), pp. 026803-026807, ISSN 1079-7114.

Stupca, M.; Alsalhi, M.; Al Saud, T.; Almuhanna, A. & Nayfeh, M.H. Enhancement of polycrystalline silicon solar cells using ultrathin films of silicon nanoparticle. Applied Physics Letters, Vol.91, No.6, (August 2007), pp. 063107-1-063107-3, ISSN 1077-3118.

Švrcek, V.; Slaoui, A. & Muller, J.-C. Silicon nanocrystals as light converter for solar cells. *Thin Solid Films*, Vol.451-452, (March 2004), pp. 384-388, ISSN 0040-6090.

Švrcek, V.; Mariotti, D.; Shibata, Y. & Kondo, M. A hybrid heterojunction based on fullerenes and surfactant-free, self-assembled, closely packed silicon nanocrystals. *Journal of Physics D: Applied Physics*. Vol.43, No.41, (October 2010), pp. 415402, ISSN 1361-6463.

Sychugov, I.; Juhasz, R.; Valenta, J. & Linnros, J. Narrow Luminescence Linewidth of a Silicon Quantum Dot. *Physical Review Letters*, Vol.94, No.8, (March 2005), pp. 087405-087409, ISSN 1079-7114.

Takeda, Y.; Ito, T.; Motohiro, T.; König, D.; Shrestha, S. & Conibeer, G. Hot carrier solar cells operating under practical conditions. *Journal of Applied Physics*, Vol.105, No.7, (July 2009), pp. 074905-074915, ISSN 1089-7550.

Teske, S. & Bitter, M. (2008). Solar Generation V-Solar electricity for over one billion people and two million jobs by 2020, Date of access: 10/09/11, Available from http://www.epia.org/S.

Trupke, T.; Green, M.A. & Würfel, P. Improving solar cell efficiencies by down-conversion of high-energy photons. *Journal of applied physics*, Vol.92, No.3, (August 2002), pp. 1668-1675, ISSN 1089-7550.

Tsakalakos, L.; Balch, J.; Fronheiser, J.; Korevaar, B.A.; Sulima, O. & Rand, J. Silicon nanowire solar cells. *Applied Physics Letters*, Vol.91, No.23, (December 2007), pp. 233117-1-233117-3, ISSN 1077-3118.

Tsybeskov, L.; Vandyshev, Ju.V. & Fauchet, P.M. Blue emission in porous silicon: Oxygen related photoluminescence. *Physical Review B*, Vol.49, No.11, (March 1994), pp. 7821-7824, ISSN 1550-235X.

Uhlir, A. Electrolytic Shaping of Germanium and Silicon. *Bell system Technical Journal*, Vol.35, No.2, (March 1956), pp. 333-347, ISSN 0005-8580.

Umezu, I.; Yoshida, K.; Sugimura, A.; Inokuma, T.; Hasegawa, S.; Wakayama, Y.; Yamada, Y. & Yoshida, T. A Comparative study of the photoluminescence properties of a-SiOx:H film and silicon nanocrystallites. *Journal of Non-Crystalline Solids*, Vol.266-269, No.2, (May 2000), pp. 1029-1032, ISSN 0022-3093.

Vinciguerra, V.; Franzò, G.; Priolo, F.; Iacona, F. & Spinella, C. J. Quantum confinement and recombination dynamics in silicon nanocrystals embedded in Si/SiO$_2$ superlattices. *Applied Physics*, Vol.87, No.11, (June 2000), pp. 8165-8174, ISSN 1089-7550.

Wadia, C.; Alivisatos, A.P. & Kammen, D.M. Materials Availability Expands the Opportunity for Large-Scale Photovoltaics Deployment. *Environmental Science and Technology*, Vol.43, No.6, (March 2009), pp. 2072-2077, ISSN 1520-5851.

Wan, Z.; Patterson, R.; Huang, S.; Green, M.A. & Conibeer, G. Ultra-thin silicon nitride barrier implementation for Si nano-crystals embedded in amorphous silicon carbide matrix with hybrid superlattice structure. *Europhysics letters*, Vol.95, No.6, (September 2011), pp. 67006p1-67006p5, ISSN 1286-4854.

Wei, D.; Andrew, P. & Ryhänen, T. Electrochemical photovoltaic cells—review of recent developments. *Journal of Chemical Technology & Biotechnology*, Vol.85, No.12, (December 2010), pp. 1547-1552, ISSN 0142-0356.

Westland, D.J.; Ryan, J.F.; Scott, M.D.; Davies, J.I. & Riffat, J.R. Hot carrier energy loss rates in GaInAs/InP quantum wells. *Solid State Electronics*, Vol.31, No.1-2, (March-April 1988), pp. 431-434, ISSN 0038-1101.

Wilson, W.L.; Szajowski, P.F. & Brus, L.E. Quantum confinement in Size-selected, Surface-Oxidized Silicon Nanocrystals. *Science*, Vol.262, No.5137, (November 1993), pp. 1242-1244, ISSN 1095-9203.

Wolfsegger, C. & Stierstorfer, J. (2007). Solar Generation IV 2007 – solar electricity for over one billion people and two million jobs by 2020, date of access 29/09/11, Available from: http://www.epia.org/fileadmin/EPIA_docs/publications/epia/EPIA_SG_IV_fin al.pdf.

Würfel, P. Solar energy conversion with hot electrons from impact ionisation. *Solar Energy Materials and Solar Cells*, Vol.46, No.1, (April 1997), pp. 43-52, ISSN 0927-0248.

Würfel, P.; Brown, A.S.; Humphrey, T.E. & Green, M.A. Particle conservation in the hot-carrier solar cell. *Progress in Photovoltaics*, Vol.13, No.4, (June 2005), pp. 277-285, ISSN 1099-159X.

Yang, C.S.; Lin, C.J.; Kuei, P.Y.; Horng, S.F.; Ching-Hsiang Hsu, C. & Liaw, M.C. Quantum size effects on photoluminescence from Si nanocrystals in PECVD silicon-rich-oxide. *Applied Surface Science*, Vol.113/114, (April 1997), pp. 116-120, ISSN 0169-4332.

Zatryb, G.; Podhorodecki, A.; Hao, X.J.; Misiewicz, J.; Shen, Y.S. & Green, M.A. Quantitative evaluation of boron-induced disorder in multilayers containing silicon nanocrystals in an oxide matrix designed for photovoltaic applications. *Optics express*, Vol.18, No.21, (October 2010), pp. 22004-22009, ISSN 1094-4087.

Zatryb, G.; Podhorodecki, A.; Hao, X.J.; Misiewicz, J.; Shen, Y.S. & Green, M.A. Correlation between stress and carrier nonradiative recombination for silicon nanocrystals in oxide matrix. *Nanotechnology*, Vol.22, No.33, (August 2011), pp. 335703-335708, ISSN 1361-6528.

3

Capping of InAs/GaAs Quantum Dots for GaAs Based Lasers

Alice Hospodková
Institute of Physics AS CR, v. v. i., Prague
Czech Republic

1. Introduction

Nowadays, self-assembled InAs quantum dots grown on GaAs substrates are intensively studied due to their potential applications: quantum dot ensembles are employed in laser active regions and mid-infrared detectors, single quantum dots have a potential for application in memories, two-level information processing or single photon sources (Henini& Bugajski, 2005; Mowbray& Skolnick, 2005; Battacharya, 2007).

This chapter is concentrated on quantum dots, which are prepared with the aim to become a base for low cost, low threshold, high output and heat sink free lasers for the 1.3 or even 1.55 µm communication band. To be embedded in a functional structure, dots must be covered by a capping layer. Capping of quantum dots is currently a very important topic of research (Bozkhurt et al., 2011; Dasika et al., 2009; Haxha et al., 2009; Kong et al., 2008). The overgrowth process is essential to obtain dot parameters necessary for the realization of novel optoelectronic devices. This chapter is a kind of review report on the structural and optical changes during the process of capping of self-assembled InAs quantum dots. The presented results are measured on samples grown by metal-organic vapour phase epitaxy which is the growth technique of choice for high volume laser production.

Growth conditions of InAs/GaAs quantum dot layer determine such parameters as density, homogeneity and original size of quantum dots. However, the final properties of quantum dots are significantly determined by quantum dot capping, since during the capping process the redistribution of In atoms takes place. This decreases the height of dots and changes their shape. The published and our results show that the GaAs capping layer transforms originally circular lens shaped quantum dots into elongated objects and their height is significantly decreased. The strain field surrounding dots is also changed during the capping process and can be influenced by the composition and the thickness of the layers grown above quantum dots.

The aim of the chapter is to elucidate the processes during the dot capping by reflectance anisotropy spectroscopy in situ measurements and to discuss the influence of the capping layer composition and thickness on the structural and optical properties of InAs/GaAs quantum dots. Three most common types of capping layers are studied: simple GaAs capping layer, InGaAs and GaAsSb. The emission wavelength of quantum dots capped by GaAs is usually around 1200 nm, which is not interesting for application in telecommunication devices, since they need to be operated at wavelengths 1300 or 1550 nm to minimize the losses in optical waveguides. One way to redshift the emission wavelength is to increase the size of dots by increasing the amount of deposited InAs (Krzyzewski et al., 2002), by controlling growth conditions (especially increasing temperature) (Johanson et al., 2002), by growth on high index surfaces (Akijama et al., 2007) or by growing multiple quantum dot structures (Hospodkova et al., 2008). Another way to shift emission wavelength is to cap dots by so called strain reducing layers. The lattice constant of the strain reducing material is bigger than that for GaAs substrate but smaller than lattice constant of the dot material (InAs or InGaAs). The redshift is then caused by combined effect of reduced strain inside quantum dots, increased dot aspect ratio and decreased bandgap of the capping barrier material. Materials most often used for the strain reduction are InGaAs and GaAsSb. Advantages and disadvantages of both types of strain reducing layers are discussed in section 3. The longest achieved electroluminescence wavelength of 1520 nm was recently reported by (Majid et al., 2011, a, b). This unique result was achieved by combination of bilayer structure with InGaAs strain reducing capping layer. The structure was grown by molecular beam epitaxy.

More than one quantum dot layer are required in lasers to increase the gain. In such a case the capping layer serves also as a separating layer between two quantum dot layers. During the growth of the separation layers, surfactant atoms like In have to be removed from epitaxial surface, so that the In is not transported from one quantum dot layer to the following one. Different types of In flushing method during the growth of capping layers and their importance are discussed at the end of the chapter in section 4.

The photoluminescence or electroluminescence and reflectance anisotropy spectroscopy as in situ monitoring of the technological process are the main methods which help us to evaluate the quantum dot properties, supporting methods are atomic force microscopy of quantum dot surface, transmission electron microscopy, X-ray diffraction used to determine the composition of the ternary capping layers and X-ray reflection used for capping layer thickness evaluation.

2. GaAs capping

Combination of InAs and GaAs layers forms highly mismatched heterostructure (7% mismatch of lattice constants). This mismatch is a driving force for Stranski-Krastanow quantum dot formation. Dots are formed to decrease the strain in deposited InAs material. The lowest strain occurs at the apex of quantum dot. On the other hand the strain in the structure complicates the quantum dot preparation, since accumulated strain energy can lead to the formation of dislocations in the structure for oversized dots or in the case of more quantum dot layers.

The compressive strain inside quantum dots also obstructs achieving longer emission wavelength. When formed InAs dots are subsequently capped by GaAs, the compressive strain inside dots is further increased. The increased strain after the capping blue shifts the wavelength of quantum dot photoluminescence maximum. The dependence of photoluminescence maximum energy on the capping layer thickness is shown in Fig. 1, where experimental and simulated data are compared (Hazdra et al., 2009). It is apparent from the calculated band diagram (see inset of the Fig. 1) that not only increased compressive strain inside InAs dots but also decreased tensile strain in GaAs barriers is responsible for a considerable blue shift of ground state transition energies with increasing capping thickness. More detailed study of the impact of the capping layer thickness on the electronic structure of quantum dots can be found in (Wu et al., 2006).

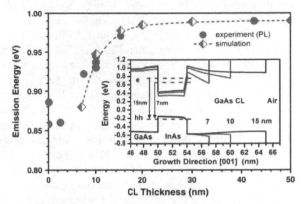

Fig. 1. Comparison of measured (photoluminescence - circles) and calculated (diamonds) room temperature transition energies originating from InAs quantum dots covered with GaAs capping layers with different thickness. The corresponding band diagrams (taken in the quantum dot centre along the [001] axis) are shown in the inset (Hazdra et al., 2009, a).

The increased strain in InAs material during the capping process not only modulates the quantum dot electronic structure, but also the shape and the composition of quantum dots is considerably changed during capping (Eisele et al.,2008; Costantini et al., 2006; Songmuang et al., 2002). Since the GaAs has smaller lattice constant than InAs, the GaAs capping layer is at first preferably deposited between quantum dots or near a dot base, leaving uncovered a quantum dot apex. The increased strain during the growth of the GaAs capping layer is the driving force for a quantum dot dissolution and migration of In atoms from the uncovered dot apex to the dot base, transforming thus the originally lens shaped InAs quantum dots with the circular base (see Fig. 2a) to elongated InGaAs dots (see Fig. 2b). The alloyed interface between InAs core of quantum dot and GaAs capping together with elongated dot shape causes lower energy separation between ground state and excited state emission (Hospodková, 2007, b), which is disadvantageous for laser design.

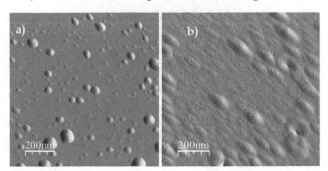

Fig. 2. Atomic force microscopy images of a) uncapped lens shaped InAs quantum dots on GaAs substrate with the circular base. b) InAs dots after capping by 10 nm of GaAs, dots are elongated in [-110] direction.

The capping process of InAs quantum dots can be monitored in situ by time resolved reflectance anisotropy spectroscopy, which enables one to observe in situ the mechanism of In atoms migration from quantum dots and surfactant behaviour of In atoms during GaAs capping layer growth (Steimetz, 1998). To monitor processes during the capping of quantum dots by GaAs layer we grew a set of samples A1, A2 and A3 using Triethyl gallium (TEGa), Trimethyl indium (TMIn) and Tertiary-butyl arsin (TBAs). The V/III ratio during the GaAs capping growth was gradually changing from a very low value at the beginning to 3.4 after 30 s of capping layer growth. The V/III ratio for InAs growth was 0.5, 1 and 2 for samples A1, A2 and A3. Such a low V/III ratio was used to improve quantum dot size homogeneity (Hoglund et al., 2006). However, very low V/III ratio during InAs growth and at the beginning of capping layer growth enabled us to achieve In rich (4×2) surface reconstruction. The gradual increase of V/III ratio during capping layer growth helped us to distinguish between In rich (4×2) InAs and β2(2×4) GaAs surfaces, since both surfaces have opposite development of reflectance anisotropy signal intensity at 4.2 eV with increasing V/III ratio. The signal intensity increases with increasing V/III ratio for InAs epitaxial surface, while it decreases for GaAs one. The reason of this behaviour is the change of dominant surface reconstruction for InAs from (4×2) to (2×4) with increasing V/III ratio, the relevant reflectance anisotropy spectra were reported in (Goletti et al., 2001; Knorr et al., 1997).

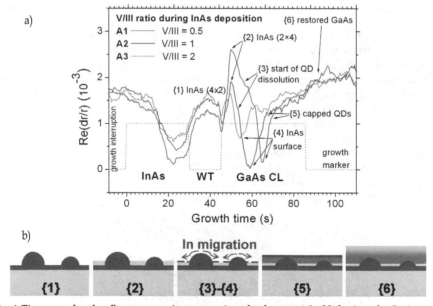

Fig. 3. a) Time resolved reflectance anisotropy signal taken at 4.2 eV during the InAs and GaAs capping layer growth. Samples differ only in the partial TBAs pressure during the InAs growth and during the waiting time for quantum dot formation. The beginning of InAs deposition was set as a time 0. b) Proposed model of the growth {1}–{6} according to reflectance anisotropy data (Kita, 2000).

This behaviour could be checked during the InAs deposition (see Fig. 3), sample A1 with the lowest V/III ratio has also the lowest intensity of the reflectance anisotropy signal at 4.2 eV.

The GaAs epitaxial surface reconstructions of our samples changed with increasing V/III ratio from β2(2×4) to c(4×4), the relevant reflectance anisotropy spectra were published for example in (Richter, 2002; Hingerl et al., 1993). Knowing this, we can disclose the processes during the growth of GaAs capping layer.

The surface processes according to the reflectance anisotropy signal are shown in Fig. 3. We can see that the surface reconstructions differ considerably during several tens of seconds of capping layer growth and the reflectance anisotropy signals of all three samples align approximately after 30 s when identical V/III ratio was achieved. Different stages of surface processes are marked in Fig. 3 by numbers in braces {n}. Stranski–Krastanow quantum dot formation takes place partly during the InAs growth and partly during the waiting time and causes the signal intensity increase {1}. Two well-distinguished extremes, maximum and minimum, can be recognized during GaAs capping layer growth {2}–{6} in reflectance anisotropy signal. The maximum {2} occurs approximately at the same time for all three samples. It is connected with the introduction of triethyl gallium into the reactor. According to the signal intensity we can ascertain when the β2(2×4) GaAs reconstruction is established (Richter, 2002), which is also proved by the dependence of reflectance anisotropy intensity on concentration of As atoms on the surface (the lowest As concentration corresponds with the highest intensity of reflectance anisotropy signal – opposite to InAs surface behaviour – see Fig. 3). The following decrease of signal intensity can also be expected for β2(2×4) GaAs surface because As atom concentration is gradually increasing. But surprisingly the signal intensity reaches its minimum {4} and then increases again. The increase of signal intensity with As concentration on the surface can be explained only by the presence of InAs layer on the epitaxial surface. This means that the decrease of signal intensity {3} is not caused only by the increase of As atom concentration on the epitaxial surface, but also by the presence of In atoms migrating from strained, uncovered dots. It is clearly seen from reflectance anisotropy signal especially of samples A2 and A3 that the slope of signal intensity decrease suddenly changes. It happens between 7 and 9 s of the GaAs capping layer growth when approximately 2.5 monolayers of GaAs should be deposited. We suppose that at this time In atoms start to migrate from strained dots. This start of In migration is driven by the strain caused by the deposited GaAs capping layer. The dissolution of dots starts at a certain thickness of the deposited GaAs. Since it starts according to reflectance anisotropy measurement at somewhat different time for each sample, we suppose that the growth rate of GaAs probably slightly decreases with As concentration on the surface which was also observed in (Reinhardt, 1994). At the minimum {4} the InAs covers the dominant part of the epitaxial surface. The migration of In atoms from dots continues also over the time when the signal intensity steeply increases. The termination of steep increase of the reflectance anisotropy signal {5} is caused by achieving the V/III ratio of 3.4 and stabilization of As atom concentration in the reactor. The slow increase of the signal intensity after the point {5} represents decreasing In atom concentration on the epitaxial surface, In migration from quantum dots decreases and the GaAs surface restores again {6}.

Conclusion from reflectance anisotropy monitoring is that first the thin GaAs layer is formed (approximately 2.5 monolayers) and then the In migration from quantum dots follows. These two processes do not start at the same time; the In dissolution is delayed (Hospodková et al., 2010). This confirms the hypothesis that the increased strain during capping is the driving force for quantum dot dissolution, which was also published in (Takehama, 2003).

To study the effect of the growth rate of the capping layer on quantum dot photoluminescence a set of three samples B1, B2 and B3 was grown. The samples differ in capping layer growth rates which were 6.4, 10.2 and 18.8 nm/min, respectively. The capping layer thickness was 7.5 nm for all three samples. We have expected that the dot dissolution would be higher and consequently the quantum dot size would be smaller for slow capping. Surprisingly, the photoluminescence results of this set of samples (Fig. 4) do not confirm this expectation. On the contrary, the photoluminescence maximum energy decreases with slower capping layer growth rate which means either that the quantum dot size is increased (not probable) or that the strain in quantum dots is decreased. To explain these results atomic force microscopy images of all three samples were compared (Fig. 5).

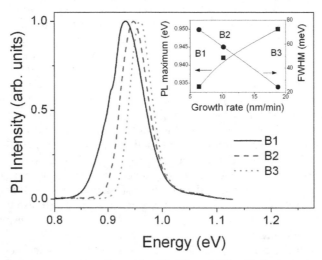

Fig. 4. Photoluminescence spectra of samples B1, B2 and B3 with the capping layer growth rates 6.4, 10.2 and 18.8 nm/min. The dependence of the photoluminescence maximum energy and full width at half maximum on the capping layer growth rate is shown in the inset; the lines are a guide for the eye.

It is apparent from the atomic force microscopy images that the dot dissolution during capping depends not only on the capping layer growth rate but also on the size of quantum dots. The strong dissolution with formation of the depression in the dot centre was observed only on the biggest dots from the dot ensemble. Samples B1, B2 and B3 differ significantly in the morphology of the big dots after capping (Fig. 5). The size of these depressions is increased when the capping process is slower (see Fig. 5.d). The InAs material from partly dissolved dots is subsequently incorporated into the capping layer (as was demonstrated by in situ reflectance anisotropy measurement), which becomes InGaAs instead of GaAs. This causes red shift of the photoluminescence maximum of smaller quantum dots and also increases the full width at half maximum (inset of Fig. 4). The In segregation near smaller quantum dots cannot be excluded. These results are in agreement with the work of (Takehama et al., 2003) who has studied the influence of the GaAs capping layer thickness on quantum dot morphology. The stronger dissolution of bigger dots can be also deduced from reflectance anisotropy results presented in (Steimetz et al., 1998).

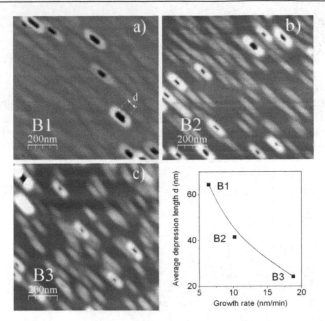

Fig. 5. a) - c) Atomic force microscopy images of samples B1, B2 and B3. d) The dependence of the average length d of depressions on the capping layer growth rate, the line is a guide for the eye.

In conclusion of this section the GaAs capping layer proved to have many disadvantages for a quantum dot laser design. GaAs capping strongly compresses InAs quantum dots formed by the Stranski-Krastanow growth mode, which causes the undesirable blue shift of photoluminescence maximum and also dissolution of quantum dots during capping. The InAs dissolution starts after the deposition of critical thickness of the capping layer which is according to our in situ measurements around 2.5 monolayers of GaAs. The dissolution is stronger for bigger dots. The quantum dots shape transforms during capping from the lens to the elongated boat-like shape, which decreases the energy difference between ground and excited state transitions (Hospodkova et al., 2007, a).

3. Strain reducing capping layers

To overcome the problems during quantum dot capping summarized in the previous section, other capping materials, which could decrease the strain in quantum dots, were studied. The most common capping materials used for the reduction of strain inside dots are InGaAs (Dasika et al., 2009; Liu et al., 2003; Kim et al., 2003) and GaAsSb (Haxha et al., 2009; Bozkhurt et al., 2011; Ulloa et al., 2007; Akahane et al., 2004), sometimes InAlAs or some combination of these materials is used (Liu et al., 2005; Kong et al., 2008). Strain reducing layers covering InAs dots are used to shift the emission wavelength to 1.3 μm and 1.55 μm, typical of optical fibre communication. Both types of strain reducing layers, InGaAs and GaAsSb, red shift the emission wavelength by similar mechanism: they decrease the dot dissolution during capping, relieve the strain inside quantum dots and also decrease the barrier energy for electrons and holes.

3.1 InGaAs strain reducing capping layer

InGaAs layers are most often used for the strain reduction in so-called "dot in a well" structures (Liu et al., 2003). InGaAs decreases the strain and not only preserves, but also it can even effectively increase the quantum dot size during InGaAs growth due to the In gathering in dot vicinity (Hazdra et al. 2009, b), so it is easy to shift the quantum dot photoluminescence maximum significantly to longer wavelengths. Set C of four samples with quantum dots capped by 10 nm of $In_xGa_{1-x}As$ with different composition, x ranging from 0 to 0.24, was prepared. The photoluminescence spectra of samples C1 –C4 are presented in Fig. 6.

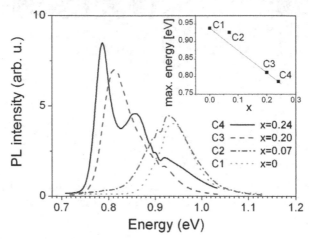

Fig. 6. Photoluminescence spectra of samples with quantum dots capped by 10 nm thick $In_xGa_{1-x}As$ with different composition: x=0 (GaAs) for C1, x=0.07 for C2, x=0.20 for C3 and x=0.24 for C4. The inset shows the dependence of the energy of the ground state emission on the quantum dot capping layer composition.

The strong red shift of the quantum dot photoluminescence maximum with an increasing In content in the capping layer is not caused only by the strain reducing effect, but in the case of samples C3 and C4 (with higher In content) also by increasing quantum dot size due to the In gathering in quantum dot vicinity. Intensive ground state photoluminescence maximum of sample C4 with $In_{0.24}Ga_{0.76}As$ strain reducing layer on the 1.578 μm wavelength is a promise for an application of InAs quantum dots at 1.55 μm telecommunication wavelength.

However, the lower energy barrier for electrons in quantum dots is, on the other hand, a disadvantage of this type strain reducing layer for an application in laser structures. First, the lower energy barrier together with increased dot size causes smaller energy separation between the ground and excited states and lasing then starts usually from the excited state transitions with higher density of states and with shorter wavelength emission. Second, the lower energy barrier for electrons and additionally diffused heterojunction between quantum dots and InGaAs quantum well weakens the localization of electrons in dots and electroluminescence from InGaAs quantum well becomes dominant in the case of capping layer with higher In content (Hospodková et al., 2011). Comparison of photoluminescence and electroluminescence of quantum dot structures with different In content in InGaAs strain reducing layer is presented in Fig. 7. Two sets of samples with different composition of In$_x$Ga$_{1-x}$As capping layers were prepared by low pressure metal-organic vapour phase epitaxy. In the first set, prepared for photoluminescence measurement, the quantum dot structure was embedded in the undoped GaAs. In the second set the identically prepared quantum dot structures were embedded in the p-i-n structure of electroluminescent diode.

The results measured on these two sets of samples are presented in Fig. 7. and demonstrate that even if the photoluminescence evinces a significant red shift with increasing In content in the quantum dot capping layer (see Fig. 7a), the electroluminescence main maximum is blue shifted (see Fig. 7b-d). The ground state electroluminescence was achieved as the most intensive maximum only in the case of capping by GaAs (Fig. 7b). The main electroluminescence maximum became from the first excited state transition in the case that In$_{0.13}$Ga$_{0.87}$As strain reducing layer was used for quantum dot capping (Fig. 7c). When In content was further increased to x=0.23, only electroluminescence from the quantum well complex of InAs wetting layer with InGaAs strain reducing layer was achieved (Fig. 7d) due to the poor localization of electrons in quantum dots.

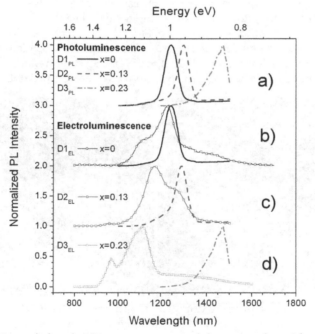

Fig. 7. a) Comparison of photoluminescence spectra of three samples with quantum dots capped by 5 nm of GaAs, In$_{0.13}$Ga$_{0.87}$As or In$_{0.23}$Ga$_{0.77}$As , resp. and subsequently covered by 60 nm thick GaAs. b-d) Comparison of electroluminescence and photoluminescence spectra of samples with quantum dots capped by GaAs (b), In$_{0.13}$Ga$_{0.87}$As strain reducing layer (c) and In$_{0.23}$Ga$_{0.77}$As (d). Samples for electroluminescence were embedded in p-i-n diode structure.

Electroluminescence and lasing from the ground state at 1300 nm was achieved on quantum dot structures with InGaAs strain reducing layer prepared by molecular beam epitaxy (Liu, 2004) since this technology allows one to grow structures at lower temperature which decrease the In segregation and diffusion and improves localization of electrons in quantum dots. The localization of electrons in dots can be further improved by graded composition of InGaAs (Kim, 2003) or by combining the InGaAs strain reducing layer with InAlAs (Kong, 2008). Unfortunately, graded InGaAs composition is difficult to achieve when metal-organic vapour phase epitaxy is used for structure growth due to high In segregation during capping (Steimetz, 1998). Also the combination of InGaAs with InAlAs cannot be used by this technology because of the high decomposition temperature of available Al precursors. So the InGaAs is not suitable strain reducing layer for quantum dot structures prepared by metal-organic vapour phase epitaxy.

3.2 GaAsSb strain reducing capping layer

GaAsSb strain reducing layer offers several promising advantages compared to InGaAs. Firstly, it provides higher energy barrier for electrons, so it seems to be more suitable for laser structures, where high carrier density is needed. Secondly, it considerably decreases the In segregation during dot capping, which has consequence not only in the sharper interfaces between quantum dot and strain reducing layer, but also in a more circular shape of resulting quantum dot after capping (see Fig. 8). The circular shape preserves higher energy separation between ground and excited state emission which is required for lasing wavelength stability. The important advantage of GaAsSb strain reducing layer is also the increased photoluminescence intensity in comparison to GaAs capped dots.

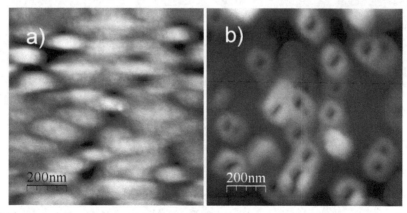

Fig. 8. a) Elongated shape of quantum dots capped by 5 nm of $In_{0.23}Ga_{0.77}As$ strain reducing layer. b) Circular shape of quantum dots capped by 5 nm of $GaAs_{0.81}Sb_{0.19}$. The hollows in the dot centre are caused by the strain induced decrease of GaAsSb growth rate on quantum dot apex, they were also observed by scanning tunnelling microscopy (Ulloa, 2007; Bozkurt, 2011)

Due to these advantages GaAsSb seems to be promising strain reducing material for quantum dot structures prepared by metal organic vapour phase epitaxy. However, this material does not allow such a strong red shift of emission wavelength as in the case of InGaAs strain reducing layer. There are two reasons for lower red shift. First, GaAsSb capping layer can only preserve the original size of dots because this material does not contain any In atoms which would be gathered near quantum dots and second, the applicable composition of GaAsSb strain reducing layer is only up to 14% of Sb, since for higher Sb content the heterostructure becomes type II with holes confined in GaAsSb quantum well and electrons localized in quantum dots (see Fig. 11a). The emission intensity decreases significantly for type II heterostructure due to the worse overlap of electron and hole wave functions.

Our effort was to obtain emission wavelength of quantum dots capped by GaAsSb as long as possible, but keeping type I heterojunction between quantum dots and strain reducing layer. Since in the case of GaAsSb capping quantum dots cannot be increased by In atoms gathering around dots and because the Sb content in the capping layer is limited, it is important to cover quantum dots with the biggest achievable original size. An effective way to increase the quantum dot size is to increase the growth temperature for InAs deposition. The set of four samples with quantum dots covered by 3 monolayers of GaAsSb was prepared under different temperatures ranging from 480 °C to 530 °C. The dot size was checked by atomic force microscopy. The clear increase in quantum dot size with increasing growth temperature is evident (Fig. 9).

The quantum dot photoluminescence efficiency decreased significantly for growth temperatures bellow 480 °C and above 520 °C. The optimal temperature for quantum dot formation was set to 510 °C. To increase the dot size we have also lowered the InAs growth rate, which increased the photoluminescence intensity and also slightly red shifted the ground state emission maximum.

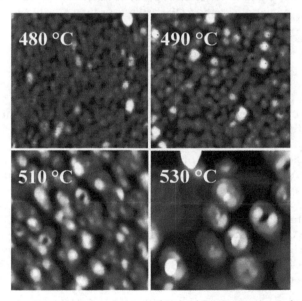

Fig.9. Atomic force microscopy images 1×1μm of InAs/GaAs quantum dot surface prepared at different growth temperatures, revealing the quantum dot size increase with increasing growth temperature. Quantum dots were capped by 3 monolayers of GaAsSb to preserve the surface morphology during the cooling process.

Comparison of photoluminescence spectra of quantum dots capped by 50 nm of GaAs before and after the InAs growth optimization can be found in the Fig. 10, corresponding red shift with increased growth temperature and decreased growth rate is 62 meV. The achieved wavelength of the ground state photoluminescence maximum was 1270 nm.

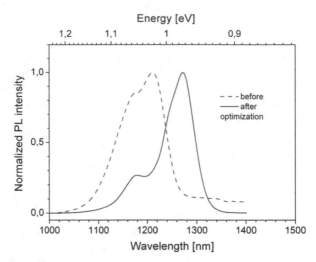

Fig. 10. Comparison of photoluminescence spectra of quantum dots capped by 50 nm of GaAs before InAs growth optimization (490 °C, InAs growth rate 0.1 ML/s) and under optimized growth conditions (510 °C, InAs growth rate 0.05 ML/s).

After increasing the dot size the second task was to achieve a maximal red shift of emitted wavelength by capping quantum dots with a GaAsSb layer, but to maintain type I heterojunction between the InAs quantum dots and the GaAsSb layer. A new structure of GaAsSb layer with graded Sb concentration was proposed to accomplish this goal (see Fig. 11b). GaAsSb with lower Sb concentration immediately caps quantum dots then the Sb concentration gradually increases during GaAsSb growth. The part of GaAsSb layer with lower Sb concentration represents the barrier for holes localized in InAs quantum dots and, on the other side also the barrier for triangular GaAsSb valence band quantum well formed by increased Sb concentration. To preserve type I heterojunction it is necessary to propose the structure parameters (composition and thickness of GaAsSb, size of quantum dots) in such a way that the ground state in GaAsSb triangular quantum well has a lower energy than ground state in InAs quantum dots, so that the holes remain localized inside dots (Fig. 11b). With this GaAsSb structure we have achieved intensive luminescence with the record ground state emission at 1391 nm maintaining type I heterojunction (see Fig. 12a). The Sb concentration in GaAsSb strain reducing layer was increased from 5 to 25% according to X-ray diffraction. The thickness of GaAsSb layer was 5 nm.

The same combination of quantum dots with graded GaAsSb strain reducing layer was embedded in p-i-n structure of light emitting diode. The electroluminescence main maximum of this diode originates from the ground state transitions and is only slightly blue shifted in comparison to the photoluminescence of corresponding structure (compare Fig. 12a and b). The strong electroluminescence of the ground state maximum with increasing current density is a sign of a good localization of electrons and holes in quantum dots using this type of strain reducing layer.

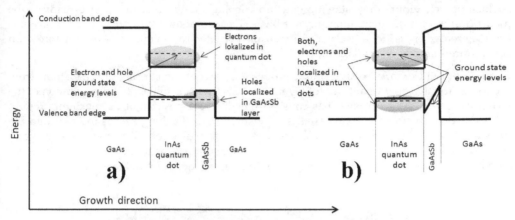

Fig. 11. Schematic illustration of band structure alignment of a) InAs quantum dot with conventional GaAsSb strain reducing layer forming type II heterojunction between InAs and GaAsSb semiconductors. Electrons are localized in quantum dots while holes in GaAsSb quantum well. b) InAs quantum dots capped by GaAsSb layer with graded composition. Both, electrons and holes are localized in quantum dots.

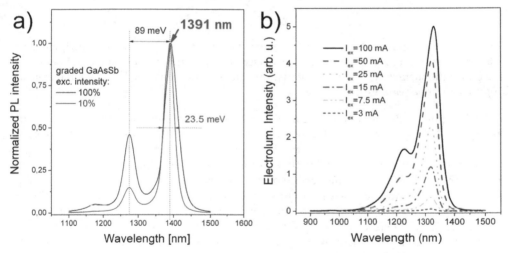

Fig. 12. a) Photoluminescence spectra measured at two different excitation intensities of quantum dots capped by 5 nm thick GaAsSb layer with graded composition. b) Electroluminescence spectra of the same quantum dot structure embedded in the p-i-n structure of light emitting diode.

4. In-flushing methods

To obtain lasing and to increase the quantum efficiency of quantum dot laser it is necessary to incorporate several quantum dot layers into the laser active region, the most common numbers are from three to ten layers of quantum dots. Since In and Sb atoms have strong surfactant behaviour, they stay flowing on capping GaAs epitaxial surface ready to be incorporated into the next quantum dot layer. The size of quantum dots in upper layers is then increased as is demonstrated by transmission electron microscopy of structure with seven quantum dot layers (Fig. 13)

Even 50 nm thick GaAs separation layer does not prevent transferring of these atoms from the lower to the upper quantum dot layer. This usually leads to the fatal deterioration of the structure or at least each layer has considerably different position of PL maximum due to different dot size in each layer. That is why so called In-flushing step is inserted into the growth of quantum dot separation layers when multiple quantum dot structure is grown to remove surfactant atoms from epitaxial surface (Tatebayashi, 2004) and to prevent transportation of InAs material from a lower to an upper quantum dot layer.

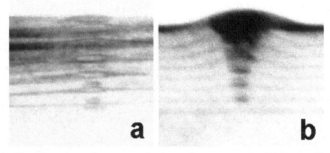

a   **b**

Fig. 13. Transmission electron microscopy image of seven stacked layers of InAs quantum dots separated by 10 nm of GaAs. The increased size of quantum dots is predominantly caused by the surfactant behaviour of In atoms transporting thus the InAs material from a lower to an upper quantum dot layer.

In-flushing methods are based on growth interruption and increasing temperature during the separation layer growth. The increased temperature helps to remove effectively the surfactant In atoms from the epitaxial surface. How effective the method is can be recognized during the growth by time resolved reflectance measurement or after the growth from the photoluminescence spectra of the structure. We have prepared three samples E1, E2 and E3 with multiple quantum dot structures by three different technological procedures. All three samples consist of six layers of InAs quantum dots capped by GaAsSb strain reducing layer with graded Sb concentration separated by 35 nm thick GaAs layers. Sample E1 was prepared without any In-flushing step. The structure was so deteriorated that no photoluminescence was detected on sample E1. Also the reflectance anisotropy signal was strongly increased during the growth of the structure indicating that the surface quality was deteriorated (Fig. 14b). The In-flushing step was inserted into the growth of GaAs separation layers of sample E2. After the growth of 10 nm of GaAs at 510 °C the growth was interrupted, temperature was increased to 600 °C for 7 minutes then decreased back to 510 °C and the growth of remaining 25 nm of GaAs continued. The reflectance anisotropy signal did not increase during the E2 structure growth indicating good surface quality (Fig. 14b); however, presence of In atoms on the GaAs epitaxial surface was recognized in the signal. The photoluminescence spectrum of sample E2 also revealed transport of InAs or Sb atoms between quantum dot layers, the photoluminescence intensity was very low and several broad peaks originating from different quantum dot layers were recognized in the spectrum (see Fig. 14). The In-flushing method with the GaAs growth at elevated temperature according to (Liu, 2004) was used to improve the E3 sample quality. After the growth interruption and temperature elevation (similarly to E2), 8 nm of GaAs were grown at elevated temperature, then the temperature was decreased to 510 °C and remaining 17 nm of GaAs were grown. The photoluminescence of sample E3 improved considerably compared with sample E2 (Fig. 14a), the maximal intensity increased by three orders of magnitude and only one ground state maximum was present in the spectrum. The peak intensity increased eight times in comparison to sample with one quantum dot layer and position of ground state maximum remained on the same wavelength. We can conclude from these experiments that a growth interruption with temperature increase to 600 °C may not be a sufficient method in case, when GaAsSb SRL is used, and additionally the GaAs growth at higher temperature helps to prevent transferring of surfactants from lower to upper dot layers.

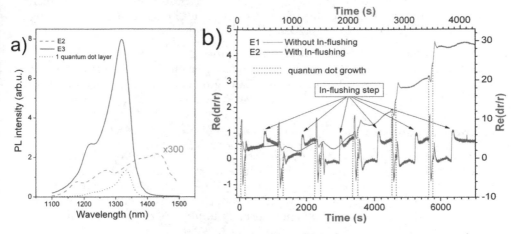

Fig. 14. a) Photoluminescence spectra of samples E2 and E3 containing six quantum dot layers with graded GaAsSb strain reducing layers separated by 35 nm of GaAs. Different In flushing methods were used for samples E2 and E3. Photoluminescence of sample with one quantum dot layer with the same GaAsSb strain reducing layer is presented for comparison. b) Comparison of time resolved reflectance anisotropy data of sample E2 with In-flushing steps during GaAs separation layer growth and of sample E1 without In-flushing step. Increased signal of sample E1 after the growth of 3rd quantum dot layer is a sign of surface quality deterioration. To compare signals in the same graph different scales were required distinguished by the title colour.

5. Conclusions

Three different types of quantum dot capping layer were studied and compared. The GaAs capping layer significantly blue shifts the quantum dot emission because the size of quantum dots is decreased during capping and the strain inside InAs is increased. Processes during quantum dot capping by GaAs are explained with the help of in situ time resolved reflectance anisotropy measurement and atomic force microscopy. Conclusion from in situ monitoring is that first the thin GaAs layer (approx. 2.5 monolayers of GaAs) is formed and then the In migration from dots follows.

Applying InGaAs strain reducing layer on InAs/GaAs quantum dots causes strong red shift of photoluminescence maximum. Possibility to obtain intensive luminescence at 1.55 μm from quantum dots with InGaAs strain reducing layer is demonstrated. However, InGaAs does not seem to be suitable strain reducing layer for quantum dot structures prepared by metal-organic vapour phase epitaxy due to higher growth temperature and consequently weaker localization of electrons in quantum dots.

The GaAsSb SRL is presented as capping layer of choice for 1.3 μm quantum dot lasers prepared by metal-organic vapour phase epitaxy due to more suitable conduction band alignment and due to suppression of In segregation from quantum dots during capping process. New design of GaAsSb strain reducing layer with graded Sb concentration is presented and the record wavelength of intensive ground state photoluminescence maximum 1391 nm was achieved with graded GaAsSb strain reducing layer maintaining type I heterostructure. Intensive electroluminescence at 1340 nm is demonstrated using the GaAsSb strain reducing layer with graded Sb concentration.

Different surfactant flushing methods are compared. A growth interruption with temperature increase to 600 °C may not be a sufficient method in case that GaAsSb strain reducing layer is used and additionally, the GaAs growth at higher temperature helps to prevent transferring of surfactants from the lower to the upper quantum dot layer.

6. Acknowledgements

This work was supported by Czech Science Foundation Projects 202/09/0676, P102/10/1201, and the research program of Institute of Physics AV0Z 10100521. I would like to thank to Jiří Oswald for photoluminescence and electroluminescence measurements, to Ondřej Caha for X-ray measurements, to Aliaksei Vetushka and Pavel Hazdra for atomic force microscopy and especially to Jiří Pangrác for technical assistance and structure growth.

7. References

Akahane, K., Yamamoto, N., & Ohtani, N. (2004). Long-wavelength light emission from InAs quantum dots covered by GaAsSb grown on GaAs substrates. *Physica E - Low dimensional systems & Nanostructure*, Vol. 21, No. 2-4, (March 2004), pp. (295 – 299), ISSN: 1386-9477 , DOI: 10.1016/j.physe.2003.11.016

Akiyama, Y., & Sakaki, H. (2007). MBE growth of self-assembled InGaAs quantum dots aligned along quasi-periodic multi-atomic steps on a vicinal (111)B GaAs surface. *Journal of Crystal Growth*, Vol. 301-2, (April 2007), pp (697 – 700), ISSN: 0022-0248, doi:10.1016/j.jcrysgro.2006.09.021

Bhattacharya, P., & Mi, Z. (2007). Quantum-dot optoelectronic devices. *Proc. of IEEE*, Vol. 95, No. 9, (September 2007), pp. (1723 – 1740), ISSN 0018-9219

Bozkurt, M., Ulloa, J. M., & Koenraad, P. M. (2011). An atomic scale study on the effect of Sb during capping of MBE grown III-V semiconductor QDs. *Semiconductor Science and Technology*, Vol. 26, No. 6, (April 2011), pp. (064007-1 – 064007-11), ISSN: 0268-1242, DOI: 10.1088/0268-1242/26/6/064007

Dasika, V. D., Song, J. D., Choi, W. J., Cho, N. K., Lee, J. I., & Goldman, R. S. (2009). Influence of alloy buffer and capping layers on InAs/GaAs quantum dot formation. *Applied Physics Letters*, Vol. 95, No. 16, (October 2009), pp. (163114-1 – 163114-3), ISSN: 0003-6951, DOI: 10.1063/1.3243688

Goletti, C., Arciprete, F., Almaviva, S., Chiradia, P., Esser, N., & Richter W. (2001). Analysis of InAs (001) surfaces by reflectance anisotropy spectroscopy. *Physical Review B*, Vol. 64, No. 19, (November 2001), pp. (193301-1 – 193301-4), ISSN: 1098-0121, DOI: 10.1103/PhysRevB.64.193301

Haxha, V., Drouzas, I., Ulloa, J. M., Bozkurt, M., Koenraad, P. M., Mowbray, D. J., Liu, H. Y., Steer, M. J., Hopkinson, M., & Migliorato, M. A. (2009). Role of segregation in InAs/GaAs quantum dot structures capped with GaAsSb strain-reduction layer. *Physical Review B*, Vol. 80, No. 16, (October 2009), pp. (165331-1 – 165331-6), ISSN: 1098-0121, DOI: 10.1103/PhysRevB.80.165334

Hazdra, P., Oswald, J., Komarnitskyy, V., Kuldová, K., Hospodková, A., Hulicius, E., & Pangrác, J. (2009). Influence of capping layer thickness on electronic states in self assembled MOVPE grown InAs quantum dots in GaAs. *Superlattices and Microstructures*, Vol. 46, No.1-2, (July - August 2009), pp. (324 – 327), ISSN: 0749-6036, DOI: 10.1016/j.spmi.2008.12.002

Hazdra, P., Oswald, J., Komarnitskyy, V., Kuldová, K., Hospodková, A., Vyskočil, J., Hulicius, E., & Pangrác, J. (2009). InAs/GaAs quantum dot structures emitting in the 1.55 µm band, *IOP Conference Series: Materials Science and Engineering*, Vol. 6, pp. (012007-1 – 012007-4), doi:10.1088/1757-899X/6/1/012007

Henini, M., & Bugajski, M. (2005). Advances in self-assembled semiconductor quantum dot lasers. *Microelectronics Journal*, Vol. 36, No. 11, (May 2005), pp. (950 – 956), ISSN: 0026-2692, DOI: 10.1016/j.mejo.2005.04.017

Hingerl, K., Aspnes, D. E., Kamiya, I., & Florez L. T. (1993). Relationship among reflectance-difference spectroscopy, surface photoabsorption, and spectroellipsometry. *Applied Physics Letters*, Vol. 63, No. 7, (August 1993), pp. (885 – 887), ISSN: 0003-6951, DOI: 10.1063/1.109890

Hoglund, L., Petrini, E., Asplund, C., Malm, H., Andersson, J. Y., & Holtz, P. O. (2006). Optimising uniformity of InAs/(InGaAs)/GaAs quantum dots grown by metal organic vapor phase epitaxy. *Appied Surface Science*, Vol. 252, No. 15, (May 2006), pp. (5525 – 5529), ISSN: 0169-4332, DOI: 10.1016/j.apsusc.2005.12.128

Hospodková, A, Křápek, V., Kuldová, K., Humlíček, J.,. Hulicius, E, Oswald, J., Pangrác, J., & Zeman, J., (2007). Photoluminescence and magnetophotoluminescence of vertically stacked InAs/GaAs quantum dot structures, *Physica E-Low-Dimensional Systems & Nanostructures*, Vol. 36, No. 1, (January 2007), pp. 106-113, ISSN: 1386-9477, DOI: 10.1016/j.physe.2006.09.010

Hospodková, A., Křápek, V., Mates, T., Kuldová, K., Pangrác, J., Hulicius, E., Oswald, J., Melichar, K., Humlíček J., & Šimeček, T. (2007). Lateral-shape of InAs/GaAs quantum dots in vertically correlated structures. *Journal of Crystal Growth*, Vol. 298,

Special Issue, (January 2007), pp. (570 – 573), ISSN: 0022-0248, DOI: 10.1016/j.jcrysgro.2006.10.156

Hospodková, A., Hulicius, E., Pangrác, J., Oswald, J., Vyskočil, J., Kuldová, K., Šimeček, T., Hazdra, P., & Caha, O. (2010). InGaAs and GaAsSb strain reducing layers covering InAs/GaAs quantum dots. *Journal of Crystal Growth*, Vol. 312, No. 8, (January 2010), pp. (1383 – 1387), ISSN: 0022-0248, DOI: 10.1016/j.jcrysgro.2009.10.057

Hospodková, A., Vyskočil, J., Pangrác, J., Oswald, J., Hulicius, E., & Kuldová, K. (2010). Surface processes during growth of InAs/GaAs quantum dot structures monitored by reflectance anisotropy spectroscopy. *Surface Science*, Vol. 604, No. 3-4, (January 2010), pp. (318 – 321), ISSN: 0039-6028, DOI: 10.1016/j.susc.2009.11.023

Hospodková, A., Pangrác, J., Oswald, J., Hazdra, P., Kuldová, K., Vyskočil, J., & Hulicius, E., (2011) Influence of strain reducing layers on electroluminescence and photoluminescence of InAs/GaAs QD structures. *Journal of Crystal Growth*, Vol. 315, No.1 (January 2011) pp. (110–113), ISSN 0022-0248

Johansson, J., & Seifert, W. (2002). Kinetics of self-assembled island formation: Part II - Island size. *Journal of Crystal Growth*, Vol. 234, No. 1, (January 2002), pp.(139 – 144), ISSN: 0022-0248, DOI: 10.1016/S0022-0248(01)01675-X

Kim, J. S., Lee, J. H., Hong, S. U., Han, W. S., Kwack, H.-S., Lee, C. W., & Oh D. K. (2003). Manipulation of the structural and optical properties of InAs quantum dots by using various InGaAs structures. *Journal of Applied Physics*, Vol. 94, No. 10, (November 2003), pp. (6603 – 6606), ISSN: 0021-8979, DOI: 10.1063/1.1621714

Kita, T., Tachikawa, K., Tango H., Yamashita K., & Nishino T. (2000). Self-assembled growth of InAs-quantum dots and postgrowth behavior studied by reflectance-difference spectroscopy. *Appied Surface Scence*, Vol. 159, (June 2000) pp. (503 – 507), ISSN: 0169-4332, DOI: 10.1016/S0169-4332(00)00132-X

Knorr, K., Pristovsek, M., Resch-Esser, U., Esser, N., Zorn, M., & Richter. W. (1997). In situ surface passivation of III–V semiconductors in MOVPE by amorphous As and P layers. *Journal of Crystal Growth*, Vol. 170, No. 1 – 4, (January 1997), pp. (230 – 236), ISSN: 0022-0248, DOI: 10.1016/S0022-0248(96)00629-X

Kong, L. M., Feng, Z. C., Wu, Z. Y., & Lu, W. J. (2008). Emission dynamics of InAs self-assembled quantum dots with different cap layer structures. *Semiconductor Science and Technology*, Vol. 23, No. 7, (May 2008), pp. (075044-1 – 075044-5), ISSN: 0268-1242, DOI: 10.1088/0268-1242/23/7/075044

Krzyzewski, T. J., Joyce, P. B., Bell, G. R., & Jones, T. S. (2002). Role of two- and three-dimensional surface structures in InAs-GaAs (001) quantum dot nucleation. *Physical Review B*, Vol.66, No. 12, (September 2002), pp. (121307-1 – 121307-4), ISSN: 1098-0121, DOI: 10.1103/PhysRevB.66.121307

Liu, H. Y., Hopkinson, M., Harrison, C. N., Steer, M. J., Frith, R., Sellers, I. R., Mowbray, D. J., & Skolnick, M. S. (2003). Optimizing the growth of 1.3 μm InAs/InGaAs dots-in-a-well structure, *Journal of Aplied Physics*, Vol. 93 No. 5, (March 2003), pp. (2931 – 2936), ISSN: 0021-8979, DOI: 10.1063/1.1542914

Liu, H. Y., Sellers, I. R., Hopkinson, M., Harrison, C. N., Mowbray, D. J., & Skolnick, M. S. (2003). Engineering carrier confinement potentials in 1.3-μm InAs/GaAs quantum dots with InAlAs layers: Enhancement of the high-temperature photoluminescence intensity. *Applied Physics Letters*, Vol. 83, No. 18, (November 2003), pp. (3716 – 3718), ISSN: 0003-6951, DOI: 10.1063/1.1622443

Liu, H. Y., Sellers, I. R., Badcock, T. J., Mowbray, D. J., Skolnick, M. S., Groom, K. M., Gutiérrez, M., Hopkinson, M., Ng, J. S., David, J. P. R., & Beanland, R. (2004). Improved performance of 1.3 mm multilayer InAs quantum-dot lasers using a high-growth-temperature GaAs spacer layer, *Applied Physics Letters*, Vol. 85, No. 5, (August, 2004), pp. (704 – 705), DOI: 10.1063/1.1776631

Majid, M. A., Childs, D. T. D., Shahid, H., Airey, R., Kennedy, K., Hogg, R. A., Clarke, E., Spencer, P.,& Murray, R. (2011). 1.52 μm electroluminescence from GaAs-based quantum dot bilayers. *Electronics letters*, Vol. 47, No. 1, (January 2011) pp. 44-45, ISSN: 0013-5194, DOI: 10.1049/el.2010.3203

Majid, M. A., Childs, D. T. D., & Shahid, H. (2011). 1.52 μm electroluminescence from GaAs-based quantum dot bilayers. *Electronics letters*, Vol. 47, No. 4, (February 2011) pp. 289-289, ISSN: 0013-5194, DOI: 10.1049/el.2011.9015

Mowbray, D. J., & Skolnick, M. S. (2005). New physics and devices based on self-assembled semiconductor quantum dots. *Journal of Physics D: Applied Physics*, Vol. 38, No. 13, (June 2005) pp. (2059 – 2076), ISSN: 0022-3727, DOI: 10.1088/0022-3727/38/13/002

Reinhardt, F., Jonsson, J., Zorn, M., Richter, W., Ploska, K., Rumberg, J., & Kurpas, P. (1994). Monolayer growth oscillations and surface-structure of GaAs (001) during metalorganic vapor-phase epitaxy growth. *Journal of Vacuum Science and Technology B*, Vol. 12, No. 4, (July 1994), pp. (2541 – 2546), ISSN: 1071-1023, DOI: 10.1116/1.587798

Richter, W. (2002). In-situ observation of MOVPE epitaxial growth. *Applied Physics A - Materials Science & Processing*, Vol. 75, No. 1, (April 2002), pp.(129 – 140), ISSN: 0947-8396, DOI: 10.1007/s003390101061

Steimetz, E., Wehnert, T., Haberland, K., Zettler, J. T., Richter W. (1998). GaAs cap layer growth and In-segregation effects on self-assembled InAs-quantum dots monitored by optical techniques. *Journal of Crystal Growth*, Vol. 195, No. 1-4, (December 1998), pp. (530 – 539), ISSN: 0022-0248, DOI: 10.1016/S0022-0248(98)00650-2

Takehana, K., Pulizzi, F., Patanè, A., Henini, M., Main, P. C., Eaves, L., Granados, D., & Garcia J. M. (2003). Controlling the shape of InAs self-assembled quantum dots by thin GaAs camping layers, *Journal of Crystal Growth*, Vol. 251, No. 1-4, (April 2003), pp. (155 – 160), ISSN: 0022-0248, DOI: 10.1016/S0022-0248(02)02407-7

Tatebayashi j., Arakawa Y., Hatori N., Ebe H., Sugawara M., Sudo H., & Kuramata A. (2004). InAs/GaAs self-assembled quantum-dot lasers grown by metalorganic chemical vapor deposition - effects of postgrowth anealing on stacked InAs quantum dots. *Applied Physics Letters*, Vol. 85, No. 6, (August 2004), pp. (1024-1026)

Ulloa, J. M., Drouzas, I. W. D., Koenraad, P. M., Mowbray, D. J., Steer, M. J., Liu, H. Y., & Hopkinson, M. (2007). Suppression of InAs/GaAs quantum dot decomposition by the incorporation of a GaAsSb capping layer. *Applied Physics Letters*, Vol. 90, No. 21, (May 2007), pp. (213105-1 – 213105-3), ISSN: 0003-6951, DOI: 10.1063/1.2741608

Wu, H. B., Xu, S. J., & Wang, J. (2006). Impact of the cap layer on the electronic structure and optical properties of self-assembled InAs/GaAs quantum dots. *Physical Review B*,

Vol. 74, No. 20, (November 2006), pp. (205329-1 – 205329-6), ISSN: 1098-0121, DOI: 10.1103/PhysRevB.74.205329

Wasilewski, Z. R., Fafard, S., & McCaffrey, J. P. (1999). Size and shape engineering of vertically stacked self-assembled quantum dots. *Journal of Crystal Growth*, Vol. 201 - 202, (May 1999), pp. (1131 – 1135), ISSN: 0022-0248, DOI: 10.1016/S0022-0248(98)01539-5

Factors Affecting the Relative Intensity Noise of GaN Quantum Dot Lasers

Hussein B. AL-Husseini

Nassiriya Nanotechnology Research Laboratory (NNRL)
Physics Department, Science College, Thi-Qar University, Nassiriya
Iraq

1. Introduction

The motivation for writing this chapter is at least threefold: the extremely promising laser characteristics of GaN quantum dot semiconductor lasers; their vast application potential in high-power and high-temperature optoelectronic devices like optical communications, industrial manufacturing, medicine, military uses, sensing and optical coherent tomography. Low noise is one of the requirements in the wide range of applications that one hopes to arrive at in the use of optoelectronic devices that deals with QD structures. When the size of an electron system reaches the nanometer scale, noise becomes a very interesting problem [2]. Its origin lies in spontaneous emission and shot noise; the last of which is a nonequilibrium fluctuation, caused by the discreteness of the charge carriers. From the investigation of shot noise, one can learn additional information on electronic structure and transport properties. It is directly related to the degree of randomness in carrier transfer [2,3]. In most experimental studies, it is assumed [4] that the frequency fluctuation processes are either infinitely fast, leading to homogeneous broadening, or infinitely slow, leading to inhomogeneous broadening. However, effects of a spectral fluctuation on a picosecond time scale have been observed in experiments, showing that the frequency fluctuation [4] is not always infinitely fast or slow. The random forces have no memory and the correlation of their product function is a delta function. The memory effects [6,7] arise because the wave functions of the particles are smeared out so that there is always some overlap of wave functions. As a result, the particle retains some memory of the collisions experienced through its correlation with other particles in the system.

Quantum dot semiconductor lasers is the subject of intense investigations for the last and this decades owing to their unique properties that arise from 3-D quantum confinement. This is due to the expected characteristics like: Low threshold current density, high characteristic temperature and small linewidth enhancement factor (LEF) [1]. This latter property should lead to an increased tolerance to optical noise with the perspective of achieving isolator-free lasers, of particular importance for low-cost lasers in metropolitan and local area networks [8]. The peculiar property of the semiconductor QDs grown by self-assembling are: 1) the strong compressive strain inside the QDs which leads to a significant deformation of the potentials respect to the bulk case [9], 2) the inhomogeneous distribution of the QD size [10], [11] and 3) the presence of more than one confined energy state inside the QDs [12]. The origin of laser fluctuations in intensity and phase lies in the quantum nature of lasing process itself. These fluctuations are of great importance since they introduce errors in optical communications. In fact, every spontaneous emission event in the oscillating mode, varying the phase of the electromagnetic field (quantum noise) is responsible for the carrier density variation [13].

In this chapter, we report on the RIN performance of GaN/Al$_{0.25}$Ga$_{0.75}$N/AlN QD laser. The influence of power and relaxation time, as well as the intensity noise properties of the input signals, is investigated. It will be shown that improvement in the RIN performance is possible under certain operating conditions. This paper is organized as follows. In Section II, the general aspects of our investigation are presented, as well as the derivation of the RIN in QDs. The results are presented and discussed in Section III. Finally, the current and future developments are given in Section IV.

In the recent years we have seen an ever increasing number of papers and conferences dealing with theory and experiments with QD devices. This has produced a huge amount of information and sometimes contradictory results. As a consequence it can be difficult for new starters finding the right way for getting a basic understanding of the physics of QD devices. In this chapter we have therefore introduced several references to those papers we judged to be fundamental for understanding the basic physics of QD lasers with the hope of providing a useful guideline for those who start adventuring in this field.

2. Relative intensity noise

Relative intensity noise (RIN) is studied by introducing the Langevin noise terms in the laser rate equations (REs) as a driving force where the REs of the carrier-number, or -density, for the active, barrier and SCH layers are used [14, 15]. Here, we modeled the REs by using occupation probability of carriers in the QD, WL and SCH layers. The QDs used here are assumed to be in the form of a disk shape. The laser structure studied is illustrated in Fig.(1). The energy diagram of the QD laser active region is depicted in Fig. (2). QD states are taken into account here by introducing the excited state (ES) in addition to the ground state (GS). The WL, which is in the form of a quantum well, is represented by one energy state [16]. The Pauli-blocking due to state filling is taken into account. The resulting REs system including noise then becomes,

$$\frac{df_{sch}}{dt} = -\frac{f_{sch}(1-f_{wl})}{\tau_{co,sch}} + \frac{f_{wl}}{\tau_{e,wl}} - \frac{f_{sch}}{\tau_{sr}} + F_{f_{sch}}(t) \tag{1}$$

$$\frac{df_{wl}}{dt} = \left(\frac{f_{sch}}{\tau_{co,sch}} + \frac{4f_{ES}}{\tau_{eo,ES}}\right)(1-f_{wl}) - \frac{f_{wl}}{\tau_{e,wl}} - \frac{f_{wl}}{\tau_{co,wl}}(1-f_{ES}) - \frac{f_{wl}}{\tau_{wr}}$$
$$+F_{f_{wl}}(t) \tag{2}$$

$$\frac{df_{ES}}{dt} = \left(\frac{f_{wl}}{4\tau_{co,wl}} + \frac{f_{GS}}{2\tau_{eo,GS}}\right)(1-f_{ES}) - \frac{f_{ES}}{\tau_{eo,ES}}(1-f_{wl}) - \frac{f_{ES}}{\tau_{co,ES}}(1-f_{GS})$$
$$-\frac{f_{ES}}{\tau_r} + F_{f_{ES}}(t) \tag{3}$$

$$\frac{df_{GS}}{dt} = \frac{2f_{ES}(1-f_{GS})}{\tau_{co,ES}} - \frac{f_{GS}(1-f_{ES})}{\tau_{eo,GS}} - \frac{f_{GS}}{\tau_r} - v_g\Gamma g_{GS}S_p + F_{f_{GS}}(t) \tag{4}$$

$$\frac{dS_p}{dt} = v_g\Gamma g_{GS}S_p - \frac{S_p}{\tau_p} + \beta\frac{2f_{GS}}{\tau_r} + F_{f_s}(t) \tag{5}$$

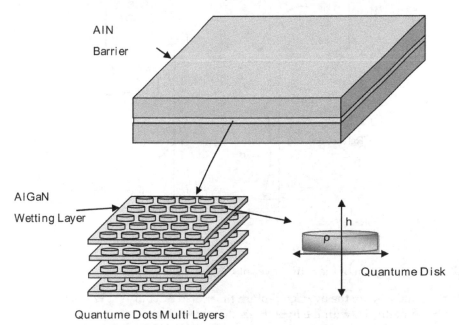

Fig. 1. Schematic illustration of the QD active layer which consists of an arrays of GaN quantum disks grown on an AlGaN wetting layer covered by AlN barrier.

The occupation probabilities for the GS and ES are defined as $f_{GS} = n_{GS}/2N_D$ and $f_{ES} = n_{ES}/4N_D$, where n_{GS} (n_{ES}) is the number of electron-hole (e-h) pairs in the GS (ES), N_D is the total number of QDs, g_{GS} is the GS gain while the SCH and WL populations and the photon occupation are described by the normalized terms: $f_{sch} = n_{sch} N_D$, $f_{wl} = n_{wl} N_D$, and $S_p = n_p N_D$, respectively. n_p is the photon number, Γ is the optical confinement factor. In Eq. (5), the last term on the right-hand side ($2\beta f_{GS}/\tau_r$) represents the rate of photons emitted by spontaneous emission coupled into the lasing mode. Note that, the size inhomogeneity of the dots is included in the expression of the gain. The capture and escape times; in Eqs. (1-4) are given by [1]

$$\left.\begin{array}{l}\tau_{c,sch}^{-1} = (1-f_{wl})\tau_{co,sch}^{-1} \\[4pt] \tau_{c,wl}^{-1} = (1-f_{ES})\tau_{co,wl}^{-1} \\[4pt] \tau_{c.ES}^{-1} = (1-f_{GS})\tau_{co,ES}^{-1} \\[4pt] \tau_{e,GS}^{-1} = (1-f_{ES})\tau_{eo,GS}^{-1} \\[4pt] \tau_{e,ES}^{-1} = (1-f_{wl})\tau_{eo,ES}^{-1}\end{array}\right\} \tag{6}$$

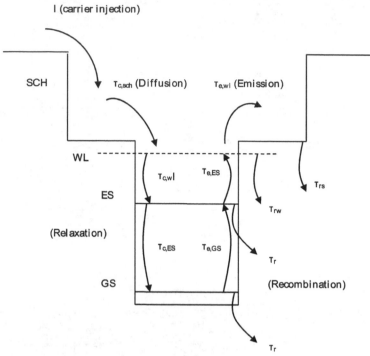

Fig. 2. Energy diagram of the active layer of the QD laser.

$\tau_{c,sch}$, $\tau_{c,wl}$, and $\tau_{c,ES}$ are the average capture times from SCH to WL, WL to ES and from QD ES to GS, respectively, with the hypothesis that the final state is empty. The emission times $\tau_{e,GS}$, and $\tau_{e,ES}$ are the escape times from the GS back to the ES and from the ES back to the WL. τ_{0-1} is the relaxation rate when the state is unoccupied, i.e., $f = 0$. As f approaches 1 the relaxation rate decreases, resulting in the occupation of the upper level. Furthermore, at room temperature and before reaching stimulated emission the system must converge to a quasi-thermal equilibrium characterized by a Fermi distribution of the carriers in all the states. The carrier transport time from the SCH to WL or QD can be expressed as [13],

$$\tau_c = \frac{d_s^2}{2D_{n.p}} \tag{7}$$

Where d_s is the thickness of the SCH layer, D_n and D_p are the diffusion coefficients of electrons and holes, respectively, which can be calculated from the Einstein relation $D_{n,p} = (K_B T / q)\mu_{n,p}$, where K_B, T, $\mu_{n,p}$, correspond to the Boltzmann's constant, lattice temperature, electron and hole mobility, respectively. The thermionic emission time in the QD can be calculated by [15]

$$\tau_e = d_w \left[\frac{2\pi m_e^*}{K_B T}\right]^{1/2} \exp\left(\frac{E_b}{K_B T}\right) \tag{8}$$

where d_w, $m_e{}^*$, E_b denote the thickness of the QD, electron effective mass and the effective barrier energy, respectively. The carrier relaxation time τ_r in general, is about 2.8-4.5 ns for typical quantum well laser [15]. The thickness of the SCH (WL) is 33nm (5nm). In Eqs (1-5) F_{fsch}, F_{fwl}, F_{fES}, F_{fGS} and F_{fs} are the Langevin noise forces. Physically F_{fs} arise from spontaneous emission, while F_{fsch}, F_{fwl}, F_{fES} and F_{fGS} have their origin in the discrete nature of the carrier generation and recombination processes in the SCH, WL, ES and GS region, respectively. The small-signal analysis [1] (along with the Fourier transform) is used to get RIN spectra for these systems. It is defined as $RIN = S_s(w)/S_0^2$ where $S_s(w) = \langle \, |\hat{s}(w)|^2 \, \rangle$ is the spectral fluctuation of photon population around its steady-state value (intensity noise). The angle brackets denotes ensemble average. The small-signal solution of the REs (1-5) yields the following expression

$$\hat{s}(w) = A_{f_{sch}}(w).\hat{F}_{f_{sch}}(w) + A_{f_{wl}}(w).\hat{F}_{f_{wl}}(w) + A_{f_{ES}}(w).\hat{F}_{f_{ES}}(w)$$
$$+ A_{f_{GS}}(w).\hat{F}_{f_{GS}}(w) + A_s(w).\hat{F}_S(w) \tag{9}$$

where $\hat{F}_f$ and $\hat{F}_s$ are the Fourier transforms of the Langevin forces. We recall, however, that these Fourier transforms do not possess a mathematical existence, where

$$A_{f_{sch}}(w) = \frac{a_{21}a_{32}a_{43}a_{54}}{D(w)} \tag{10a}$$

$$A_{f_{wl}}(w) = \frac{a_{32}a_{43}a_{54}(a_{11} + jw)}{D(w)} \tag{10b}$$

$$A_{f_{ES}}(w) = \frac{a_{43}a_{54}\left[-a_{12}a_{21} - (a_{11} + jw)(a_{22} - jw) \right]}{D(w)} \tag{10c}$$

$$A_{f_{GS}}(w) = \frac{a_{54}\left[-a_{23}a_{32}(a_{11} + jw) + a_{12}a_{21}(a_{33} - jw) + (a_{11} + jw)(a_{22} - jw)(a_{33} - jw) \right]}{D(w)} \tag{10d}$$

$$A_s(w) = \frac{\left[a_{12}a_{21}\left(-(a_{33} - jw)(a_{44} - jw) + a_{34}a_{43} \right) + a_{23}a_{32}(a_{11} + jw)(a_{44} - jw) -a_{34}a_{43}(a_{11} + jw)(a_{22} - jw) - (a_{11} + jw)(a_{22} - jw)(a_{33} - jw)(a_{44} - jw) \right]}{D(w)} \tag{10e}$$

The denominator D(w) is given by

$$D(w) = A_o + A_1(jw) + A_2(jw)^2 + A_3(jw)^3 + A_4(jw)^4 + (jw)^5$$

Here A_1, A_2, A_3, A_4 are defined in Appendices A and B. It is clear that the RIN is due to the contributions of the noise sources F_{fsch}, F_{fwl}, F_{fES}, F_{fGS} and F_{fs}. The Langevin forces satisfy the general relations [16, 17]

$$\langle \, F_i(t) \, \rangle = 0$$

$$\langle\, F_i(t)F_j(t')\,\rangle = 2D_{ij}\delta(t-t') \tag{11}$$

where D_{ij} is the diffusion coefficient associated with the corresponding noise source. According to our analysis, we found the following nonzero steady-state diffusion coefficients D_{ij} given by

$$
\left.
\begin{aligned}
D_{f_{sch}f_{sch}} &= \frac{f_{sch}}{\tau_{11}(f_{sch})} \\[4pt]
D_{f_{wl}f_{wl}} &= \frac{f_{wl}}{\tau_{22}(f_{wl})} \\[4pt]
D_{f_{ES}f_{ES}} &= \frac{f_{ES}}{\tau_{33}(f_{ES})} \\[4pt]
D_{f_{GS}f_{GS}} &= v_g\Gamma g_{GS}S_p \\[4pt]
D_{S_p f_{GS}} &= -D_{f_{GS}f_{GS}} \\[4pt]
D_{S_p S_p} &= \beta\frac{2f_{GS}}{\tau_r}S_o
\end{aligned}
\right\} \tag{12}
$$

Where $1/\tau_{ij}$ is the carrier lifetime in the SCH, WL and QD regions [1]. Using Eqs. (11) and (12), the spectral density of the intensity noise is given by the following expression

$$
\begin{aligned}
S_s(w) = &\, 2\left|A_{f_{sch}}(w)\right|^2\frac{f_{sch}}{\tau_{11}(f_{sch})} + 2\left|A_{f_{wl}}(w)\right|^2\frac{f_{wl}}{\tau_{22}(f_{wl})} + \\[4pt]
&\, 2\left|A_{f_{ES}}(w)\right|^2\frac{f_{ES}}{\tau_{33}(f_{ES})} + 2\left|A_{f_{GS}}(w)\right|^2 v_g\Gamma g_{GS}S_p + \\[4pt]
&\, 2\left|A_s(w)\right|^2\beta\frac{2f_{GS}}{\tau_r}S_o + 4\big[\ \mathrm{Re}\big(A_{f_{GS}}(w)\big).\mathrm{Re}(A_s(w)) \\[4pt]
&\, + \mathrm{Im}\big(A_{f_{GS}}(w)\big).\mathrm{Im}(A_s(w))\ \big]\big(-v_g\Gamma g_{GS}S_p\big)
\end{aligned} \tag{13}
$$

Where Re (.) and IM (.) stands respectively, for the real and imaginary parts of the complex argument.

To see the origin of this RIN model let us compare it with the earlier rate equation models. Properties including structure dependence, in this model, are to be discussed elsewhere. Its effect on the modulation response in the carrier transport model is discussed in Ref. [11]. One can see that

1. Setting is $\tau_{co,wl}=0$, $\tau_{e,wl}=\infty$, $\tau_{wr}=\infty$ equivalent to neglecting the effect of the WL. This returns it to the three-level carrier transport model as in Ref. [12].
2. After making the assumption described for (1), this returns it to the earlier one where it becomes only the active region (QD-region here) which is taken in the analysis as in Ref. [8].

Accordingly, the RIN can be split into six components by make each equation refers to only force effecting without the correlation with other forces in the system.

$$(1)\ RIN\big|_{F_{sch}.F_{sch}} = 2\big|A_{f_{sch}}(w)\big|^2 \langle F_{sch}.F_{sch} \rangle$$

$$(2)\ RIN\big|_{F_{wl}.F_{wl}} = 2\big|A_{f_{wl}}(w)\big|^2 \langle F_{wl}.F_{wl} \rangle$$

$$(3)\ RIN\big|_{F_{ES}.F_{ES}} = 2\big|A_{f_{ES}}(w)\big|^2 \langle F_{ES}.F_{ES} \rangle$$

$$(4)\ RIN\big|_{F_{GS}.F_{GS}} = 2\big|A_{f_{GS}}(w)\big|^2 \langle F_{GS}.F_{GS} \rangle \qquad (14)$$

$$(5)\ RIN\big|_{F_{sp}.F_{sp}} = 2\big|A_s(w)\big|^2 \langle F_{sp}.F_{sp} \rangle$$

$$(6)\ RIN\big|_{F_{GS}.F_{sp}} = 4\big[\ \mathrm{Re}\big(A_{f_{GS}}(w)\big).\mathrm{Re}\big(A_s(w)\big)$$
$$+ \mathrm{Im}\big(A_{f_{GS}}(w)\big).\mathrm{Im}\big(A_s(w)\big)\ \big]\langle F_{GS}.F_{sp} \rangle$$

$RIN\big|_{F_{GS}.F_{sp}}$ denotes that the RIN is due to the mixing of the two noise sources, F_{GS} and F_{SP}.

3. Results and discussion

The GaN/Al$_{0.25}$Ga$_{0.75}$N/AlN QD laser structure is taken as an example for relative intensity noise study. GaN dots studied here are assumed to have a disk shape with a radius of (14 nm) and (2nm) height. Its maximum gain (1800 cm^{-1}) appears at (λ=345nm(3.6 eV)). The differential linear gain and the transparency surface carrier density (the carrier density at zero gain) values are calculated, respectively, as (3.35×10^{18} cm^2 and 1.54×10^{12} cm^{-2}). The RIN for GaN/Al$_{0.25}$Ga$_{0.75}$N/AlN QD lasers has been computed using Eq. (13). The parameters used in the calculation are listed in Table 1. Fig. 3 shows contributions of RIN components results due to effect of: SCH (F_{SCH}), WL (F_{WL}), ES (F_{ES}), GS (F_{GS}) and that part results from photon density (F_{SP}). These components are listed in Eq. (14). It is shown that the main noise contribution comes from WL and ES. This also with our conclusion about the chaotic behavior of QD structures [18]. RIN pronounced peak from relaxation oscillations at resonance frequency of 5 GHz, Fig. 4 shows ES contribution to noise (F_{ES}) at two ES recombination times compared with RIN spectrum. Longer recombination time reduces the noise. Fig. 5 shows the recombination time effect on the WL contribution to noise (F_{WL}). It is shown that this time has more pronounced effect on reducing WL contribution. Noise reduction due to recombination can be attributed to reduction of threshold current which can reduce noise. Both figures (4 and 5) show a different situation when the recombination time (τ_{33}) in the QD region becomes the variable parameter. The larger effect of τ_{33} on RIN spectra is obvious . At (τ_{33} 2.8 ps) the relaxation oscillation frequency begins to retard.

Laser parameters
Active region length, L = 900 nm
Active region width, W = 0.1 μm
Number of QD layers, N_w = 10
QD density per unit area, N_d = 5×10^{12} cm^{-2}
Internal loss, α_{int} = 3 cm^{-1}
Spontaneous emission factor, β = 10^{-4}

Table 1. Parameters of the QD material and laser [1]

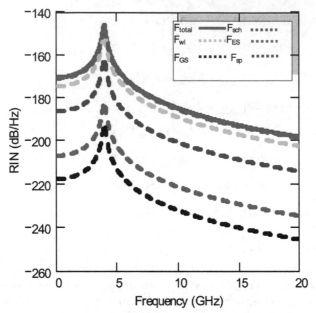

Fig. 3. The RIN of GaN/Al$_{0.25}$Ga$_{0.75}$N/AlN QD lasers with different contributions of RIN parts: SCH part (F_{SCH}), WL part (F_{WL}), ES part (F_{ES}), GS part (F_{GS}). The RIN spectrum for the structure is referred to by F_{total} in the inset.

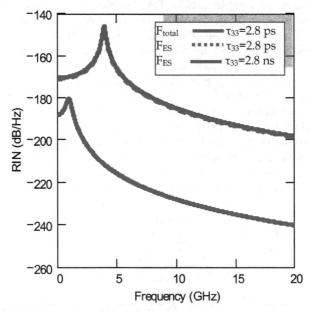

Fig. 4. The RIN of GaN/Al$_{0.25}$Ga$_{0.75}$N/AlN QD lasers compared with the contribution of RIN from ES (F_{ES}) at two QD recombination times. The RIN spectrum for the structure is referred to in the inset as F_{total}.

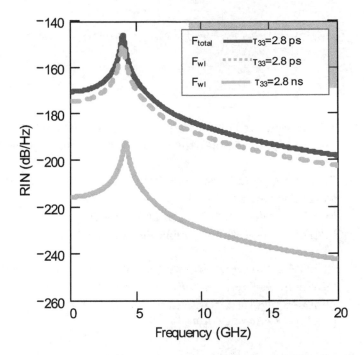

Fig. 5. The RIN of GaN/Al$_{0.25}$Ga$_{0.75}$N/AlN QD lasers compared with the contribution of RIN from WL (F_{WL}) at two QD recombination times. The RIN spectrum for the structure is referred to in the inset as F_{total}.

Appendix A

All the partial derivatives coefficients a_{ij} are real quantities evaluated at steady-state values f_{scho}, f_{wlo}, f_{ESo}, f_{GSo} and S_{pos}.

$$a_{11} = \frac{1}{\tau_{11}(f_{sch})} + \frac{1}{\tau_{12}}(1-f_{wl}), \ a_{12} = -\frac{1}{\tau_{21}}, \ a_{23} = \frac{4}{\tau_{32}}(1-f_{wl})$$

$$a_{21} = \frac{(1-f_{wl})}{\tau_{12}}, \ a_{22} = -\frac{1}{\tau_{22}(f_{wl})} - \frac{1}{\tau_{21}} - \frac{1}{\tau_{23}}(1-f_{ES}), \ a_{32} = \frac{1}{4\tau_{23}}(1-f_{ES})$$

$$a_{34} = \frac{(1-f_{ES})}{2\tau_{43}}, \ a_{33} = -\frac{1}{\tau_{33}(f_{ES})} - \frac{1}{\tau_{32}}(1-f_{wl}) - \frac{1}{\tau_{34}}(1-f_{GS})$$

$$a_{43} = \frac{2}{\tau_{34}}(1 - f_{GS}), \ a_{44} = -\frac{1}{\tau_{44}(f_{GS})} - \frac{1}{\tau_{43}}(1 - f_{ES}) - G_{f_{GS}}S_0$$

$$a_{45} = -G_0 - G_S S_0, \ a_{55} = G_0 + G_S S_{po} - \frac{1}{\tau_p}, \ a_{54} = G_{f_{GS}}S_{po} + \beta\frac{2}{\tau_r}$$

Appendix B

$$A_o = -a_{11}a_{23}a_{32}a_{44}a_{55} - a_{11}a_{22}a_{43}a_{34}a_{55} + a_{11}a_{23}a_{32}a_{45}a_{54} + a_{11}a_{22}a_{33}a_{44}a_{55}$$
$$- a_{12}a_{21}a_{34}a_{43}a_{55} + a_{12}a_{21}a_{33}a_{44}a_{55} - a_{12}a_{21}a_{33}a_{45}a_{54} - a_{11}a_{22}a_{33}a_{45}a_{54}$$

$$A_1 = a_{11}a_{23}a_{32}a_{55} + a_{11}a_{22}a_{43}a_{34} - a_{11}a_{22}a_{33}a_{44} - a_{11}a_{22}a_{33}a_{55} + a_{11}a_{23}a_{32}a_{44}$$
$$- a_{44}a_{11}a_{22}a_{55} + a_{11}a_{22}a_{45}a_{54} + a_{55}a_{34}a_{11}a_{43} - a_{55}a_{44}a_{33}a_{11} + a_{33}a_{11}a_{45}a_{54}$$
$$+ a_{55}a_{44}a_{23}a_{32} + a_{45}a_{23}a_{32}a_{54} - a_{55}a_{34}a_{22}a_{43} + a_{55}a_{44}a_{33}a_{22} - a_{33}a_{22}a_{45}a_{54}$$
$$+ a_{12}a_{21}a_{34}a_{43} - a_{55}a_{12}a_{21}a_{33} - a_{12}a_{21}a_{33}a_{44} - a_{55}a_{44}a_{12}a_{21} + a_{12}a_{21}a_{45}a_{54}$$

$$A_2 = -a_{11}a_{23}a_{32} + a_{11}a_{22}a_{33} + a_{44}a_{11}a_{22} + a_{55}a_{11}a_{22} - a_{34}a_{11}a_{43} + a_{44}a_{33}a_{11}$$
$$+ a_{55}a_{33}a_{11} + a_{55}a_{11}a_{44} - a_{11}a_{45}a_{54} - a_{44}a_{23}a_{32} + a_{55}a_{23}a_{32} + a_{34}a_{22}a_{43}$$
$$- a_{44}a_{33}a_{22} - a_{55}a_{33}a_{22} - a_{55}a_{44}a_{22} + a_{22}a_{45}a_{54} - a_{55}a_{44}a_{33} + a_{55}a_{34}a_{43}$$
$$+ a_{33}a_{45}a_{54} + a_{12}a_{21}a_{33} + a_{55}a_{12}a_{21} + a_{44}a_{12}a_{21}$$

$$A_3 = -a_{11}a_{22} - a_{33}a_{11} - a_{11}a_{44} - a_{55}a_{11} - a_{23}a_{32} + a_{33}a_{22} + a_{44}a_{22} + a_{55}a_{22}$$
$$- a_{45}a_{54} - a_{34}a_{43} + a_{44}a_{33} + a_{55}a_{33} + a_{55}a_{44} - a_{12}a_{21}$$

$$A_4 = a_{11} - a_{22} - a_{33} - a_{44} - a_{55}$$

4. Conclusions

Relative intensity noise characteristics in QD structure are studied using The resulting REs system including noise. This enables us to study relaxation, recombination and emission processes in the QD region. This is impossible with other models. GaN QDs are taken as an example for the study. The following points can be stated:

1. Most of the contributions due to carriers in both wetting layer and excited state of QD have a main contribution in noise. This gives the importance of carrier dynamics in the QD and WL regions .

2. τ_{33} governs the effects of emission from — and recombination outside QD region, then it can reduce the intensity noise at some short value.

From these results one can conclude the important effect of the phonon-bottleneck on intensity noise behavior of these structures. Crystal quality and smaller recombination and relaxation times inside the dot reduce the noise.

5. Acknowledgments

The author express his deep gratitude for the support provided by the Proof. Dr. Amin H. AL-Khursan, head of (NNLR) for project control and guidance.

6. References

[1] H. Al-Husseini, Amin H. Al-Khursan, S. Y. Al –Dabagh, "III-Nitride QD lasers", Open Nanosci J. 3, 1-11, 2009.

[2] H.K. Zhao, "Shot noise in the hybrid systems with a quantum dot coupled to normal and superconducting leads", Phys. Lett. A 299 (2002) 262.

[3] K.Y. Lau, in: P.S. Zory (Ed.), Quantum Well Lasers, Dynamics of Quantum Well Lasers, Academic Press, New York, 1993, p. 252 (Chapter 5).

[4] H.J. Bakker, K. Leo, J. Shah, K. Kohler, Phys. Rev. B 49 (1994) 8249.

[5] G.P. Agrawal, N.K. Dutta, Long Wavelength Semiconductor Lasers, Van Nostrand Reinhold, New York, 1986 (Chapter 9).

[6] D. Ahn, IEEE J. Select. Top. Quantum Electron. 1 (1995) 301.

[7] D. Ahn, Phys. Rev. B 51 (1995) 2159.

[8] T. Steiner, Semiconductor Nanostructures for Optoelectronic Applications. British Library Cataloguing in Publication Data London: Boston, 2004.

[9] M. Grundman, O. Stier, and D. Bimberg, "InGs/GaAs pyramidal quantum dots: Strain distribution, optical phonons, and electronic structure," Phys. Rev. B 52, vol. 52, no. 16, pp. 11 969 – 11 981, October 1995.

[10] O. Qasaimeh, "Efffect of inhomogeneous line broadening on gain and differential gain of quantum dot lasers," IEEE Tran. Electron. Dev., vol. 50, no. 7, pp. 1575–1581, July 2003.

[11] M. Gioannini, "Numerical modelling of the emission characteristics of semiconductor quantum dash materials for lasers and optical amplifiers," IEEE J. Quantum Electron., vol. 40, no. 4, April 2004.

[12] A. Markus, J. Chen, O. G.-L. J. Provost, C. Paranthoen, and A. Fiore, "Impact of intraband relaxation on the performance of a quantum-dot laser," IEEE J. Select. Top. Quantum Electron., vol. 9, no. 5, pp. 1308–1314, September/October 2003.

[13] C. Z. Tong , S. F. Yoon, C. Y. Ngo, C. Y. Liu, W. K. Loke, "Rate equations for 1.3-μm dots-under-a-well and dots-in-a-well self-assembled InAs-GaAs quantum-dot lasers", IEEE J. Quantum Electron., 42, 1175-83, 2006.

[14] Amin H. Al-Khursan, "Intensity noise characteristics in quantum-dot lasers: Four-level rate equations analysis" J. Lumin, 113, 129-36, 2005.

[15] S. Schulz, G. Czycholl, "Spin-orbit coupling and crystal-field splitting in the electronic and optical properties of nitride quantum dots with a wurtzite crystal structure", Condens. Matter Mater. Sci., 1, 2436, 2008.

[16] U. Goesele, S. Senz, V. Schmidt, Y. Wang, C. Hangzhou, "Process for group III-V semiconductor nanostructure synthesis and compositions made using same" US7557028 (2009).

[17] M. Gioannini, Amin H. Al-Khursan, G. A.P. Thé and I. Montrosset, "Simulation of quantum dot lasers with weak external optical feedback", Dynamics Days Conference, Delft-Netherlands, 2008.

[18] H. Al-Husseini, Amin H. Al-Khursan, "Relative Intensity Noise for Self-Assembled III-Nitrides Quantum-Dot Lasers", Recent Patents on Electrical Engineering J, 3, 211-217,2010.

Quantum Dots as a Light Indicator for Emitting Diodes and Biological Coding

Irati Ugarte[1], Ivan Castelló[2],
Emilio Palomares[2] and Roberto Pacios[1]
[1]*Ikerlan S. Coop*
[2]*ICIQ Institut Català d'Investigació Química*
Spain

1. Introduction

Quantum dots (QDs) are inorganic semiconductor particles that exhibit size and shape dependent optical and electronic properties (Alivisatos, 1996; Smith & Nie, 2010). Due to the typical dimension in the range of 1-100 nm, the surface-to-volume ratios of the materials become large and their electronic states become discrete. Moreover, due to the fact that the size of the semiconductor nanocrystal is smaller than the size of the exciton, charge carriers become spatially confined, which raises their energy (quantum confinement). Thus, the size and shape-dependent optoelectronic properties are attributed to the quantum confinement effect. Because of this effect, light emission from these particles can be tuned, throughout the ultraviolet, visible and near infrared spectral ranges.

Quantum dots possess many advantages that make them interesting for several applications:

- They show symmetrical and **narrow emission** spectra and **broad absorption** spectra, enabling that a single light source can be used to excite multicolour quantum dots simultaneously without signal overlap.
- They have a **brighter emission** and a **higher signal to noise ratio** compared with organic dyes.
- Their **stability** is due to its inorganic composition which reduces the effect of photobleaching compared to organic dyes.
- The **lifetime** of the excited states that give rise to fluorescence in quantum dots is about 10 to 40ns, which is longer than the few nanoseconds observed for organic dyes.
- The **large Stokes shift** (difference between peak absorption and peak emission wavelengths) reduces autofluorescence, which increases sensitivity to detect particular wavelengths.
- QDs have **high quantum yields** in the visible range (0.65-0.85 for CdSe) as well as for the NIR (0.3-0.7 for PbS), while organic dyes are moderate in the NIR (0.05-0.25).

Figure 1 shows an example of different size CdSe based nanoparticles narrow emission and broad absorption spectra. Both were excited at the same wavelength of 390 nm to record the emission.

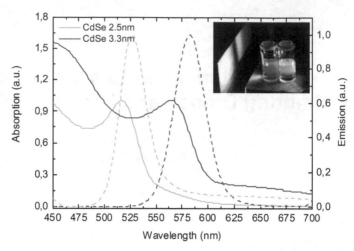

Fig. 1. Emission and absorption spectra of different size CdSe nanocrystals.

All these features make QDs excellent candidates for fluorescent probes. Such multicolour emission enables multiplexed optical coding of biomolecules for biomedical applications, for molecular and cellular imaging and ultrasensitive detection (Bruchez eta al., 1998; Han et al., 2001).

Moreover, QDs are not only used for biological detection technologies but diverse applications. Intensive research is carried out for their use in optoelectronic devices, such as light emitting diodes (Coe eta al., 2002; Colvin et al., 1994), thermoelectric devices (Harman et al., 2002) and photovoltaic devices (Milliron et al., 2005).

In addition, QDs inserted in an electroluminiscent polymer matrix can modify the intrinsic emission of the polymer, very interesting for many industry applications.

Taking into account the advantageous properties of colour tunability of these inorganic semiconductor particles, different research technologies along with the latest experimental results on both topics –biological detection and electroluminiscent devices- will be discussed throughout this chapter.

2. Quantum dots for electroluminescence devices

A new technology based on organic light emitting diodes (OLEDs) is revolutionizing the display industry that has been dominated in the past three decades by liquid-crystal displays (LCDs). OLEDs are no longer an exclusively research activity but approaching industrial applications. However, QD-based light emitting diodes (QD-LEDs) have emerged as a competitive choice and they have attracted intense research and commercialization efforts over the last decade. Colloidal quantum dots, which consist of nanometer diameter inorganic semiconductor crystals surrounded by surface passivating organic ligands, are particularly attractive because they can be solution processed and are compatible with low cost, OLED printing techniques. Although this QD-LED technology is still relatively new, it presents promising advantages compared to its organic counterpart. These are as follows:

- Full width at half maximum (FWHM) of the emission peak from QDs is only 20-30nm, depending on the monodispersity achieved during the colloidal synthesis. Besides, organic electroluminescent materials' FWHM is more than 50nm. Therefore, QD-LEDs present better potential for high quality image and colour contrast and saturation. Attending to the Commission International de l'Eclairage (CIE) chromaticity diagram which defines different colour saturation and hues, the triangle representing possible QD emission is larger than the HDTV (High Definition Television) standard. In figure 2, both HDTV and QD emission triangles are represented in the CIE chromaticity diagram.

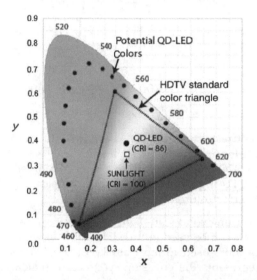

Fig. 2. CIE chromaticity diagram showing that the spectral purity of QDs enables a colour gamut larger than the HDTV standard. Source: (Wood & Bulovic´, 2010).

- Under operating at high brightness and/or high current, *Joule* heat could turn towards device degradation problems. Inorganic materials' thermal stability is usually higher than organic ones. Therefore, inorganic materials based devices are expected to exhibit longer lifetimes.
- Different colours for OLEDs are achieved by different materials, involving different degradation rates and lifetimes for each one. This means that display colour of these devices generally changes with time. However, in QD-LEDs all of the three primary colours can be obtained with the same composition changing the particle size. Due to the same composition and thus, similar lifetime, the display colour variation in quantum-dot-based devices could be less pronounced.
- The emission spectra for QD-LEDs can reach the IR range, while is rather difficult to overcome the 650nm peak emission of OLEDs. With QDs, for example, CdSe of different sizes provide emission from the blue through the red, while QDs made of a smaller band gap material, such as PbSe, PbS, or CdTe, offer spectral tunability in the NIR spectral region. In figure 3, different size and composition QD emission spectra are depicted. It demonstrates the possibility to achieve a really wide range of different emission wavelengths.

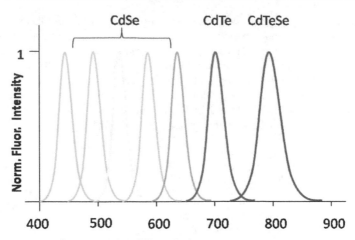

Fig. 3. Different size and composition QD emission spectra.

- Spin statistics are not restrictive for QDs. In organic fluorescent materials, where the singlet to triplet state formation ratio is 1:3, the efficiency is limited to 25% because only the singlet state recombination results in fluorescence (Carvelli et al., 2011). Therefore, theoretically, using QDs higher external quantum efficiency (EQE) could be achieved comparing to organic electroluminescent materials.

On the other hand, concerning solid state lighting applications, warmer white is preferred over the bluish white. QDs can be used to tune the quality of white lighting; by adding QD lumophores to the phosphors used to fabricate the LED it is possible to achieve warmer white light. Figure 4 represents the emission spectra of a standard LED bulb and a QD added LED bulb.

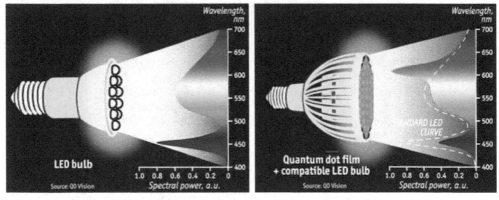

Fig. 4. Standard and QD added LED bulb emission spectra. Source: QD Vision.

The basic structure of QD-LED is similar to the OLED arquitecture. Basically, a QD layer is sandwiched between electron and hole transporting materials. An applied electric field causes electrons and holes to move into the QD layer where they recombine forming excitons and emitting photons.

Figure 5 represents the basic structure and the electron-hole recombination that occurs in a QD-LED device.

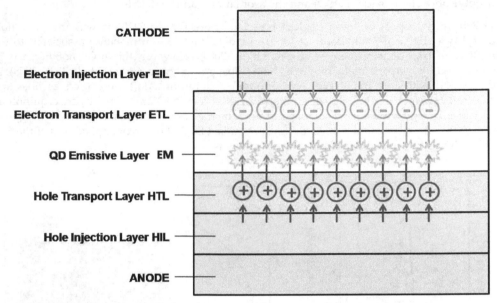

CATHODE

Electron Injection Layer EIL

Electron Transport Layer ETL

QD Emissive Layer EM

Hole Transport Layer HTL

Hole Injection Layer HIL

ANODE

Fig. 5. QD-LED basic structure and exciton generation.

Although the QD-LED potential is impressive, pure organic LEDs are still brighter, they have longer lifetimes and show much higher efficiency than QD-LEDs. OLED devices are already on the market and are widely used in several applications. However, QD-LED will surely find a niche in the market owing to their high colour purity, as the emission linewidth is in the order of 30 nm and the wavelength can, in principle, be tuned continuously from UV to infrared. Moreover, the combination of both organic materials and inorganic QDs in optoelectronic devices can be an interesting approach to achieve efficient and high-throughput products.

3. Polymer-QD blend light emitting diodes

The integration of organic and inorganic materials at the nanometre scale into hybrid light emitting devices (HYLEDs) combines the ease of processability of organic materials with the high-performance electronic and optical properties of inorganic nanocrystals.

The first QD-polymer-based hybrid electroluminescent (EL) device was demonstrate in 1994 (Colvin et al., 1994), showing relatively low EQE (0.01%). They used spin-coated poly-(phenylene vinylene) (PPV) polymer as hole transport layer and a multilayer film of cadmiun selenide nanocrystals as emitting as well as electron transport layer. PPV was also used in many other early QD-LED devices (Mattoussi et al., 1998; Schlamp et al., 1997) but the EQE was still low (<1%).

In 2002, a QD-LED structure only consisting on a single monolayer of QDs, sandwiched between two organic thin films demonstrated considerably better device performance (Coe et al., 2002). This sandwiched structure was achieved making use of the material phase segregation effect that takes place between the QD aliphatic capping groups (in this case trioctyl phosphine oxide, TOPO) and the aromatic organic materials.

The limiting aspect of these devices based on pure CdSe QD films was the poor hole mobility ($\sim 10^{-12}$ cm^2/V in contrast to $\sim 10^{-4}$ cm^2/V for electron mobility) associated to the high hole trap density (Ginger & Greenham, 2000). However, this hole mobility can be compensated by using metal oxides as hole transporting layers (Caruge et al., 2006, 2008). In this work NiO and tris-(8-hydroxyquinoline) aluminium (Alq$_3$) were used as hole and electron transporting layers, respectively. In this work NiO and tris-(8-hydroxyquinoline) aluminium (Alq$_3$) were used as hole and electron transporting layers, respectively. Figure 6 represents the electroluminescence (EL) spectra and EQE at different applied bias of the first QD-LED using metal oxide charge transport layers.

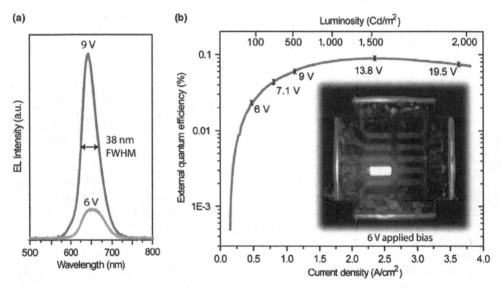

Fig. 6. a) EL spectra and b) external quantum efficiency at different applied bias of the QD-LED developed by Caruge et al.

Besides metal oxides, different charge transport layers have been used for QD-LEDs. Usually organic materials present faster hole mobility than QDs. Therefore, as it was mentioned before, first QD-LEDs were PPV-based devices. Other several research studies carried out in this field used N-N'-diphenyl-N,N'-bis(3-methylphenyl)-1-1'biphenyl-4,4'diamine (TPD) as hole transport layer (Coe-Sullivan et al., 2005; Coe et al., 2002; Sun et al., 2007). Blue emitting electroluminescent poly(9,9-dioctylfluorene) (PFO), for example, allows hole mobility in the range of $\sim 10^{-3}$-10^{-5} cm^2/V depending on the processing technique (Kreouzis et al., 2006). A single active layer consisting of PFO doped with CdSe QD showed around 0.5% EQE (Campbell & Crone, 2008).

Further optimization was obtained by multilayer QD-LEDs, where the thicknesses of the QD and organic layers could be varied independently (Zhao et al., 2006). By optimizing the thicknesses of the constituent QD layers of devices, improved electroluminescent efficiency (around 2-3 cd/A), low turn-on voltages (3–4 V) and long operation lifetimes were developed (Sun et al., 2007).

Table 1 summarizes the performance evolution for different QD-LED device arquitectures reported in literature.

Emitting λ (nm)	$V_{turn-on}$ (V)	L_{max} (cd/m²)	Efficacy (cd/A)	Efficiency (lm/W)	EQE (%)	Device structure	Citation
610	4	100	--	--	> 0.01	ITO / CdSe / PPV / Mg & ITO / PPV / CdSe / Mg	(Colvin et al., 1994)
560	4	600			0.2	ITO / PPV / CdSe – CdS / Mg:Ag	(Schlamp et al., 1997)
600	3-3.5				0.1	ITO / PPV (multilayer) / CdSe / Al	(Mattoussi et al., 1998)
560	3.5	2000	1.9		0.52	ITO / TPD / CdSe-ZnS / Alq₃ / Mg:Ag & ITO / TPD / CdSe-ZnS / TAZ / Alq₃ / Mg:Ag	(Coe et al., 2002)
615	3	7000	2.0	1.0	2	ITO / TPD / CdSe / Alq₃ / Mg:Ag / Ag	(Coe-Sullivan et al., 2005)
625		3000			0.18	ITO / NiO / CdSe-ZnS / Alq₃ / Ag:Mg / Ag	(Caruge et al., 2006)
517 (G) 546 (Y) 589 (O) 600 (R)	4 (G) 5 (Y) 3 (O) 3 (R)	3700 (G) 4470 (Y) 3200 (O) 9064 (R)	1.1-2.0 (G-R)	<1.1		ITO / PEDOT:PSS / poly-TPD / CdS-ZnS / Alq₃ / Ca:Al	(Sun et al., 2007)
600	1.9	12380		2.41		ITO / PEDOT:PSS / TFB / CdSe-CdS-ZnS / TiO₂ / Al	(Cho et al., 2009)
470 (B) 540 (G) 600 (R)		4200 68000 31000		0.17 8.2 3.8		ITO / PEDOT:PSS / poly-TPD / CdS-ZnS / ZnO / Al	(Lei Qian et al., 2011)

Table 1. Performance evolution of different QD-organic LED devices.

Technological development and the accumulation of fundamental knowledge, such as designing new device structures and formulating the composition and structure of the QDs, has led to rapidly improve the performance of these devices. As a result, combining an organic hole transport layer and an inorganic metal oxide electron transport layer with the emissive QD films (Cho et al., 2009; Lei Qian et al., 2011) devices showed maximum luminance and power efficiency values over 4200cd/m² and 0.17lm/W for blue emission, 68000cd/m² and 8.2lum/W for green emission, and 31000 cd/m² and 3.8lm/W for red emission. Moreover, with the incorporation of ZnO nanoparticles, these devices exhibit high environmental stability and longer operating lifetimes.

On the other hand, the integration of QDs into a full-colour LED structure requires successful individual red-green-blue (RGB) QD patterning for a pixelated active display panel. The most widely used method to fabricate monochromatic devices, spin coating, results in mixing the pixel colours. More recently, inkjet (Tekin et al., 2007) and contact printing (L. Kim et al., 2008; T.H. Kim et al., 2011) processes have been proposed as alternative methods for QD patterning. These techniques demand intensive efforts to optimize film uniformity and flat surface morphology in order to ensure the good performance of devices.

Although improving the performance is the key requisite in all optoelectronic devices, other characteristics, as colour tunability of a single emissive layer or ease of processing, could give QD-LEDs an extra added value to be used in specific applications.

3.1 Light tunability of hybrid polymer-QD-based LEDs. Applications as light indicators

Light tunability is an attractive characteristic for a diverse amount of application fields. In this context, QD-LEDs have attracted great attention because they are able to tune the outcomig light by not only varying the composition but also by controlling the size and shape of the QD.

In patterned matrix display systems, a variety of colours are possible by changing the relative intensities of the red-green-blue (RGB) subpixels. Moreover, combining QDs with organic emitting materials can provide a wide range colour emitting devices (Anikeeva et al., 2009).

In addition to displays, different colour emission from the same device would be highly desirable for some applications like ambient light at wish and distinctive indicators of a systems state (polarity, voltage, pressure indicators, etc.). Therefore, all RGB components together in the same film are required.

Unlike green and red components, the blue component (440-490nm) of the visual spectrum is characterized by low luminous efficacies. Besides, the diminutive dimension QDs needed to achieve the blue component adds more difficulty to enhance the efficiency of bright-blue QD-LEDs. Therefore, the synergic combination of organic and inorganic semiconducting materials can lead to obtain a single layer that contains all RGB components. As it was mentioned before, PFO, whose hole mobility is much higher than CdSe, can be the blue emitter while green and red emission can be achieved from quantum dots. In figure 7, all RGB components from both organic and inorganic materials are represented: the blue emission corresponds to PFO while green and red emissions correspond to different size CdSe QDs.

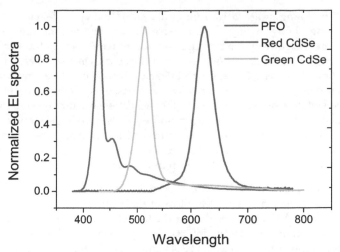

Fig. 7. All RGB components from blue emitting PFO (organic), green emitting 2.2nm CdSe particles and red emitting 5.8 nm CdSe particles.

Once all RGB components can be achieved, white emission is also feasible by adjusting the balance between these three constituents. White emission from polymer-QD based LEDs is very attractive for mainly lighting purposes. For general room lighting, the challenge to obtain low consumption devices combined with the ability to vary the colour temperature of a lamp is the major advantage offered by this type of light emitting devices. Moreover, due to the mechanical properties of the organic films, the possibility to achieve very thin and even flexible devices could allow new and modern lighting designs.

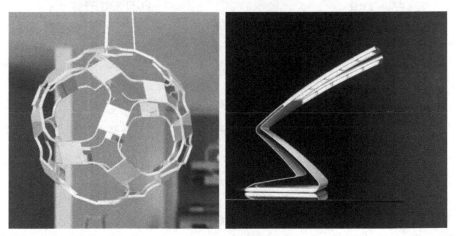

Fig. 8. Ultrathin and flexible character of OLED and HYLEDs could allow new and modern lighting designs. Different OLED lamp prototype from Novaled.

Several research studies demonstrate that it is possible to reach white light in organic or hybrid polymeric-inorganic diodes by blending different materials with emission in RGB

(Chou et al., 2007; Huang et al., 2010; Luo et al., 2007). Luo et al. developed high efficiency white-light emitting diodes using a single copolymer. Fluorescent red, green and blue chromophores were introduced in a conjugated polymer backbone as emissive layer, resulting in a stable white light with 6.2 cd/A luminance efficiency. On the other hand, in the work developed by Chou et al. the fabrication and characterization of white-emitting HYLEDs was reported. The emissive layer of these devices was based on PFO as the blue emitter and CdSe/ZnS as the yellow emitter. Contrary to the multilayer structure used in this work, in the activity performed by Huang et al. the green and red QDs were incorporated into the PFO as the single emissive layer. Minimizing the number of layers that have to be used involves simplifying the device structure and the fabrication process, an important requisite in order to be potentially introduced in the market as high throughput and low cost devices.

Green emission from oxidized PFO has been in detail studied mainly by means of photoluminescence (Ferenczi et al., 2008; Sims et al., 2004). These works demonstrate that a green emission band at 535nm arises due to the formation of oxidation-induced fluorenone defects. The generation of fluorenone defects during diode operation has been documented in different experiments (Ouisse et al., 2003; Ugarte et al., 2009). These studies conclude that the spectral shift of polyfluorene-based LEDs was enhanced by the current passing through the device. One could benefit from this characteristic in order to obtain combined blue and green emission from a single polymer emissive layer.

In figure 9, deliberately oxidized PFO and non-oxidized PFO based OLEDs EL spectra are depicted. The polymer solution oxidation was achieved by annealing in excess the original polyfluorene solution.

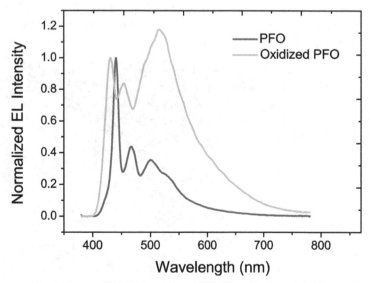

Fig. 9. EL spectra for different OLEDs based on PFO (blue line) and deliberately oxidized PFO solution (green line).

Additionally, blending the polymer solution with red emitting colloidal CdSe QDs, all three RGB components are possible in the same emitting film.

Blending two different materials could give rise to Förster Resonance Energy Transfer (FRET) if certain conditions are fulfilled. Basically, the absorption of the guest has to overlap the emission of the host. If additionally, these different materials (polymer chains and QD nanoparticles) are within a minimum distance (Förster radius) the emitted energy by the host is absorbed and re-emitted by the guest (Buckley et al., 2001; Cabanillas-Gonzalez et al., 2005). Therefore, in order to make use of emission from both materials a very precise control of the nanoscale phase-segregation is necessary.

In this way, by varying polymer:QD ratio, colour tunable light emitting devices can be achieved. In figure 10, different ratio PFO-CdSe devices EL emission is depicted.

For a certain PFO:QD ratio, emission from RGB components are balanced and the desired white emission is also feasible (green line in figure 9).

On the other hand, the observed tunability does not only depend on the polymer:QD ratio but also on the applied bias. Voltage tunable emission has been reported in both polymer-QD (Chou et al., 2007; Ugarte et al., 2009) and polymer-polymer (Berggren et al., 1994) devices. Materials with different composition and miscibility properties that consequently tend to phase separate, can simultaneously emit if they are excited at once. Due to the different charge injection properties in both phases, light emission from these micrometre-sized domains depends on the turn on applied voltage. In this case, for low applied bias emission from the polymer is predominant, while increasing the voltage, emission from both materials allows to redshift the spectra.

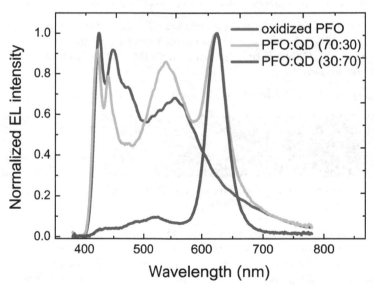

Fig. 10. Oxidized PFO-based device (blue line) and different ratio PFO-QD-based device (red and green lines) emission.

Figure 11 shows the voltage dependent EL spectrum of a HYLED containing PFO:CdSe (70:30) electro-active film. The same device can emit towards either blue or red depending on the externally applied voltage.

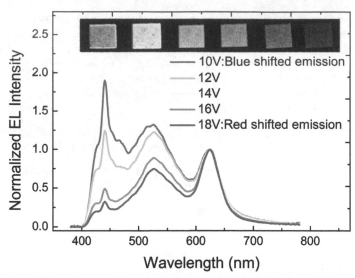

Fig. 11. Colour tunability varying the voltage in a PFO-QD (70-30) based emissive film.

This interesting characteristic, the voltage tunability that covers the whole visible spectrum, will give an extra added value in optoelectronic devices for many applications; from general lighting to distinctive indicators of a system state (voltage, pressure, pH etc.).

Recently, other studies demonstrate a monolithic, full-colour, tunable inorganic LED (Hong et al., 2011). This inorganic device is based on multifacetted gallium nitride (GaN) nanorod arrays with $In_xGa_{1-x}N$/GaN multiple quantum wells (MQWs) anisotropically formed on the nanorod tips and sidewalls.

In figure 12 is shown the LED developed by Hong et al. at different applied bias.

Fig. 12. Color tunability varying the voltage in a PFO-QD (70-30) based emissive film. Source: (Hong et al., 2011).

Although lower voltage values are needed for these purely inorganic devices, they require more complex fabrication processes that could limit their use to economic and low consumption devices. Therefore, QD-polymer LEDs present the simplicity and advantages of a monolithic and single-layer device with the thin, flexible and ease of processing characteristics provided by organic optoelectronic technology.

4. Different QD blends for biological coding

Biomedicine and bioengineering are research areas where most important and revolutionary advances have been carried out during the last decades. Analysis of biological molecules is one of the most important activities in these fields, where a large number of proteins and nucleic acids have to be rapidly studied and/or screened. Thus, the development of a fast and efficient coding procedure for biomolecules that could allow high-throughput analysis would be desirable. In this context, the possibility to achieve new labels and barcodes that can be miniaturized is one of the major challenges. Up to date, a variety of multiplex encoding techniques (chemical, physical, electronic, spectrometric etc.) have been developed. One of the most attractive strategies, the spectrometric encoding, provides a large quantity of codes combined by operational simplicity. Amongst the different available spectra, as Raman or FTIR, fluorescence-based methods are probably the simplest in decoding. Moreover, due to the variety of fluorescence techniques stablished on the market, fluorescence is one of the most important encoding method.

Although so far organic dyes have mostly been used as fluorescent codes, the use of QD nanoparticles has attracted great interest in the last years. Compared with traditional organic dyes, fluorescent QDs offer unique photochemical and photophysical properties (Resch-Genger et al., 2008). Moreover, they show longer photostability and lower photobleaching compared to organic dyes (Bruchez Jr et al., 1998; Wu et al., 2003). Figure 13 represents the comparison of the fluorescence intensity evolution in time between organic fluorophores (Alexa 488) and inorganic quantum dots (QD 608).

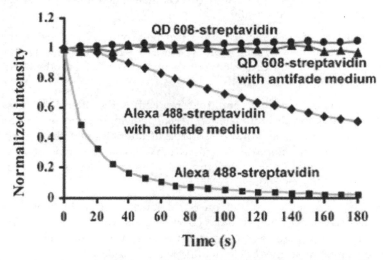

Fig. 13. Fluorescence intensity evolution in time of QD 608 and organic dye Alexa 488, measured every 10 s for 3 minutes. QD fluorescence intensity remains stable while the organic dye is degraded after few seconds. Source: (Wu et al., 2003).

Their principal advantage over their organic counterparts is that due to their symmetrical and narrow emission spectra and broad absorption spectra it is possible to use a single light source to simultaneously excite different quantum dots. Therefore, exciting different QD blends at a certain wavelength, even small amounts of QD are clearly detectable in the fluorescence spectra. Figure 14 represents the PL spectra of different blends obtained by mixing, in different ratios, individual CdSe populations. One of them emits at 532nm while the other emits at 582nm.

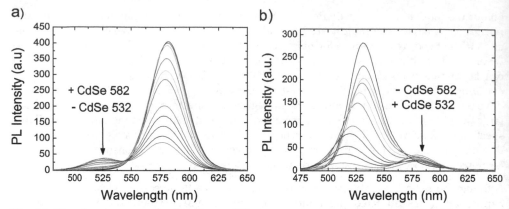

Fig. 14. Photoluminance intensity of two different size quantum dot blends. a) Addition of small amounts of 532nm emitting CdSe nanoparticles to 582nm emitting population are detectable by photoluminance measurements. b) In the same way, small amounts of 582nm emitting QD in 532nm population are detectable.

In addition, due to the larger stokes shift of QDs their autofluorescence is considerably reduced leading to an increased sensitivity.

Therefore, in the last decade these robust and bright light emitters have opened the door to a variety of biological applications, such as fluorescent labels for cellular labeling, deep-tissue and tumor imaging agents and drug-delivery nanoparticles amongst others (Alivisatos et al., 2005; Wu et al., 2003).

Concerning fluorescence labelling, multiplexing with QDs offers the possibility of simultaneous excitation and discriminated emission, allowing more compact and less step required detection systems. The working principle of multiplexed coding is based on the combination of different emission wavelength (colour) QDs and different intensity levels. In this way, combining, for example, 6 colours and 10 intensity levels would theoretically enable one million of different possible codes.

In figure 15, the PL spectra and the measured intensity as a function of the blend composition are represented. The particular composition of each blend can be identified by both QD intensity levels and emission wavelength peaks. At the same time, each blend can be associated to a target molecule for bioidentification applications.

Embeding the nanoparticles into microbeads is crucial in order to allow binding to biomolecules by functionalization of these fluorescent labels. Moreover, in case of non-water soluble QDs, it is vital to ensure the biocompatibility of QDs with the biological environment.

Therefore, the ability to incorporate multicolour QDs into micron-size polymer beads at precisely controlled ratios is the critical step preparation for efficient, homogeneous and biocompatible encoding labels. Currently, there are two basic technologies mostly used for multiplexed QD encoded beads:

1.   Trapping of different colour QDs directly into the beads.

2.   Direct *in situ* encapsulation of QDs during the bead synthesis and later shell formation of different colour QDs on the surface of the bead.

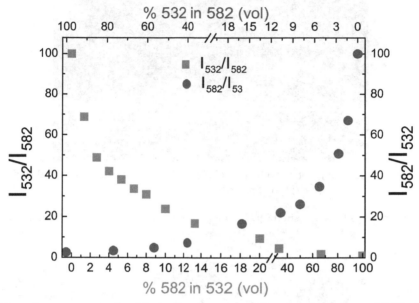

Fig. 15. Intensity pattern obtained from blending two different populations of CdSe QD in different ratios.

Trapping of QDs directly into polymeric beads was first developed by Nie and coworkers (Han et al., 2001). Dissolving the polymeric beads (in this case polystyrene) in chloroform/alcohol QDs solution, the inorganic nanoparticles were embedded into the spheres due to the swelling of the beads. Not only single colour QDs were inserted by this method but also three colour QDs in different ratios were combined increasing the number of possible encoding labels.

Figure 16 represents a fluorescence image obtained from a mixture of CdSe/ZnS QD-tagged beads emitting different single-colour signals (Han et al., 2001). First, different colour QD populations were synthesized and embedded into polystyrene beads; then, different beads were blended altogether in a single dispersion and finally these beads were spread on a glass substrate.

The direct *in situ* encapsulation of QDs consist of embedding the nanoparticles into the spheres during the polymerization of the beads. In order to achieve multiplexed colour encoded tags, a different colour QD shell has to be formed on the surface of previously synthesized bead (Ma et al., 2011). In the work developed by Palomares and coworkers multiplexed colour encoded nanospheres were fabricated encapsulating different colour QD/silica layers and result in highly efficient photoluminescent nanospheres with monodisperse, photostable, and excellent luminescence properties. Moreover, this technique based on a "layer by layer" QD-bead system ensures to avoid any possible electron transfer between different QD populations due to the physical silica barrier separation.

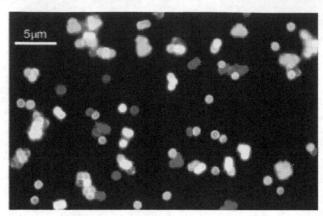

Fig. 16. Fluorescence image of a mixture of CdSe/ZnS tagged beads emitting different single-colour signals. Source: (Han et al., 2001).

Figure 17 represents the core-shell QD/silica procedure to prepare three different colour QD beads.

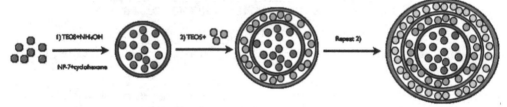

Fig. 17. Illustration of the procedure used to prepare of multiplexed colour encoded silica nanospheres encapsulating QDs multilayers. Source: (Ma et al., 2011).

The combination of different size-colour QDs for core and shell fabrication with the variation of different QD ratios enables multiple different labels or codes that could allow high-throughput analysis of biomolecules.

5. Conclusion

Due to their unique electrical and optical properties, quantum dots are excellent candidates to be used as indicators in very diverse applications. Not only size, shape and composition dependent colour but also high quantum yield and stability against photobleaching are the key of their potential success in many areas from electronics to biology and medical diagnostics.

In the field of light emitting diodes, they can be combined either with other different size QDs or organic electroluminiscent materials in order to fabricate tunable light emitting devices.

In the same way, embedding and coating different size QDs onto microbeads can lead to create multiple optical codes for biomedical applications.

6. References

Alivisatos, A. P. (1996). Semiconductor clusters, nanocrystals, and quantum dots. *Science,* Vol. 271, No. 5251, (February 1996), pp. (933-937), 00368075.

Alivisatos, A. P., Gu, W., & Larabell, C. (2005). Quantum dots as cellular probes. *Annual review of Biomedical Engineering,* Vol. 7, (April 2005), pp. (55-76), 15239829.

Anikeeva, P.O., Halpert, J. E., Bawendi, M. G., & Bulovic', V. (2009). Quantum dot light-emitting devices with electroluminescence tunable over the entire visible spectrum. *Nano Letters,* Vol. 9, No. 7, (July 2009), pp. (2532-2536), 15306984.

Berggren, M., Inganas, O., Gustafsson, G., Rasmusson, J., Andersson, M. R., Hjertberg, T., & Wennerstrom, O. (1994). Light emitting diodes with variable colours from polymer blends. *Nature,* Vol. 372, No. 6505, (December 1994), pp. (444-446), 00280836.

Bruchez Jr, M., Moronne, M., Gin, P., Weiss, S., & Alivisatos, A. P. (1998). Semiconductor nanocrystals as fluorescent biological labels. *Science,* Vol. 281, No. 5385, (September 1998), pp. (2013-2016), 00368075.

Buckley, A. R., Rahn, M. D., Hill, J., Cabanillas-Gonzalez, J., Fox, A. M., & Bradley, D. D. C. (2001). Energy transfer dynamics in polyfluorene-based polymer blends. *Chemical Physics Letters,* Vol. 339, No. 5-6, (May 2001), pp. (331-336), 00092614.

Cabanillas-Gonzalez, J., Virgili, T., Lanzani, G., Yeates, S., Ariu, M., Nelson, J., et al. (2005). Photophysics of charge transfer in a polyfluorene/violanthrone blend. *Physical Review B - Condensed Matter and Materials Physics,* Vol. 7, No. 1, (January 2005), pp. (014211-1-014211-8), 10980121.

Campbell, I. H., & Crone, B. K. (2008). Efficient, visible organic light-emitting diodes utilizing a single polymer layer doped with quantum dots. *Applied Physics Letters,* Vol. 92, No. 4, (2008), Article number (043303), 00036951.

Caruge, J. M., Halpert, J. E., Bulovic, V., & Bawendi, M. G. (2006). NiO as an inorganic hole-transporting layer in quantum-dot light-emitting devices. *Nano Letters,* Vol. 6, No. 12, (December 2006), pp. (2991-2994), 15306984.

Caruge, J. M., Halpert, J. E., Wood, V., Bulovic, V., & Bawendi, M. G. (2008). Colloidal quantum-dot light-emitting diodes with metal-oxide charge transport layers. *Nature Photonics,* Vol. 2, No. 4, (April 2008), pp. (247-250), 17494885.

Carvelli, M., Janssen, R. A. J., & Coehoorn, R. (2011). Determination of the exciton singlet-to-triplet ratio in single-layer organic light-emitting diodes. *Physical Review B - Condensed Matter and Materials Physics,* Vol. 83, No. 7, (February 2011), Article number (075203), 10980121.

Coe-Sullivan, S., Steckel, J. S., Woo, W. K., Bawendi, M. G., & Bulovic', V. (2005). Large-area ordered quantum-dot monolayers via phase separation during spin-casting. *Advanced Functional Materials,* Vol. 15, No. 7, (July 2005), pp. (1117-1124), 1616301X.

Coe, S., Woo, W. K., Bawendi, M., & Bulovic', V. (2002). Electroluminescence from single monolayers of nanocrystals in molecular organic devices. *Nature,* Vol. 420, No. 6917, (December 2002), pp. (800-803), 00280836.

Colvin, V. L., Schlamp, M. C., & Alivisatos, A. P. (1994). Light-emitting diodes made from cadmium selenide nanocrystals and a semiconducting polymer. *Nature,* Vol. 370, No. 6488, (August 1994), pp. (354-357), 00280836.

Cho, K. S., Lee, E. K., Joo, W. J., Jang, E., Kim, T. H., Lee, S. J., Kwon, S. -J., Han, J. Y., Kim, B. -K., Choi, B. L., Kim, J. M. (2009). High-performance crosslinked colloidal quantum-dot light-emitting diodes. *Nature Photonics*, Vol. 3, No. 6, (June 2009), pp. (341-345), 17494885.

Chou, C. H., Yang, C. H., Hsu, C. S., & Chen, T. M. (2007). Hybrid white-light emitting-LED based on luminescent polyfluorene polymer and quantum dots. *Journal of Nanoscience and Nanotechnology*, Vol. 7, No. 8, (August 2007), pp. (2785-2789), 15334880.

Ferenczi, T. A. M., Sims, M., & Bradley, D. D. C. (2008). On the nature of the fluorenone-based emission in oxidized poly(dialkyl-fluorene)s. *Journal of Physics Condensed Matter*, Vol. 20, No. 4, (January 2008), Article number (045220), 09538984.

Ginger, D. S., & Greenham, N. C. (2000). Charge injection and transport in films of CdSe nanocrystals. Journal of *Applied Physics*, Vol. 87, No. 3, (February 2000), pp. (1361-1368), 00218979.

Han, M., Gao, X., Su, J. Z., & Nie, S. (2001). Quantum-dot-tagged microbeads for multiplexed optical coding of biomolecules. *Nature Biotechnology*, Vol. 19, No. 7, (July 2001), pp. (631-635), 10870156.

Harman, T. C., Taylor, P. J., Walsh, M. P., & LaForge, B. E. (2002). Quantum dot superlattice thermoelectric materials and devices. *Science*, Vol. 297, No. 5590, (September 2002), pp. (2229-2232), 00368075.

Hong, Y. J., Lee, C. -H., Yoon, A., Kim, M., Seong, H. -K., Chung, H. J., Sone, C., Park, Y. J. & Yi, G. -C. (2011). Visible-color-tunable light-emitting diodes. *Advanced Materials*, Vol. 23, No. 29, (August 2011), pp. (3284-3288), 09359648.

Huang, C. Y., Huang, T. S., Cheng, C. Y., Chen, Y. C., Wan, C. T., Madhava Rao, M. V., Su, Y. K. (2010). Three-band white light-emitting diodes based on hybridization of polyfluorene and colloidal CdSe-ZnS quantum dots. *IEEE Photonics Technology Letters*, Vol. 22, No. 5, (March 2010), pp. (305-307), 10411135.

Kim, L., Anikeeva, P. O., Coe-Sullivan, S. A., Steckel, J. S., Bawendi, M. G., & Bulovic', V. (2008). Contact printing of quantum dot light-emitting devices. *Nano Letters*, Vol. 8, No. 12, (December 2008), pp. (4513-4517), 15306984.

Kim, T. H., Cho, K. S., Lee, E. K., Lee, S. J., Chae, J., Kim, J. W., Kim, D. H., Kwon, J.-Y., Amaratunga, G., Lee, S.Y., Choi, B. L., Kuk, Y., Kim, J. M., & Kim, K. (2011). Full-colour quantum dot displays fabricated by transfer printing. *Nature Photonics*, Vol. 5, No. 3, (March 2011), pp. (176-182), 17494885.

Kreouzis, T., Poplavskyy, D., Tuladhar, S. M., Campoy-Quiles, M., Nelson, J., Campbell, A. J., Bradley, D. D. C., (2006). Temperature and field dependence of hole mobility in poly(9,9- dioctylfluorene). *Physical Review B - Condensed Matter and Materials Physics*, Vol. 73, No. 23, (2006), Article number (235201), 10980121.

Lei Qian, Y. Z., Jiangeng Xue & Paul H. Holloway (2011). Stable and efficient quantum-dot light-emitting diodes based on solution-processed multilayer structures. *Nature photonics*, (August 2011), 17494885.

Luo, J., Li, X., Hou, Q., Peng, J., Yang, W., & Cao, Y. (2007). High-efficiency white-light emission from a single copolymer: Fluorescent blue, green, and red chromophores on a conjugated polymer backbone. *Advanced Materials*, Vol. 19, No. 8, (April 2007), pp. (1113-1117), 09359648.

Ma, Q., Castelló Serrano, I., & Palomares, E. (2011). Multiplexed color encoded silica nanospheres prepared by stepwise encapsulating quantum dot/SiO2 multilayers. Chemical Communications, Vol. 47, No. 25, (2011), pp. (7071-7073), 13597345.

Mattoussi, H., Radzilowski, L. H., Dabbousi, B. O., Thomas, E. L., Bawendi, M. G., & Rubner, M. F. (1998). Electroluminescence from heterostructures of poly(phenylene vinylene) and inorganic CdSe nanocrystals. *Journal of Applied Physics*, Vol. 83, No. 12, (June 1998), pp. (7965-7974), 00218979.

Milliron, D. J., Gur, I., & Alivisatos, A. P. (2005). Hybrid organic-nanocrystal solar cells. *MRS Bulletin*, Vol. 30, No. 1, (January 2005), pp. (41-44), 08837694.

Ouisse, T., Stephan, O., & Armand, M. (2003). Degradation kinetics of the spectral emission in polyfluorene light-emitting electro-chemical cells and diodes. *EPJ Applied Physics*, Vol. 24, No. 3, (December 2003), pp. (195-200), 12860042.

Resch-Genger, U., Grabolle, M., Cavaliere-Jaricot, S., Nitschke, R., & Nann, T. (2008). Quantum dots versus organic dyes as fluorescent labels. *Nature Methods*, Vol. 5, No. 9, pp. (763-775), 15487091.

Schlamp, M. C., Peng, X., & Alivisatos, A. P. (1997). Improved efficiencies in light emitting diodes made with CdSe(CdS) core/shell type nanocrystals and a semiconducting polymer. *Journal of Applied Physics*, Vol. 82, No. 11, (December 1997), pp. (5837-5842), 00218979.

Sims, M., Bradley, D. D. C., Ariu, M., Koeberg, M., Asimakis, A., Grell, M., Lidzey, D. G. (2004). Understanding the origin of the 535 nm emission band in oxidized poly(9,9-dioctylfluorene): The essential role of inter-chain/inter-segment interactions. *Advanced Functional Materials*, Vol. 14, No. 8, (August 2004), pp. (765-781), 1616301X.

Smith, A. M., & Nie, S. (2010). Semiconductor nanocrystals: Structure, properties, and band gap engineering. *Accounts of Chemical Research*, Vol. 43, No. 2, (February 2010), pp. (190-200), 00014842.

Sun, Q., Wang, Y. A., Li, L. S., Wang, D., Zhu, T., Xu, J., Yan, C., Li, Y. (2007). Bright, multicoloured light-emitting diodes based on quantum dots. *Nature Photonics*, Vol. 1, No. 12, (December 2007), pp. (717-722), 17494885.

Tekin, E., Smith, P. J., Hoeppener, S., Van Den Berg, A. M. J., Susha, A. S., Rogach, A. L., et al. (2007). InkJet printing of luminescent CdTe nanocrystal-polymer composites. *Advanced Functional Materials*, Vol. 17, No. 1, (January 2007), pp. (23-28), 1616301X.

Ugarte, I., Cambarau, W., Waldauf, C., Arbeloa, F. L., & Pacios, R. (2009). Precisely voltage tunable polymeric light emitting diodes by controlling polymer chemical oxidation and adding inorganic semiconducting nanoparticles. From blue to red stopping at white in the same device. *Organic Electronics*, Vol. 10, No. 8 (December 2009), pp. (1606-1609), 15661199.

Wood, V., & Bulovic', V. (2010). Colloidal quantum dot light-emitting devices. *Nano Reviews*, Vol. 1, (July 2010), 5202.

Wu, X., Liu, H., Liu, J., Haley, K. N., Treadway, J. A., Larson, J. P., Ge, N., Peale, F., Bruchez, M.P. (2003). Immunofluorescent labeling of cancer marker Her2 and other cellular targets with semiconductor quantum dots. *Nature Biotechnology*, Vol. 21; No. 1, (January 2003), pp. (41-46), 10870156.

Zhao, J., Bardecker, J. A., Munro, A. M., Liu, M. S., Niu, Y., Ding, I. K., Luo, J., Chen, B., Jen, A. K.-Y., Ginger, D. S., (2006). Efficient CdSe/CdS quantum dot light-emitting diodes using a thermally polymerized hole transport layer. *Nano Letters*, Vol. 6, No. 3, (March 2006), pp. (463-467), 15306984.

Ultrafast Nonlinear Optical Response in GaN/AlN Quantum Dots for Optical Switching Applications at 1.55 μm

S. Valdueza-Felip[1], F. B. Naranjo[1], M. González-Herráez[1],
E. Monroy[2] and J. Solís[3]

[1]*Photonics Engineering Group (GRIFO), Electronics Dept., University of Alcalá,*
[2]*CEA-Grenoble, INAC / SP2M / NPSC,*
[3]*Instituto de Óptica, C.S.I.C.,*
[1,3]*Spain,*
[2]*France*

1. Introduction

Multi-terabit optical time division multiplexing (OTDM) networks require semiconductor all-optical switches and wavelength converters operating at room temperature. These devices should be characterized by AN ultrafast response capable of sustaining high repetition rates with low switching energy and high contrast ratio. These features lead to consider the use of resonant nonlinearities for the applications envisaged (Wada, 2004).

There is a particular interest in the development of these devices for the C-band of optical fibers, ≈ 1.55 μm, where erbium-doped fiber amplifiers (EDFAs) are widely available. Highly nonlinear fibers and waveguides have been used for generating nonlinear interactions in this spectral region (Almeida et al., 2004; Bloembergen, 2000). However, due to the low values of their nonlinear coefficients ($n_2 \sim 2.4 \times 10^{-16}$ cm^2/W at 1.55 μm), their implementation often require long interaction distance (even kilometers in the case of nonlinear fibers) or high power levels (González-Herráez, 2011), making them cumbersome for real systems.

An interesting alternative comes through the use of semiconductors. Semiconductor third-order nonlinear optical phenomena, like nonlinear absorption, self- and cross-phase modulation, self- and cross-gain modulation, and four-wave mixing, find direct application in the all-optical control of data streams in fiber-optic networks (Boyd, 2008; Radic & McKinstrie, 2005). An important effort for the development of InGaAs and InGaAsP-based nonlinear optical devices at 1.55 μm has been performed in recent years (Wada, 2004): FWM wavelength conversion in resonant devices based on InGaAsP-InP waveguides ($\chi^{(3)} \sim 4.2 \times 10^{-10}$ esu) has been reported (Donnelly et al., 1996), as well as all-optical demultiplexing of 160–10 Gbit/s signals with Mach-Zehnder interferometric switch using intraband transitions in InGaAs/AlAs/AlAsSb QWs (Akimoto et al., 2007). However, new materials with larger $\chi^{(3)}$ values are required in order to improve the performance of all-optical-based devices.

From this point of view, III-nitride semiconductors are very good candidates since the asymmetry of their crystalline structure contributes to enhance nonlinear phenomena. There are two different approaches to cover the near-infrared (NIR) spectral range using nitride semiconductors. A first strategy is the use of InN which presents a room temperature direct bandgap ~0.7 eV (~1.77 μm). Naranjo *et al.* explored the potential of InN-based structures for slow-light generation by analyzing the third-order susceptibility and the nonlinear absorption at 1.55 μm (Naranjo et al., 2007; 2011; Valdueza-Felip, 2011).

An alternative approach consists on using intraband transitions in GaN/AlN quantum well (QW) or quantum dot (QD) heterostructures, taking advantage from their large conduction band offset (~1.8 eV (Tchernycheva et al., 2006)). Using this material system, the resonance frequency can be tuned to 1.55 μm by controlling the size and composition of the nanostructures. A wide range of intraband devices based on GaN/AlN QWs and QDs have been developed for telecommunication applications in the last decade. In particular intraband QW and QD emitters (Nevou et al., 2008), electro-optical modulators (EOM) (Kheirodin et al., 2008), and different kind of detectors such as: quantum well infrared photodetectors (QWIPs) (Hofstetter et al., 2009), quantum dot infrared photodetectors (QDIPs) (Hofstetter et al., 2010) and quantum cascade detectors (QCD) (Vardi et al., 2008). Furthermore, thanks to the strong electro-phonon interaction, intraband relaxation is faster than interband recombination. As an example, intraband relaxation times in the range of 150–400 fs have been reported in GaN/AlN QWs (Hamazaki et al., 2004; Heber et al., 2002; Iizuka et al., 2006; 2000; Rapaport et al., 2003), confirming the potential of these materials for ultrafast all-optical devices.

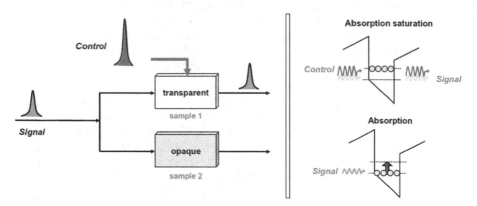

Fig. 1. GaN/AlN-based heterostructure application: ultrafast all-optical switching. The high-intensity pump beam saturates the absorption of the sample making it transparent to the input data beam.

Finally, it is also important to notice that quantum confinement in semiconductor nanostructures enhances the nonlinear response enabling nonlinear interactions at lower power levels, as already demonstrated by second-harmonic generation in GaN/AlN QWs (Nevou et al., 2006). In fact, saturation intensities in the order of 140 MW/cm^2 in GaN/AlN QD structures have been reported (Nevou et al., 2009), which correspond to control-pulse switching energies in the order of 10 pJ for optimized waveguide structures (Li et al., 2008; Valdueza-Felip, 2011).

2. Samples under study

Due to the rather large electron effective mass of GaN ($m^* = 0.2m_0$), layers as thin as 1–1.5 nm are required to achieve intraband absorptions at 1.3–1.55 μm. Plasma-assisted molecular-beam epitaxy (PAMBE) is the most suitable growth technique for this application, due to the low growth temperature, which hinders GaN-AlN interdiffusion and provides atomically abrupt interfaces (Sarigiannidou et al., 2006). Furthermore, in situ monitoring of the surface morphology by reflection high energy electron diffraction (RHEED) makes possible to control the growth at the atomic layer scale.

The samples described in this work consist of Si-doped GaN/AlN QWs and QDs superlattices (SLs) deposited by plasma-assisted molecular beam epitaxy. The substrates used for the growth of the SLs are 1-μm-thick AlN-on-sapphire templates grown on the c-axis growth direction. Figure 2 shows an schematic description of GaN QWs (a) and QDs (b) analyzed samples. The AlN templates were chosen in order to have crack-free growth (the SL is grown under compressive strain) and a low refractive index in order to optimize the optical guiding in GaN/AlN-based waveguides (Valdueza-Felip, 2011).

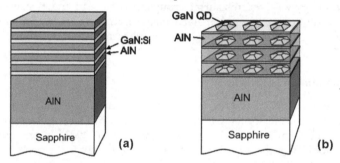

Fig. 2. Structure of the GaN/AlN QW (a) and QD (b) samples under study.

2.1 Growth conditions

One fascinating aspect of PAMBE-grown III-V nitride semiconductors is the possibility to control their growth mode by varying parameters such as the substrate temperature or the metal/nitrogen ratio value. As growth mode is directly related with the strain relaxation mechanism of nitride heterostructures, a simple tuning of growth parameters allows growing either two-dimensional (2D) quantum wells (QWs) or three-dimensional (3D) quantum dots (QDs), depending on the desired application.

2.1.1 GaN/AlN quantum wells growth

For the growth of GaN quantum wells embedded in AlN barriers, Ga-rich conditions are needed. In particular, GaN optimal growth conditions are under 2 ML of Ga excess for obtaining flat interface surfaces (Adelmann et al., 2003; Kandaswamy et al., 2009; 2008). A substrate temperature above 700°C is required to have a Ga desorption rate high enough to stabilize the Ga bilayer on the growing surface. However, for $T_{sub} \geq 750°C$, the GaN decomposition becomes important. In the case of AlN, the Al flux is fixed to the Al/N stoichiometric value and an additional Ga flux is introduced to stabilize the surface. Since the Al-N binding energy is much higher than the Ga-N binding energy, Ga segregates on the surface and it is not incorporated into the AlN layer.

The quality of the GaN/AlN heterostructures was found to be particularly sensitive to the Ga/N ratio. The strain fluctuations induced by Si doping and by the presence of the AlN barriers favor the formation of V-shaped pits, even in layers grown in the regime of 2 ML Ga-excess (Hermann et al., 2004; Nakamura et al., 2002). The suppression of these defects has been achieved by an enhancement of the Ga-flux so that growth is performed at the limit of Ga-accumulation. Regarding the growth temperature, it has been observed that overgrowth of GaN quantum wells with AlN at high temperatures results in an irregular thinning of the quantum well thickness due to exchange of Ga-atoms in the GaN layer with Al (Gogneau et al., 2004). This is a thermally activated phenomenon that becomes relevant for AlN growth temperatures above 730°C.

In conclusion, best interface results were achieved at a substrate temperature $T_s = 720°C$, growing AlN at the Al/N stoichiometry without growth interruptions and using also Ga as a surfactant during the growth of AlN barriers. Under these conditions, the samples present a flat surface morphology with an rms surface roughness of 1.0–1.5 nm measured in an area of 5×5 μm^2. High-resolution transmission electron microscopy (HRTEM) shows homogeneous QWs, with an interface roughness of $\sim$1 ML (see interface analysis in ref. (Sarigiannidou et al., 2006)).

2.1.2 GaN/AlN quantum dots growth

The synthesis of polar GaN/AlN QDs can be performed by two methods: either by GaN deposition under N-rich conditions (Guillot et al., 2006) or by GaN deposition under Ga-rich conditions followed by a growth interruption (Gogneau et al., 2003). N-rich growth implies a reduction of the mobility of the adsorbed species during growth that results in a high density (10^{11}-10^{12} cm^{-2}) of small QDs (1–2 nm high). On the contrary, Ga-rich conditions enhance the adatom mobility, leading to a lower QD density (10^{10}-10^{11} cm^{-2}) and bigger QDs (2–5 nm high). The difference between these growth techniques is illustrated by the atomic force microscopy (AFM) images in Fig. 3, which present QDs resulting from the deposition of 4 MLs of GaN (a) under N-rich conditions and (b) under Ga-rich conditions.

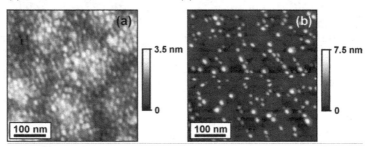

Fig. 3. AFM image of GaN/AlN QDs synthesized by deposition of 4 MLs of GaN under (a) N-rich and (b) Ga-rich conditions. Note that N-rich conditions lead to higher density of smaller QDs whereas Ga-rich conditions lead to lower density of bigger QDs (after Gacevic *et al.* (Gacevic et al., 2011)).

Whatever the growth method, GaN QDs are hexagonal truncated pyramids with $1\bar{1}03$ facets (Chamard et al., 2004), as illustrated in Fig. 4, and no Ga-Al interdiffusion has been observed (Sarigiannidou et al., 2005). They are connected to the underlying AlN by a 2 ML thick defect-free GaN wetting layer.

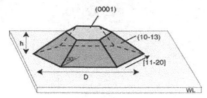

Fig. 4. Schematic representation of a GaN/AlN quantum dot (h = height, D = diamtere, WL = wetting layer).

Si-doping (n-type) is required to populate the first confined electronic level of the nanostructures, enabling efficient intraband absorption. Silicon does not modify the structural quality of the SLs (Guillot et al., 2006). The QDs can be n-type doped by incorporating a silicon flux during the GaN deposition, without modification of the QD growth kinetics or morphology (Guillot et al., 2006).

In order to tune the absorption wavelength within the telecommunication range (1.3–1.55 μm), the GaN QD height must be reduced down to approximately 1 nm, i.e. ~4 ML. Thus, this QDs must be synthesized using N-rich growth conditions. The QD size can be tuned by modifying the amount of GaN in the QDs, the growth temperature, or the growth interruption time (Ostwald Ripening) (Guillot et al., 2006).

2.2 Design of the heterostructures

It is well known that III-nitride materials exhibit spontaneous polarization (Bernardini et al., 1997). Furthermore, depending on the strain state of the layers, an additional (–) or (+) contribution to the total polarization appears, due to the piezoelectric effect. The polarization discontinuity in the well/barrier interface generates an internal electric field that has major consequences for the design of III-nitride intraband devices.

2.2.1 Quantum wells design

Kandaswamy et al. discussed the specific features of intraband active region design using GaN/AlN materials and showed that the $e_1 - e_2$ intraband transition can be tailored in a wide spectral range from near- to mid-infrared wavelengths by engineering the quantum confinement and the internal electric field by changing the alloy compositions and layer thicknesses (Kandaswamy et al., 2008). In particular, the $e_1 - e_2$ transition energy of GaN/AlN QWs covers the spectral range used for fiber-optic telecommunication networks for a GaN well thickness comprised between 4 and 6 ML (1–1.5 nm). For this work, the GaN/AlN QWs were designed with a well GaN thickness of 5 ML in order to provide a $e_1 - e_2$ intraband transition ~0.8 eV, which corresponds to ~1.55 μm (see Fig. 5).

2.2.2 Quantum dots design

Andreev et al. have calculated the electronic structure of GaN/AlN QDs using the k×p model and taking the internal electric field into account (Andreev, 2003; Andreev & O'Reilly, 2000;

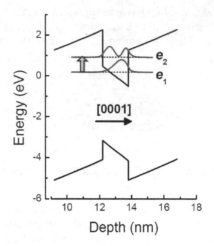

Fig. 5. GaN/AlN QWs band structure. The QWs were designed for having a resonant intraband transition ~1.55 μm.

2001). These calculations have been complemented by Ranjan *et al.* (Ranjan et al., 2003) using the strong-binding method and a self-consistent treatment to take carrier screening of the electric field into account. From the experimental point of view, infrared absorption peaked at 1.27–2.4 μm has been observed in GaN/AlN QDs (Guillot et al., 2006; Moumanis et al., 2003; Tchernycheva et al., 2005), associated to the $s - p_z$, intraband transition (z = growth axis), *i.e.* a transition with an optical dipole oriented perpendicular to the (0001) layer plane.

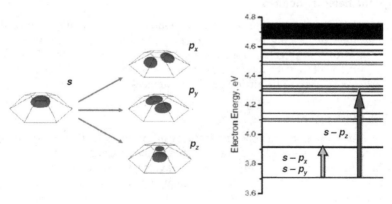

Fig. 6. Electronic states in nitride QDs with transitions $s - p_x$, $s - p_y$ and $s - p_z$ (Andreev, 2003).

Along this chapter, three different samples (S1-S3) are investigated, whose characteristics are summarized in Table 1. Samples S1 and S2 consist of a GaN/AlN QD superlattice containing 20 and 200 QD periods, respectively. The dots present a height of 1.1±0.7 nm and a diameter of 8±2 nm. The samples show a density of QDs in the range of 10^{11} cm^{-2}, as measured by atomic force microscopy and high-resolution TEM. The GaN QD layers are separated by

3-nm-thick AlN barriers. Sample S3, consists of a 100 period GaN/AlN (1.5 nm / 1.5 nm) QW superlattice. Further details on the growth of the heterostructures can be found in Refs. (Guillot et al., 2006; Sarigiannidou et al., 2006).

Sample	Structure	Height (nm)	Barrier (nm)	Number of periods
S1	QD	1.1	3	20
S2	QD	1.1	3	200
S3	QW	1.5	1.5	100

Table 1. Series of analyzed GaN/AlN heterostructures.

3. Structural characterization

The development of devices with a narrow bandwidth depends not only on the in-plane homogeneity of the layers but also on the fluctuations from QW (or QD) to QW (or QD) in the periodic structure. The parameters that might cause vertical inhomogeneities are strain-induced vertical alignment and strain relaxation along the structure by the introduction of misfit dislocations (Guillot et al., 2006).

In order to analyze the strain state of the QD stack, high-resolution x-ray diffraction (HRXRD) measurements were performed. Figure 7 displays the θ-2θ scan around the (0002) x-ray reflection of a representative QD sample compared to theoretical calculations. The simulation superimposed on the experimental scan was performed assuming that the structure is fully strained on the AlN buffer. It fits the experimental measurement, which indicates that the QD stack is strained on the AlN substrate within the resolution of the measurement (Guillot et al., 2006). Furthermore, the period thickness of the SL is confirmed by measuring the intersatellite (SL0, SL1, SL-1,...) angle in the diffractogram.

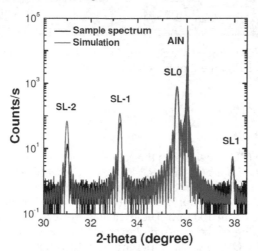

Fig. 7. HRXRD θ-2θ scan around the (0002) x-ray reflection of a representative QD sample compared to theoretical calculations. SL0, SL1, SL-1, etc. stands for intersatellite diffraction peaks associated with the periodicity of the heterostructure.

The samples were also analyzed by high-resolution transmission electron microscopy (HRTEM) measurements in order to assess the vertical arrangement of the layers. The HRTEM image of the QW sample (S3) of Fig. 8(a) shows abrupt GaN/AlN interfaces, and confirms the narrow quantum wells width (~1.5 nm) needed for telecommunication applications. No vertical correlation of the QDs is observed. The absence of vertical alignment is explained by the small size and diameter of the QDs (Guillot et al., 2006).

On the other hand, a gradual relaxation of the structure might take place in samples with high density of QDs. Dot coalescence should lead to the generation of misfit dislocations in the GaN wetting layer when its thickness exceeds the critical thickness (~2 ML) (Guillot et al., 2006). From the HRTEM images, the size of the QDs appears homogeneous along the whole structure, which suggests a negligible relaxation.

 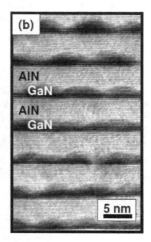

Fig. 8. High-resolution TEM image of a Si-doped GaN/AlN quantum well (a) and quantum dot (b) stack. No vertical alignment of the QDs can be observed.

4. Linear optical properties

4.1 Photoluminescence

For photoluminescence (PL) experiments the samples were mounted in a cold-finger cryostat with temperature control from $T = 7$ K to RT. PL was excited with a frequency-doubled argon laser ($\lambda = 244$ nm), and collected into a Jobin-Yvon HR460 monochromator equipped with an ultraviolet-enhanced charge-coupled device (CCD) camera. The diameter of the excitation spot on the sample was about 100 μm. The excitation power was kept around 100 μW, low enough to avoid screening of the internal electric field. The low-temperature ($T = 7$ K) PL spectra of the GaN/AlN QW and QD superlattices with various sizes are presented in Fig. 9(a). By comparison of QW and QD samples, we attribute the broadening of the PL line to the dispersion of the QD size. In the case of the GaN/AlN QWs, the spectral structure of the emission is due to monolayer thickness fluctuations in the QWs, as described elsewhere (Tchernycheva et al., 2006). In the PL from GaN/AlN QDs, the broader line width makes it possible to observe the superimposition of a Fabry-Perot interference pattern related to the total nitride thickness.

The evolution of the integrated PL intensity as a function of temperature, normalized by the integrated PL intensity at low temperature ($T = 7$ K), is presented in Fig. 9(b) for GaN/AlN QWs and QDs. Keeping in mind that the emission intensity remains stable below 25 K for all the samples, the values presented in Fig. 9(b) should correspond directly to the internal quantum efficiency (IQE) at different temperatures. These results confirm the improved thermal stability of QDs over QWs, as a result of the 3D carrier confinement (Adelmann et al., 2000; Damilano et al., 1999; Gacevic et al., 2011; Guillot et al., 2006; Sénés et al., 2007)

To probe the competition between radiative and nonradiative recombination processes, we now focus on the PL decay time (Renard et al., 2009). Time-resolved PL experiments were performed using a frequency-tripled Ti-Sapphire laser ($\lambda = 270$ nm) with a pulse width around 200 fs. The repetition rate was tuned between 100 KHz and 76 MHz depending on the temporal dynamics studied. The diameter of the excitation spot on the sample was about 50 μm. The luminescence was collected into a 0.32 m monochromator with a 150 grooves/mm grating, and detected by a streak-camera. The temporal resolution of the setup could reach 5 ps. The samples were mounted in a He-flow cryostat with temperature control from 5 K to room temperature.

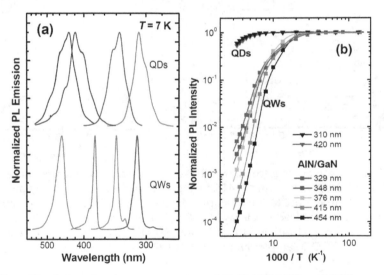

Fig. 9. (a) Normalized photoluminescence spectra of GaN/AlN QD and QW superlattices. The spectra are vertically shifted for clarity. (b) Temperature evolution of the integrated PL emission of GaN/AlN QW and QD samples emitting at different wavelengths. Details about the solid lines fits can be founded in Ref. (Gacevic et al., 2011).

Figure 10(a) displays the PL decay measured in a GaN/AlN QW sample emitting at 415 nm at room temperature and at 5 K. In the inset, the temperature dependence of the $1/e$ decay times measured on the four QW samples emitting at 310, 365, 415 and 455 nm is illustrated. The PL decay times are roughly constant up to 75 K, which strongly suggests that the decay is dominated by radiative recombination at low temperature. However, the decay times decrease significantly for higher temperatures -typically at least a factor of 2 is already lost at 150 K and at least a factor 10 at room temperature. This is valid for all samples, even for the ones presenting radiative decay times below 1 ns.

In contrast, Fig. 10(b) displays the behavior of a GaN/AlN QD sample emitting at 440 nm, showing thermal insensitivity in all the temperature range. The decay times remain constant as an indication of the inefficiency of nonradiative recombination paths in such structures. The most striking feature is that this is true even for QDs presenting radiative decay times of 500 ns as illustrated in the figure.

Comparing QWs (Fig. 10(a)) and QDs (Fig. 10(b)), nonradiative recombination of photogenerated carriers in the QDs is suppressed in the 4–300 K range. This is a consequence of the 3D confinement in the QDs which, combined with the huge band offset in GaN/AlN, prevents the carrier escape towards nonradiative centers. We can estimate a lower limit for the nonradiative recombination lifetime in the long-lived GaN/AlN QDs of 10 μs. Conversely, the absence of lateral carrier confinement in QWs favors the thermal diffusion towards nonradiative recombination centers such as dislocations.

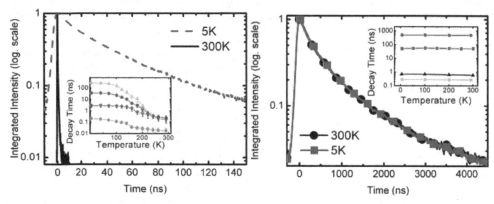

Fig. 10. (a) Comparison of the normalized PL decay at 5 K and room temperature for a sample of GaN/AlN QWs emitting at 415 nm. Inset: Evolution of the decay time of the PL as a function of the temperature for QW samples emitting at 310, 365, 410 and 455 nm. (b) Comparison of the normalized PL decay at 5 K and room temperature for a sample of GaN/AlN QDs emitting at 440 nm. Inset: Evolution of the decay time of the PL as a function of the temperature for QD samples emitting at 310, 340, 400 and 440 nm.

4.2 Intraband absorption

In order to assess the NIR absorption, samples were mechanically polished to form a 45° multipass waveguide with 4–5 total internal reflections. The infrared absorption for p- and s-polarized light was measured at room temperature using a Fourier transform infrared (FTIR) spectrometer. The absorption spectra of samples S1-S3 are illustrated in Fig. 11. For all samples under study, an absorption peak ~1.55 μm for p-polarized light is obtained. No response to s-polarized excitation is observed within the measurement spectral range (0.9–4.1 μm), as expected from the polarization selection rules of the involved intraband transitions ($e_1 - e_2$ in the QWs and $s - p_z$ in the QDs). The broadening of the absorption in the QD superlattices remains below 150 meV, as a result of QD size fluctuations. On the other hand, the multipeak structure present in the GaN/AlN QW sample can be attributed to ML thickness fluctuations (Tchernycheva et al., 2006).

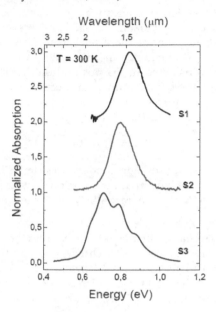

Fig. 11. Normalized infrared absorption from S1-S3 samples under p-polarized excitation. The spectra are vertically shifted for clarity. No response to s-polarized light was observed for any of the samples in the 0.9 μm $< \lambda <$ 4.1 μm spectral range.

The spectral dependence of the linear absorption evaluates the linear optical processes associated to the electronic transition. The linear absorption (α_0), refractive index, length of the sample (L_s), and effective length (L_{eff}) are essential parameters to evaluate its nonlinear response. The effective length measures the length over which a lossless material would exhibit the same nonlinear effect as the (lossy) material under study, being calculated by: $L_{eff}=(1 - e^{-\alpha_0 L_s})/\alpha_0$. When $\alpha_0 = 0$, $L_{eff} \cong L_s$ and when $\alpha_0 > 0$, $L_{eff} < L_s$. Thus L_{eff} becomes smaller as the loss increases. Note that as L_s increases, the effective length converges to the value $1/\alpha_0$ (as $L_s \to \infty$, $L_{eff} \to 1/\alpha_0$). In the case of these heterostructures, the length is given by the total length of the active region, $i.e.$ the GaN QW (or QD) thickness evaluated along the whole periodic structure. Table 2 summarizes the results obtained from the linear optical characterization at 1.5 μm.

Sample	Doping (cm^{-3})	Absorption per pass (%)	$\alpha_0 L_s$	L_s (nm)	α_0 (cm^{-1})	L_{eff} (nm)
S1	10^{20}	1.6	0.032	24	13.4×10^3	23.6
S2	5×10^{19}	10	0.23	200	11.3×10^3	179
S3	5×10^{19}	4.2	0.088	150	5.8×10^3	144

Table 2. Summary of the results obtained from linear optical measurements at 1.5 μm.

5. Nonlinear optical characterization

5.1 Third-order nonlinear effect: four-wave mixing

High-intensity lasers enable a variety of interesting nonlinear optical phenomena such as the generation of new frequencies from monochromatic light in a nonlinear medium. At the excitation intensities applied to generate these nonlinear effects, the optical parameters of the pumped material become a function of the light intensity. In nonlinear optics regime additional terms have to be considered in the polarization density to properly explain the behavior of light propagating through a nonlinear medium. For small nonlinearities $P(t)$ can be expressed as a function of the applied electric field $E(t)$ by the superposition of both linear and nonlinear polarization components (Boyd, 2008; González-Herráez, 2011) as:

$$P(t) = P_L + P_{NL} \tag{1}$$

where $P_{NL} = P^{(2)} + P^{(3)} + \cdots$ are the nonlinear polarization terms that depend on the applied electric field. The optical control of light in communication networks is performed thought the use of third-order nonlinear optical phenomena like the Kerr-effect-based four-wave mixing (FWM) process.

The optical **Kerr effect** is a change in the refractive index of a material in response to an applied electric field. This induced index change is proportional to the square of the electric field. Considering a material under a single, monochromatic wave oscillating at the fundamental frequency ω, the total polarization can be described by a scalar expression (González-Herráez, 2011; Shutherland, 2003) as:

$$P_\alpha(\omega) = \epsilon_0 \left(\chi^{(1)}(\omega) E_\alpha(\omega) + 3\chi^{(3)}_{\alpha\beta\gamma\delta}(\omega) |E|^2 E_\delta(\omega) \right) \tag{2}$$

where $|E|^2$ is the amplitude of the electric field and $\chi^{(3)}(\omega)$ denotes the third-order susceptibility parameter. In this equation, the nonlinear term $3\chi^{(3)}(\omega)|E|^2$ is treated as a perturbation (small variation) of the linear term, i.e. a correction to the linear susceptibility that depends on the input intensity ($|E|^2 \propto I$).

Assuming that the index and absorption changes are an instantaneous function of the applied electric field, the material response can be described by $n(I) = n_0 + n_2 I$ and $\alpha(I) = \alpha_0 + \alpha_2 I$, where the nonlinear refractive index n_2 and nonlinear absorption coefficient α_2 are related to the third-order nonlinear susceptibility. In weakly absorbing media the real part of $\chi^{(3)}$ drives the nonlinear refraction while the imaginary portion characterizes the nonlinear absorption (Shutherland, 2003) as:

$$n_2 = \frac{3}{4} \frac{1}{n_0^2 \epsilon_0 c} \Re e \left[\chi^{(3)} \right] \tag{3}$$

$$\alpha_2 = \frac{3\pi}{\lambda} \frac{1}{n_0^2 \epsilon_0 c} \Im m \left[\chi^{(3)} \right] \tag{4}$$

Since the nonlinear polarization depends on ω, the $\chi^{(3)}$ parameter varies when the material is in resonance with the input beam frequency. Thus, in the case of strongly absorbing media a more complex treatment must be considered following the description reported in Ref. (del Coso & Solís, 2004).

Four-wave mixing through optical phase conjugation is a useful experimental technique for the characterization of third-order nonlinear materials (Sheik-Bahae et al., 1990). FWM is an intermodulation phenomenon in optical systems, whereby interactions between 3 waves in a nonlinear medium produces a 4th wave via the third-order polarization (Hill et al., 1978; Inoue, 1992). It is possible, using nonlinear optical processes, to reverse the phase of a beam of light. The reversed beam is called conjugate beam.

Figure 12 shows an abstract drawing of the FWM process, where three coherent waves are incident on a nonlinear medium, and a fourth wave (the phase-conjugated one) is generated. In a simplified gratings formalism (not valid for beams with crossed polarizations), the generation of the conjugated signal (beam) can be viewed as follows. The conjugated beam is produced by the refractive index modulation caused by the interference of two of the beams forming a phase grating. Two of the beams ($\mathbf{E}_1$ and $\mathbf{E}_2$ or pump beams p_1 and p_2) interfere and "write" a real-time diffraction pattern or phase grating, while the third beam ($\mathbf{E}_3$) "reads" (or probe, s beam) the grating. The diffraction of the third beam when propagating through the grating generates the fourth beam ($\mathbf{E}_4 = \mathbf{E}_c^*$, the c beam), whose phase is conjugated to the three incident beams (Zeldovich et al., 1985). The diffracted phase-conjugated beam is transmitted by the material in a direction determined by the wave vectors of the interacting photons. In the degenerate case ($\omega_1 = \omega_2 = \omega_3 = \omega_4 = \omega$) energy and momentum conservations implies that that the sum of the wave vectors of $\mathbf{E}_1$ and $\mathbf{E}_2$ should be equal to the sum of the wave vectors of $\mathbf{E}_3$ and $\mathbf{E}_4$. In this work, the third-order nonlinear susceptibility $\chi^{(3)}$

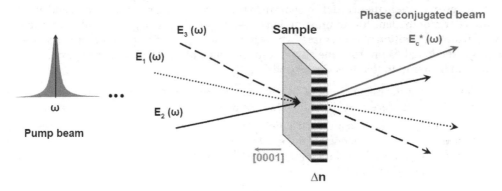

Fig. 12. Schematic of phase conjugation via the perturbation of the refractive index through the third-order nonlinear susceptibility.

of III-nitrides is studied by resonant degenerate four-wave mixing (DFWM) in the forward boxcars configuration. In this configuration, three parallel and co-propagating input pulsed laser beams are focused simultaneously in space and time on the sample with a lens. The z direction is the direction of propagation of the beams before the lens, normal to the surface of the sample. If the three input beams are placed in the three vertices of an square centered at the lens, the energy and momentum conservations ($\sum \mathbf{k}_i = 0$) is guaranteed and enables the generation of the conjugated fourth beam at the output of the sample (Shutherland, 2003).

Figure 13 illustrates the experimental setup used for DFWM measurements. The excitation source is an optical parametric amplifier (OPA)ť'which operates at 1 kHz and provides 100

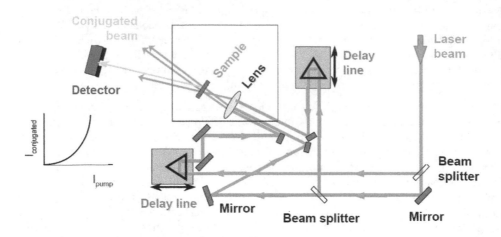

Fig. 13. Experimental DFWM setup.

fs pulses tunable in the 300–3000 nm interval. The laser pulse intensity is controlled by a $\lambda/2$ wave plate and a polarizer. Two beam splitters provide three beams with equal intensity in order to satisfy the maximum conjugate signal condition. The three beams propagate the same optical path by adjusting two delay lines. The system is aligned so that each beam lies on one vertex of a square in order to guarantee the wave vector conservation rule and at 4 mm of the center of the lens for minimizing lens aberrations. The conjugate signal is detected far away from the position of the sample, where the spatially filtered pumps and probe beams are enough separated to improve the signal-to-noise ratio. More details about the experimental setup can be found in Ref. (del Coso López, 2004).

In order to obtain the dependence of the intensity of the conjugate signal with the third-order susceptibility it is necessary to solve the nonlinear wave equation, where the nonlinear polarization $\mathbf{P}_{NL}^{(3)}(\omega)$ acts as an electric field **source term** at the given frequency. Assuming the monochromatic components of the electric field $\mathbf{E}(\omega)$ polarized in the x and y direction and propagating along the z direction, the electric field associated to the conjugate beam can be written (Shutherland, 2003) as:

$$\frac{\partial E_c(\omega)}{\partial z} = j\frac{2\omega}{\epsilon_0 n_0 c}P_c^{(3)}(\omega)e^{-jkz}$$

$$\Rightarrow \frac{\partial E_c(\omega)}{\partial z} = j\frac{3\omega}{n_0 c}\chi_{cp_1 p_2 s}^{(3)}(\omega)E_{p_1}(\omega)E_{p_2}(\omega)E_s^*(\omega)e^{-j\Delta kz} \tag{5}$$

where $\chi^{(3)}(\omega)$ denotes the three possible complex components of the third-order susceptibility tensor for generating ω and $\mathbf{k}$ is the wave vector, or change in phase per length unity along the path traveled by the wave. The last term of the previous equation leads to an exponential attenuation of the conjugate signal unless the term

$$\Delta\mathbf{k} = \mathbf{k}_{p_1} + \mathbf{k}_{p_2} - \mathbf{k}_s - \mathbf{k}_c \tag{6}$$

is equal to zero. This implies that for maximizing the conjugated signal, the phase matching condition $\sum k_i$=0 has to be fulfilled. This is straightforward in the boxcar configuration.

Taking into account the boundary conditions given by the input beams, the solution of the nonlinear wave equation can be obtained for low conversion efficiency ($I_c << I_{p_1}, I_{p_2}, I_s$) and absorbing media ($\alpha_0 \neq 0$). The intensity of the conjugate signal is then given by:

$$I_c \cong \left(\frac{3\pi L_{eff}}{\epsilon_0 n_0^2 c \lambda} \right)^2 e^{-\alpha L} \left| \chi_{cp_1p_2s}^{(3)} \right| I_{p_1} I_{p_2} I_s \tag{7}$$

From this equation we can obtain the third-order nonlinear susceptibility $\left| \chi_{cp_1p_2s}^{(3)} \right|$ by measuring the intensity of the conjugate beam as a function of the intensity of the pumps and probe beams. The maximum conjugate beam intensity is obtained for equally intense pump and probe beams from Eq. 7, hence: I_c máx. $\leftrightarrow I_{p_1} = I_{p_2} = I_s = I_0/3$. Thus, the dependence of the conjugate beam intensity on the pump beam intensity follows a cubic relationship $I_c = c I_0^3$.

It is also possible to quantify the $\left| \chi_{cp_1p_2s}^{(3)} \right|$ by using a transparent (non absorbing) reference material with a known $\chi^{(3)}$ (Shutherland, 2003). Then, the third-order susceptibility of the sample under study is expressed as:

$$\left| \chi_s^{(3)} \right| = \left(\frac{n_0^s}{n_0^r} \right)^2 \left(\frac{L^r}{L^s} \right) \left(\frac{c^s}{c^r} \right)^{1/2} \frac{\alpha_0 L^s e^{-\alpha_0 L^s/2}}{1 - e^{-\alpha_0 L^s}} \left| \chi_r^{(3)} \right| \tag{8}$$

where the subscripts r and s stands for the reference and the sample, respectively, and $\alpha_0 L$ is the linear absorption of the sample. In this experiment, the reference used is a fused silica (SiO$_2$) plate with a well known $|\chi^{(3)}| = 1.28 \times 10^{-14}$ esu (1.79×10^{-22} m^2/V^2) and $n_0 = 1.45$ (Shutherland, 2003). Throughout this chapter the cgs system of units (esu) for the quantification of $\chi^{(3)}$ is used. The conversion to SI units is given by $\chi^{(3)} \left(m^2/V^2 \right) = \frac{4\pi}{9} 10^{-8} \chi^{(3)} \left(esu \right)$.

5.2 QD nonlinear optical properties

5.2.1 Third-order susceptibility $\chi^{(3)}$

In DFWM measurements, the conjugate beam intensity I_c is plotted versus the pump intensity I_p to obtain the coefficient c of the relationship $I_c = c I_0^3$. The third-order susceptibility $|\chi^{(3)}|$ is calculated using Eq. 8. The refractive index of the active region (GaN) considered for the estimation of the third-order susceptibility is $n_0 = 2.2$.

The third-order susceptibility of the AlN-template substrate ($\sim 1.8 \times 10^{-8}$ esu) was measured at 1.55 μm in order to evaluate its potential contribution to the nonlinear signal. It should be pointed out that the laser source used in the experiment is not resonant with the bandgap of the AlN (~ 200 nm). Thus, the nonlinear contribution of the AlN to the incident electromagnetic field should be negligible compared to the contribution of light polarized in the z direction, which is absorbed by the active region of the GaN/AlN heterostructure. To avoid the nonlinear contribution of the substrate, the input power density was fixed so that the signal from the substrate remains below the experimental resolution of the setup. Thus,

it is ensured that the nonlinear response measured by DFWM is only generated by the GaN
QWs or QDs and not by the substrate.

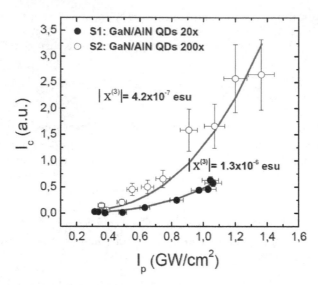

Fig. 14. I_c vs. I_p plot of GaN/AlN QDs samples at 1.55 μm. Solid lines correspond to the
cubic data fit given by $I_c = cI_p^3$.

Figure 14 shows a representative plot of the measured conjugate-beam intensity (I_c) vs. the
pump-beam intensity (I_p) for GaN/AlN QD heterostructures at 1.55 μm. The conjugate signal
follows a cubic dependence on the pump power as expected from a third-order nonlinear
process. The solid lines correspond to the cubic data fitted to a $I_c = cI_0^3$ curve. For sample S2,
a higher value of I_c is obtained due to its higher effective length since $I_c \propto L_{eff}^2$ (see Eq. 7).

Third-order susceptibility $|\chi^{(3)}|$ values of $(1.3\pm0.1)\times10^{-6}$ and $(4.2\pm0.4)\times10^{-7}$ esu were
estimated at 1.55 μm for GaN/AlN QD samples S1 and S2, respectively. The obtained
values increase with the linear absorption of the samples, being higher for the sample with
highest Si concentration (S1) due to the increase of the intraband absorption efficiency with
Si-doping. For the GaN/AlN QW sample (S3) a value of third-order susceptibility $|\chi^{(3)}|$ of
$(2.4\pm0.2)\times10^{-7}$ esu was estimated at 1.55 μm (Valdueza-Felip et al., 2008).

The obtained value of $|\chi^{(3)}|$ for the GaN/AlN QW sample is very similar to the one estimated
by Suzuki et al. of 1.6×10^{-7} esu in GaN/AlN QWs at 1.55 μm (Suzuki & Iizuka, 1997).
However, it is one order of magnitude larger than the experimental value achieved by
Hamazaki et al. in GaN/AlN multiple QWs at the same wavelength (Hamazaki et al.,
2004). This improvement of $|\chi^{(3)}|$ might be related to the structural properties of the samples
since the absorption saturation is known to be very sensitive to the structure of the material
(Iizuka et al., 2006). It must also be pointed out the use of different measurement technique
(we measure the modulus of the third-order susceptibility, while Hamazaki et al. make an
estimation of its imaginary part). The values of $|\chi^{(3)}|$ obtained in the QD samples (S1-S2) are
$\approx$ 5 times larger than the one obtained in the QW sample (S3), as expected from the higher

confinement in the QDs. It is also noteworthy that GaN/AlN-based heterostructures present an increase of 4 orders of magnitude of $\chi^{(3)}$ compared to InGaAsP-InP (Donnelly et al., 1996) and InN (Naranjo et al., 2007) material systems.

5.2.2 Population grating lifetime τ

DFWM technique allows determining the lifetime, τ, of the generated population grating by adjusting the time delay between the pump and probe beams (Adachi et al., 2000). In particular, this measurement is performed by modifying the optical path length of the probe beam using an optical delay line. The pump beams interfere and generate a spatially periodic grating that interferes with the delayed probe beam. The transient response is obtained from the evolution of the intensity of the conjugate beam as a function of the probe time delay.

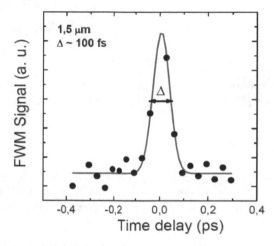

Fig. 15. DFWM signal intensity as a function of delay time between one of the pumps and the others for the GaN/AlN QD sample (S1).

Figure 15 shows the dependence of the conjugate signal intensity as a function of the time delay, obtained for the sample S1. The signal starts at zero and it increases when pump and probe beams become synchronized. It reaches its maximum when there is a full overlap in space and time of the two pump beams and the probe one in the sample. As the delay line moves away from the maximum overlap, the generated signal decays at the rhythm of the decay of the excited population in the sample. However, for the three samples under study, the measured lifetime of the population grating is smaller or comparable to the laser pulse width of 100 fs. This result is consistent with the subpicosecond intraband decay times measured in GaN QW structures, being related to a carrier relaxation process governed by the Fröhlich interaction between electrons and longitudinal-optical (LO) phonons (Hamazaki et al., 2004; Heber et al., 2002; Iizuka et al., 2006; 2000; Rapaport et al., 2003). Finally, Fig. 16 shows the figure of merit (FOM) for nonlinear optical materials: $\chi^{(3)}/\alpha$ vs. switching rate $1/\tau$, for different material systems (Hanamura et al., 2007), confirming the potential of GaN/AlN heterostructures for ultrafast all-optical devices.

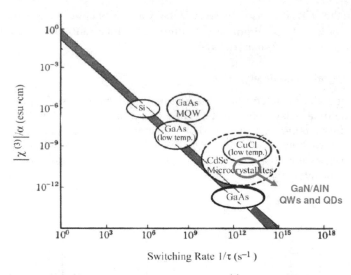

Fig. 16. Figure of merit for nonlinear optical materials: $\chi^{(3)}/\alpha$ vs. the switching rate $1/\tau$. The empirical law of constant FOM $\chi^{(3)}/\alpha$ = const. is represented by the shaded straight line (Hanamura et al., 2007).

6. Conclusions

In this chapter, the nonlinear properties of Si-doped GaN/AlN QD and QW superlattices have been evaluated at telecom wavelengths (1.55 μm). The samples under study present a near-infrared intraband absorption at ~1.55 μm wavelength assigned to the e_1-e_2 transition in the QWs and to the s-p_z transition in the QDs.

The nonlinear optical characterization of the samples was performed using the degenerate FWM technique in boxcars configuration. The conjugate signal intensity obtained for all the samples followed a cubic law with the pump power, as expected from a third-order nonlinear process. The measured conjugate-beam intensity is higher for the sample with the highest effective length and minimum linear absorption. On the other hand, the third-order susceptibility increases with the intraband absorption of the layer, achieving a maximum value of $|\chi^{(3)}|$ = (1.3±0.1)×10^{-6} esu for the GaN/AlN QD sample. This value is $\approx$ 5 times larger than the one obtained in the QW sample of $|\chi^{(3)}|$ = (2.4±0.2)×10^{-7} esu, which is consistent with results reported in the literature. This increase of $\chi^{(3)}$ for the QD sample is attributed to its higher (3D) quantum confinement.

Finally, the population grating lifetime τ was measured during the DFWM measurements, being below 100 fs for all the analyzed samples. The ultrafast nature of intraband transitions, together with the improved nonlinear optical response, render GaN/AlN heterostructures a feasible option for all-optical switching and wavelength conversion applications at 1.55 μm. The development of waveguides based on GaN/AlN heterostructures could enable new all-optical devices with potentially high speed operation, high contrast ratio and low saturation intensity suitable for future in multi-Tbit optical communication networks.

7. References

Adachi, S., Takagi, Y., Takeda, J. & Nelson, K. A. (2000). Optical sampling four-wave-mixing experiment for exciton relaxation processes, *Optics Communications* 174(1): 291.

Adelmann, C., Brault, J., Mula, G., Daudin, B., Lymperakis, L. & Neugebaue, J. (2003). Gallium adsorption on (0001) gan surfaces, *Phys. Rev. B* 67: 165419.

Adelmann, C., Simon, J., Feuillet, G., Pelekanos, N. T., Daudin, B. & Fishman, G. (2000). Self-assembled ingan quantum dots grown by molecular beam epitaxy, *Appl. Phys. Lett.* 76: 1570.

Akimoto, R., Simoyama, T., Tsuchida, H., Namiki, S., Lim, C. G., Nagase, M., Mozume, T., Hasama, T. & Ishikawa, H. (2007). All-optical demultiplexing of 160–10 gbit/s signals with mach-zehnder interferometric switch utilizing intersubband transition in ingaas/alas/alassb quantum well, *Appl. Phys. Lett.* 91(22): 221115.

Almeida, V. R., Barrios, C. A., Panepucci, R. R. & Lipson, M. (2004). All-optical control of light on a silicon chip, *Nature* 431: 1081.

Andreev, A. D. (2003). *ITQW proceedings.*

Andreev, A. D. & O'Reilly, E. P. (2000). Theory of the electronic structure of gan/aln hexagonal quantum dots, *Phys. Rev. B* 62: 15851.

Andreev, A. D. & O'Reilly, E. P. (2001). Optical transitions and radiative lifetime in gan/aln self-organized quantum dots, *Appl. Phys. Lett.* 79: 521.

Bernardini, F., Fiorentini, V. & Bosin, A. (1997). Spontaneous polarization and piezoelectric constants of iii-v nitrides, *Appl. Phys. Lett.* 70: 2990.

Bloembergen, N. (2000). Nonlinear optics: past, present, and future, *IEEE J. Selected Topics in Quantum Electronics* 6(6): 876.

Boyd, R. W. (2008). *Nonlinear optics*, 3rd edn, Academic Press.

Chamard, V., Schülli, T., Sztucki, M., Metzger, T. H., Sarigiannidou, E., Rouviére, J.-L., Tolan, M., Adelmann, C. & Daudin, B. (2004). Strain distribution in nitride quantum dots multilayers, *Phys. Rev. B* 69: 125327.

Damilano, B., Grandjean, N., Dalmasso, S. & Massies, J. (1999). Room-temperature blue-green emission from ingan/gan quantum dots made by strain-induced islanding growth, *Appl. Phys. Lett.* 75: 3751.

del Coso López, R. (2004). *Propiedades ópticas no-lineales de nanocompuestos metal-dieléctrico de $Cu:Al_2O_3$*, PhD thesis, Universidad Autónoma de Madrid.

del Coso, R. & Solís, J. (2004). Relation between nonlinear refractive index and third-order susceptibility in absorbing media, *J. Opt. Soc. Am. B* 21(3): 640.

Donnelly, J. P., Le, H. Q., Swanson, E. A., Groves, S. H., Darwish, A. & Ippen, E. P. (1996). Nondegenerate four-wave mixing wavelength conversion in low-loss passive ingaaspŨinp quantum-well waveguides, *IEEE Photon. Technol. Lett.* 8(5): 623.

Gacevic, Z., Das, A., Teubert, J., Kotsar, Y., Kandaswamy, P. K., Kehagias, T., Koukoula, T., Komninou, P. & Monroy, E. (2011). Internal quantum efficiency of iii-nitride quantum dot superlattices grown by plasma-assisted molecular-beam epitaxy, *J. Appl. Phys.* 109(10): 103501.

Gogneau, N., Jalabert, D., Monroy, E., Shibata, T., Tanaka, M. & Daudin, B. (2003). Structure of gan quantum dots grown under "modified stranski-krastanow" conditions on aln, *J. Appl. Phys.* 94: 2254.

Gogneau, N., Jalabert, G., Monroy, E., Sarigiannidou, E., Rouviére, J.-L., Shibata, T., Tanaka, M., Gérard, J.-M. & Daudin, B. (2004). Influence of aln overgrowth on structural properties of gan quantum wells and quantum dots grown by plasma-assisted molecular beam epitaxy, *J. Appl. Phys.* 96: 1104.

González-Herráez, M. (2011). *Advanced Fibre Optics: Concepts and Technology*, EPFL Press.

Guillot, F., Bellet-Amalric, E., Monroy, E., Tchernycheva, M., Nevou, L., Doyennette, L., Julien, F. H., Dang, L. S., Remmele, T., Albrecht, M., Shibata, T. & Tanaka, M. (2006). Si-doped gan/aln quantum dot superlattices for optoelectronics at telecommunication wavelengths, *J. Appl. Phys.* 100: 044326.

Hamazaki, J., Matsui, S., Kunugita, H., Ema, K., Kanazawa, H., Tachibana, T., Kikuchi, A. & Kishino, K. (2004). Ultrafast intersubband relaxation and nonlinear susceptibility at 1.55 μm in gan/aln multiple-quantum wells, *Appl. Phys. Lett.* 84(7): 1102.

Hanamura, E., Kawabe, Y. & Yamanaka, A. (2007). *Quantum Nonlinear Optics*, Springer-Verlag.

Heber, J. D., Gmachl, C., Ng, H. M. & Cho, A. Y. (2002). Comparative study of ultrafast intersubband electron scattering times at ∼1.55 μm wavelength in gan/algan heterostructures, *Appl. Phys. Lett.* 81(7): 1237.

Hermann, M., Monroy, E., Helman, A., Baur, B., Albrecht, M., Daudin, B., Ambacher, O., Stutzmann, M. & Eickhoff, M. (2004). Vertical transport in group iii-nitride heterostructures and application in aln/gan resonant tunneling diodes, *Phys. Stat. Sol. (c)* 1: 2210.

Hill, K., Johnson, D., Kawasaki, B. & MacDonald, R. (1978). Cw three wave mixing in single-mode optical fibers, *J. Appl. Phys.* 49: 5098.

Hofstetter, D., Baumann, E., Giorgetta, F. R., Théron, R., Wu, H., Schaff, W. J., Dawlaty, J., George, P. A., Eastman, L. F., Rana, F., Kandaswamy, P. K., Leconte, S. & Monroy, E. (2009). Photodetectors based on intersubband transitions using iii-nitride superlattice structures, *J. Phys.: Condens. Matter* 21: 174208.

Hofstetter, D., Francesco, J. D., Kandaswamy, P. K., Das, A., Valdueza-Felip, S. & Monroy, E. (2010). Performance improvement of alnŨgan-based intersubband detectors by using quantum dots, *IEEE Photon. Technol. Lett.* 22(15): 1087.

Iizuka, N., Kaneko, K. & Suzuki, N. (2006). Polarization dependent loss in iii-nitride optical waveguides for telecommunication devices, *J. Appl. Phys.* 99(9): 093107.

Iizuka, N., Kaneko, K., Suzuki, N., Asano, T., Noda, S. & Wada, O. (2000). Ultrafast intersubband relaxation (<150 fs) in algan/gan multiple quantum wells, *Appl. Phys. Lett.* 77(5): 648.

Inoue, K. (1992). Four-wave mixing in an optical fiber in the zero-dispersion wavelength region, *J. Lightwave Technology* 10: 1553.

Kandaswamy, P. K., Bougerol, C., Jalabert, D., Ruterana, P. & Monroy, E. (2009). Strain relaxation in short-period polar gan/aln superlattices, *J. Appl. Phys.* 106(1): 013526.

Kandaswamy, P. K., Guillot, F., Bellet-Amalric, E., Monroy, E., Nevou, L., Tchernycheva, M., Michon, A., Julien, F. H., Baumann, E., Giorgetta, F. R., Hofstetter, D., Remmele, T., Albrecht, M., Birner, S. & Dang, L. S. (2008). Gan/aln short-period superlattices for intersubband optoelectronics: A systematic study of their epitaxial growth, design, and performance, *J. Appl. Phys.* 104(9): 093501.

Kheirodin, N., Nevou, L., Machhadani, H., Crozat, P., Vivien, L., Tchernycheva, M., Lupu, A., Julien, F., Pozzovivo, G., Golka, S., Strasser, G., Guillot, F. & Monroy, E. (2008). Electrooptical modulator at telecommunication wavelengths based on gan-aln coupled quantum wells, *Photon. Technol. Lett.* 20(9): 724.

Li, Y., Bhattacharyya, A., Thomidis, C., Liao, Y., Moustakas, T. D. & Paiella, R. (2008). Refractive-index nonlinearities of intersubband transitions in gan/aln quantum-well waveguides, *J. Appl. Phys.* 104(8): 083101.

Moumanis, K., Helman, A., Fossard, F., Tchernycheva, M., Lusson, A., Julien, F. H., Damilano, B., Grandjean, N. & Massies, J. (2003). Intraband absorptions in gan/aln quantum dots in the wavelength range of 1.27Ű2.4 μm, *Appl. Phys. Lett.* 82: 868.

Nakamura, T., Mochizuki, S., Terao, S., Sano, T., Iwaya, M., Kamiyama, S., Amano, H. & Akasaki, I. (2002). Structural analysis of si-doped algan/gan multi-quantum wells, *J. Cryst. Growth* 1129: 237Ű–239.

Naranjo, F. B., González-Herráez, M., Fernández, H., Solís, J. & Monroy, E. (2007). Third order nonlinear susceptibility of inn at near band-gap wavelengths, *Appl. Phys. Lett.* 90(9): 091903.

Naranjo, F. B., Kandaswamy, P. K., Valdueza-Felip, S., Calvo, V., González-Herráez, M., Martín-López, S., Corredera, P., Méndez, J. A., Mutta, G. R., Lacroix, B., Ruterana, P. & Monroy, E. (2011). Nonlinear absorption of inn/ingan multiple-quantum-well structures at optical telecommunication wavelengths, *Appl. Phys. Lett.* 98(3): 031902.

Nevou, L., Julien, F. H., Tchernycheva, M., Guillot, F., Monroy, E. & Sarigiannidou, E. (2008). Intraband emission at $\lambda \approx 1.48$ μm from gan/aln quantum dots at room temperature, *Appl. Phys. Lett.* 92(16): 161105.

Nevou, L., Mangeney, J., Tchernycheva, M., Julien, F. H., Guillot, F. & Monroy, E. (2009). Ultrafast relaxation and optical saturation of intraband absorption of gan/aln quantum dots, *Appl. Phys. Lett.* 94(13): 132104.

Nevou, L., Tchernycheva, M., Julien, F. H., Raybaut, M., Godard, A., Rosencher, E., Guillot, F. & Monroy, E. (2006). Intersubband resonant enhancement of second-harmonic generation in gan/aln quantum wells, *Appl. Phys. Lett.* 89: 151101.

Radic, S. & McKinstrie, C. J. (2005). Optical amplification and signal processing in highly nonlinear optical fiber, *IEICE Trans. Electronics* E88-C(5): 859.

Ranjan, V., Allan, G., Priester, C. & Delerue, C. (2003). Self-consistent calculations of the optical properties of gan quantum dots, *Phys. Rev. B* 68: 115305.

Rapaport, R., Chen, G., Mitrofanov, O., Gmachl, C., Ng, H. M. & Chu, S. N. G. (2003). Resonant optical nonlinearities from intersubband transitions in gan/aln quantum wells, *Appl. Phys. Lett.* 83(2): 263.

Renard, J., Kandaswamy, P. K., Monroy, E. & Gayral, B. (2009). Suppression of nonradiative processes in long-lived polar gan/aln quantum dots, *Appl. Phys. Lett.* 95: 131903.

Sarigiannidou, E., Monroy, E., Daudin, B., Rouviére, J. & Andreev, A. (2005). Strain distribution in gan/aln quantum dots superlattices, *Appl. Phys. Lett.* 87: 203112.

Sarigiannidou, E., Monroy, E., Gogneau, N., Radtke, G., Bayle-Guillemaud, P., Bellet-Amalric, E., Daudin, B. & Rouviére, J. L. (2006). Comparison of the structural quality in ga-face and n-face polarity gan/aln multiple-quantum-well structures, *Semicond. Sci. Technol.* 21: 912.

Sénés, M., Smith, K. L., Smeeton, T. M., Hooper, S. E. & Heffernan, J. (2007). Strong carrier confinement in $in_x ga_{1-x} n$/gan quantum dots grown by molecular beam epitaxy, *Phys. Rev. B* 75: 045314.

Sheik-Bahae, M., Said, A., Wei, T., Hagan, D. & Stryland, E. V. (1990). Sensitive measurements of optical nonlinearities using a single beam, *IEEE J. Quantum Electronics* 26: 760.

Shutherland, R. L. (2003). *Handbook of Nonlinear Optics*, 2nd edn, Marcel Dekker.

Suzuki, N. & Iizuka, N. (1997). Feasibility study on ultrafast nonlinear optical properties of 1.55-μm intersubband transition in algan/gan quantum wells, *Jpn. J. Appl. Phys.* 36(Part 2, No. 8A): L1006.

Tchernycheva, M., Nevou, L., Doyennette, L., Helman, A., Colombelli, R., Julien, F. H., Guillot, F., Monroy, E., Shibata, T. & Tanaka, M. (2005). Intraband absorption of doped gan/aln quantum dots at telecommunication wavelengths, *Appl. Phys. Lett.* 87: 101912.

Tchernycheva, M., Nevou, L., Doyennette, L., Julien, F. H., Warde, E., Guillot, F., Monroy, E., Bellet-Amalric, E., Remmele, T. & Albrecht, M. (2006). Systematic experimental and theoretical investigation of intersubband absorption in gan/aln quantum wells, *Phys. Rev. B* 73(12): 125347.

Valdueza-Felip, S. (2011). *Nitride-based semiconductor nanostructures for applications in optical communications at 1.55 μm*, PhD thesis, University of Alcalá.

Valdueza-Felip, S., Naranjo, F.B., González-Herráez, M., Fernández, H., Solís, J., Guillot, F., Monroy, E., Nevou, L., Tchernycheva, M. & Julien, F. H. (2008). Characterization of the resonant third-order nonlinear susceptibility of si-doped gan/aln quantum wells and quantum dots at 1.5 μm, *IEEE Photon. Technol. Lett.* 20: 1366.

Vardi, A., Kheirodin, N., Nevou, L., Machhadani, H., Vivien, L., Crozat, P., Tchernycheva, M., Colombelli, R., Julien, F. H., Guillot, F., Bougerol, C., Monroy, E., Schacham, S. & Bahir, G. (2008). High-speed operation of gan/algan quantum cascade detectors at $\approx$1.55 μm, *Appl. Phys. Lett.* 93: 193509.

Wada, O. (2004). Femtosecond all-optical devices for ultrafast communication and signal processing, *New J. Physics* 6: 183.

Zeldovich, B. Y., Pilipetskii, N. F. & Shkunov, V. V. (1985). *Principles of Phase Conjugation*, Vol. 42 of *Springer Series in Optical Sciencies*, Springer Verlag.

Part 3

Electronics

7

Quantum Dots
as Global Temperature Measurements

Hirotaka Sakaue[1], Akihisa Aikawa[1], Yoshimi Iijima[1],
Takuma Kuriki[2] and Takeshi Miyazaki[2]
[1]Japan Aerospace Exploration Agency
[2]The University of Electro-Communications
Japan

1. Introduction

ZnS-capped CdSe semiconductor nanocrystals (quantum-dots, QDs) provide size-tunable optical properties. QDs show shifted luminescent peaks due to crystal size. They exhibit strong and stable luminescence with a 50 % quantum yield at room temperature [1]. Walker *et al.* used QDs as a global temperature probe [2]. They used a polymer support of poly(lauryl methacrylate) to create a global temperature sensor. This type of sensor, called a temperature-sensitive paint (TSP), has been widely used in aerospace measurements [3]. Conventional TSP uses a phosphorescent molecule as a temperature probe. This type of molecule has a relatively wide FWHM (full width at half maximum), which is roughly 100 nm. When applying a QD as a temperature probe, the FWHM is narrower than that of phosphorescent probes and is roughly 40 nm [1]. A low FWHM will widen the selection of probe molecules to prepare multi-color sensors in the visible wavelength range. In addition, a high quantum yield of QDs can be beneficial as an optical temperature sensing probe to increase the signal-to-noise ratio.

The material properties of polymers change the glass-transition temperature. These temperatures for TSPs are roughly 400 K [4, 5, and 6]. To use polymers as a QD support, the resultant TSP can be sprayed on a testing article. However, the measurement range is limited by these temperatures. Even for low temperature measurements, a polymer may show physical defects such as cracks at cryogenic temperatures. In aerospace engineering, anodized aluminum has been applied as a support for pressure probes. The resultant sensor, called anodized-aluminum pressure-sensitive paint, has been used in wind tunnel measurements [7]. With aluminum, which has a melting point of 930 K, as a probe support, the anodized aluminum retains its material properties at cryogenic temperatures as well as at temperatures higher than glass-transition temperatures of TSP polymers. By using anodized aluminum as a support for a temperature sensing probe, a global temperature sensor with a wide range can be created.

TSP gives global temperature information related to its luminescent output. The TSP measurement system consists of a TSP coated model, an image acquisition unit, and an image processing unit (Fig. 1). For image acquisition, an illumination source and a photo-detector are required. To separate the illumination and the TSP emission detected by a photo-detector, appropriate band-pass filters are placed in front of an illumination and a photo-detector. The image processing unit includes calibration and computation. The calibration relates the luminescent signal to temperatures. Based on these calibrations, luminescent images are converted to a temperature map using a PC.

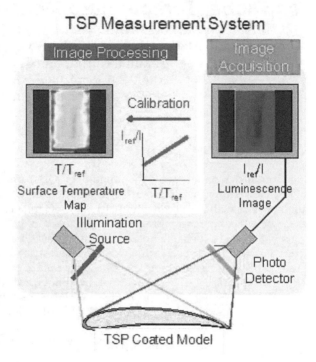

Fig. 1. Schematic of TSP measurement system

In this chapter, the development of a QD-based anodized aluminum temperature sensor is discussed. The temperature sensitivity of this sensor is characterized in the temperature range from 100 K to 500 K. An application of this sensor as a global temperature measurement is included, which is focused on a hypersonic flow where an aerodynamic heating is a critical issue.

2. Development of quantum-dot based anodized-aluminum temperature-sensitive paint (AA-TSP)

2.1 Materials and dipping deposition method

A temperature probe of QDs was applied on an anodized-aluminum surface by the dipping deposition method [8]. QDs of birch yellow from Evident Technologies (ED-C11-TOL-0580) were used. These QDs are called QD_{BY} in this chapter. The dipping deposition method requires a temperature probe of QD_{BY}, a solvent, and an anodized aluminum coating. The application procedure is schematically shown in Fig. 2. QD_{BY} was dissolved in eight different solvents, which varied according to their polarity index (Table 1). A concentration of 15 μM was adjusted to create these solutions or mixtures.

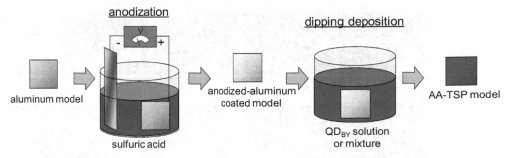

Fig. 2. Schematic description of dipping deposition method

AA-TSP	Solvent	Polarity Index
$AATSP_{ind00}$	hexane	0.1
$AATSP_{ind02}$	toluene	2.4
$AATSP_{ind03}$	dichloromethane	3.1
$AATSP_{ind04}$	chloroform	4.1
$AATSP_{ind05}$	acetone	5.1
$AATSP_{ind06}$	N,N-dimethylformamide	6.4
$AATSP_{ind07}$	dimethylsulfoxide	7.2
$AATSP_{ind10}$	water	10.2

Table 1. List of solvent conditions

Depending on the polarity index of solvents, differences in dissolution can be seen. Solvents with a lower polarity index up to 4.1 could dissolve QD_{BY}, while solvents with a higher polarity index created mixtures. These are seen in luminescence from the QD_{BY} solutions and mixtures (Fig. 3). Luminescent images were acquired using the camera system discussed in section 2.2. Solvents with a lower polarity index up to 4.1 showed relatively bright luminescence, while solvents with a higher polarity index showed dim luminescence or luminescence in spots. The anodized-aluminum coating was dipped in these solutions or mixtures for one hour at room conditions. The AA-TSPs developed were identified by the solvents used, which are listed in Table 1. We coated QD_{BY} with some of these solutions. Fig. 4 shows luminescent images of AA-TSPs, which were acquired using the camera system discussed in section 2.2. We can see uniform coatings using $AATSP_{ind00}$, $AATSP_{ind02}$, $AATSP_{ind03}$, and $AATSP_{ind04}$. Because of the effects of mixtures, QD_{BY} were coated in spots with $AATSP_{ind05}$, $AATSP_{ind06}$, $AATSP_{ind07}$, and $AATSP_{ind10}$. Except for these spots, QD_{BY} were not coated as shown in the dark luminescent signal in Fig. 4. The signal level was determined from the averaged luminescent signal of each sample. A luminescent signal from a 5-mm^2 area was averaged. The signal levels were normalized based on the signal of

AATSP$_{ind04}$, which is also shown in Fig. 4. The maximum luminescent signal was obtained from AATSP$_{ind04}$. Based on the coating uniformity and the signal level, AATSP$_{ind00}$, AATSP$_{ind02}$, AATSP$_{ind03}$, and AATSP$_{ind04}$ were selected for temperature calibration.

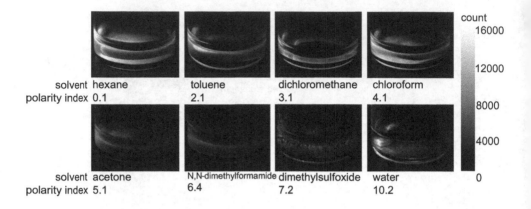

Fig. 3. Luminescence from QD$_{BY}$ solutions and mixtures. (Figures obtained from Sakaue et al. [9])

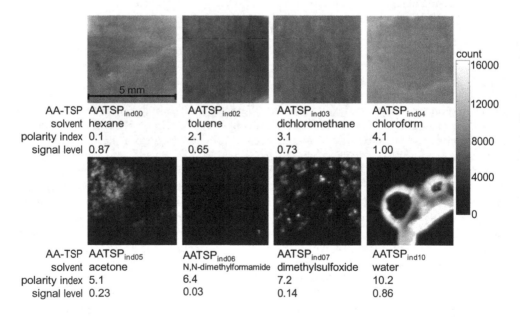

Fig. 4. Luminescent image and signal level of developed AA-TSPs. (Figures obtained from Sakaue et al. [9])

2.2 AA-TSP characterization method

Fig. 5 schematically describes the calibration setup, which combined a camera system and a spectrometer system. Both systems used a temperature-controlled chamber. The temperature could be controlled from 100 K to 500 K. The chamber was filled with dry air at 100 kPa. An AA-TSP sample was placed on the test section in the chamber. Both systems used a 407-nm laser (NEO ARK, DPS-5001) to illuminate the sample.

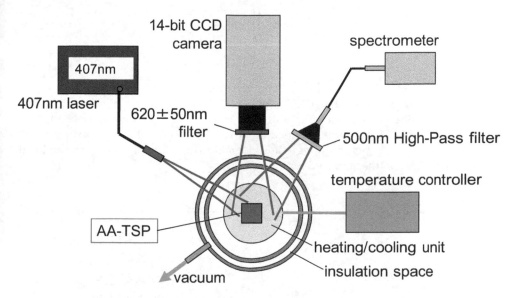

Fig. 5. Schematic of AA-TSP calibration setup

The camera system used a 14-bit CCD camera (Hamamatsu, C4880). A 620 ± 50 nm filter was placed in front of the camera lens to acquire only the luminescent signal from AA-TSPs, QD_{BY} solutions, and mixtures. The camera system was used to characterize the signal level and temperature sensitivity. For the camera system, a reference luminescent image, I_{ref}, was obtained. This image was compared with the luminescent image at a given temperature to decouple the non-temperature induced luminescence, such as illumination variation and/or coating non-uniformity. I_{ref} was used as an image at a reference temperature of 298 K. The spectrometer system used a USB4000 spectrometer from Oceanoptics. It was connected to an optical fiber with a 500-nm high-pass filter in front of the fiber end. This system was used to characterize the luminescent spectrum of the AA-TSPs.

2.3 Characterization results: Solvent dependency

Fig. 6 shows the temperature calibrations of selected AA-TSPs obtained from the camera system. The reference temperature was 298 K. The temperature calibrations decreased monotonically with increasing temperature. The temperature calibration of a relatively small temperature range, such as 30 degrees, could be assumed to be linear for engineering applications. The first-order polynomial was fitted to the calibration results.

$$\frac{I}{I_{ref}} = c_{s0} + c_{s1}T \tag{1}$$

Where c_{s0} and c_{s1} were calibration constants. The temperature sensitivity, δ_T, is defined as the slope of the temperature calibration at the reference temperature, T_{ref}, which is described as the percent change in luminescent signal, I/I_{ref}, over a given temperature in Kelvin.

$$\delta_T = \frac{d(I/I_{ref})}{dT}\bigg|_{T=T_{ref}} = c_{s1} \ (\%/K) \tag{2}$$

The δ_T is large if its absolute value is large because a high signal change over a given temperature can be obtained. Based on equations (1) and (2), the δ_Ts of selected AA-TSPs was determined: the δ_Ts are listed in Table 2. It also lists the signal levels of selected AA-TSPs. The best δ_T of -1.1 %/K was obtained for AATSP$_{ind04}$, which gave the maximum signal level. Based on these characterizations, using chloroform as a dipping solvent resulted in the application of QD$_{BY}$ on the anodized-aluminum coating.

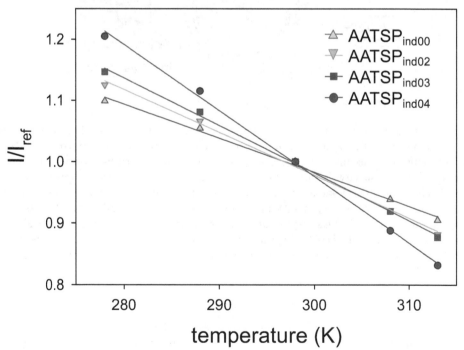

Fig. 6. Temperature calibrations of AATSP$_{ind00}$, AATSP$_{ind02}$, AATSP$_{ind03}$, and AATSP$_{ind04}$ obtained from the camera system. The reference temperature was 298 K. (Figures obtained from Sakaue et al. [9])

AA-TSP	Temperature Dependency (%/K)	Signal Level (normalized)
$AATSP_{ind00}$	-0.6	0.87
$AATSP_{ind02}$	-0.7	0.65
$AATSP_{ind03}$	-0.8	0.73
$AATSP_{ind04}$	-1.1	1.00

Table 2. Temperature sensitivity and the signal level of $AATSP_{ind00}$, $AATSP_{ind02}$, $AATSP_{ind03}$, and $AATSP_{ind04}$. The reference temperature was 298 K [9]

2.4 Characterization results: Temperature calibration from 100 K to 500 K

Based on the solvent dependency discussed in the previous section, $AATSP_{ind04}$, which used chloroform as a solvent, was calibrated for a wide temperature range. Fig. 7 shows temperature spectra of $AATSP_{ind04}$, obtained from the spectrometer system. The temperature was varied from a cryogenic temperature of 100 K up to a high temperature of 500 K. A total

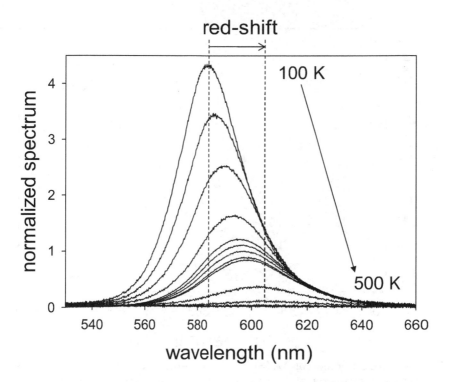

Fig. 7. The temperature spectra of $AATSP_{ind04}$ at temperature ranges from 100 K to 500 K. Each spectrum was normalized by the luminescent peak at 298 K. Thirteen temperature spectra were measured: 100 K, 150 K, 200 K, 250 K, 278 K, 288 K, 298 K, 308 K, 313 K, 350 K, 400 K, 450 K, and 500 K. (Figures obtained from Sakaue et al. [9])

of thirteen temperature spectra were measured. Each spectrum was normalized by the luminescent peak at 298 K. As temperatures increased, the luminescent spectra decreased. A red-shift of the luminescent peak occurred with increasing temperature. A maximum change of at least 20 nm was observed. Luminescent peaks at 450 K and 500 K were not clearly identified. Thus, the luminescent peak at 400 K was used as the maximum red-shift wavelength of AATSP$_{ind04}$ to determine the peak location.

Fig. 8 shows the temperature calibration of AATSP$_{ind04}$ from 100 K to 500 K, obtained using the camera system. The plot was fitted with a four-parameter sigmoid.

$$\frac{I}{I_{ref}} = c_0 + \frac{c_1}{1 + \exp\left(-\dfrac{T - c_2}{c_3}\right)} \tag{3}$$

Where c_0, c_1, c_2, and c_3 were calibration constants. The reference temperature was 298 K. The temperature calibration showed a monotonic decrease in luminescent signal with increasing temperature.

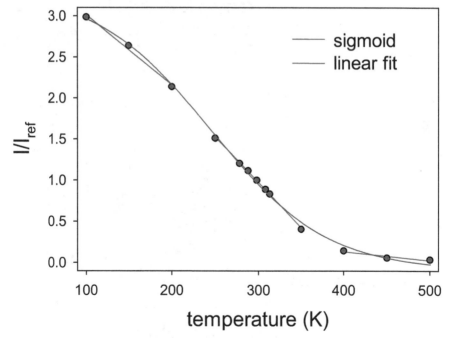

Fig. 8. Temperature calibration of AATSP$_{ind04}$ at temperature ranges from 100 to 500 K obtained from the camera system. The reference temperature was 298 K. A four-parameter sigmoid as well as the first-order polynomial was fitted. The first-order fitting was applied at three temperature regions: low-temperature region from 100 K to 200 K, middle-temperature region from 250 K to 350 K, and high-temperature region from 400 K to 500 K. (Figures obtained from Sakaue et al. [9])

The first-order fitting of equation (1) was also applied to three temperature regions: a low-temperature region from 100 K to 200 K, a middle-temperature region from 250 K to 350 K, and a high-temperature region from 400 K to 500 K. These regions are shown in Fig. 7. Although fewer calibration points were used, linear fits gave single δ_Ts that are useful for engineering application.

Because of the non-linear calibration, δ_T varied with the reference temperature, T_{ref}, for the sigmoidal fitting, which can be derived as a derivative of equation (3) at T_{ref}. For linear fittings, equation (2) can be used to determine δ_T in the three temperature regions.

$$\delta_T = \frac{d\left(I/I_{ref}\right)}{dT}\bigg|_{T=T_{ref}} = \frac{c_1}{c_3} \cdot \left(\exp\left(\frac{T_{ref}-c_2}{2c_3} \right) + \exp\left(-\frac{T_{ref}-c_2}{2c_3} \right) \right)^{-2} \quad (\%/K) \qquad (4)$$

Fig. 9 shows δ_T as a function of T_{ref}. It varied with T_{ref} when the sigmoidal fitting was used. The best δ_T of -1.3 %/K could be obtained at a T_{ref} of 245 K. By using linear fits, three δ_Ts could be obtained at the low-, the middle-, and the high-temperature regions, respectively. These fittings gave single values of δ_Ts of -0.9 %/K, -1.1 %/K, and -0.1 %/K, respectively. For a T_{ref} at 298 K, which was the representative reference temperature, δ_T was -1.1 %/K as determined from both fits.

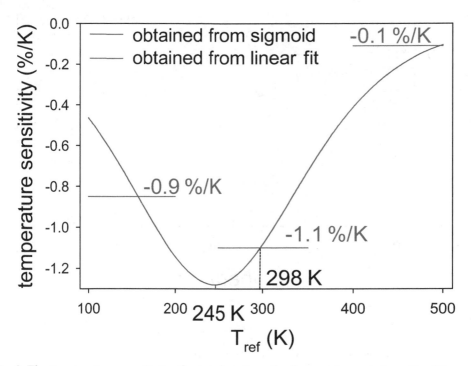

Fig. 9. The temperature sensitivity, δ_T, as a function of reference temperature, T_{ref}. (Figures obtained from Sakaue et al. [9])

2.5 Discussion: Repeatability

Fig. 10 shows the repeatability of the temperature calibration. Mean values and error bars are shown. The calibration was repeated five times and averaged for each temperature. The standard deviation is shown as an error bar. The temperature calibration showed a relatively large error at cryogenic temperatures, which may be due to the uncertainty of the temperature control of the chamber as well as the hysteresis of our AA-TSP. Because the coating provided a micro-porous structure, which was on the order of ten nanometers in diameter and ten micrometer in depth, the solvent used for dipping deposition tended to remain in the pores. This remaining solvent might have caused the hysteresis at cryogenic temperatures. The calibration results in Figures 6 and 8 are shown as mean values of five repeated calibrations.

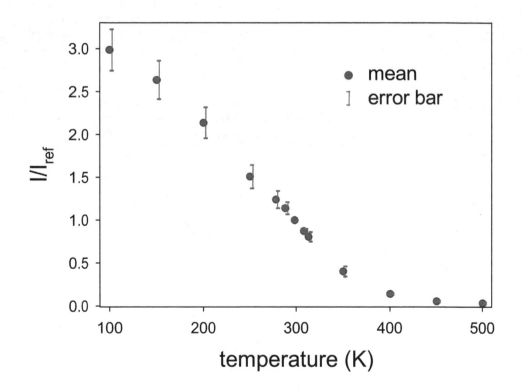

Fig. 10. Repeatability of the temperature calibration obtained from the camera system. Mean values and error bars are shown. Calibrations were repeated five times. (Figures obtained from Sakaue et al. [9])

3. Hypersonic wind tunnel application

To reveal the aerodynamic heating, hypersonic wind tunnels are used. To apply a TSP to a hypersonic flow, the heating causes a material defects to a conventional TSP, which uses a polymer as a supporting matrix. Because a supporting matrix is aluminum instead of a polymer, we can expect AA-TSP to hold its material properties at higher temperatures as that in the hypersonic application.

3.1 Compression corner model

Fig. 11 showed a photograph of wind tunnel model. Referring from Nakakita et al. [11] and Ishiguro et al. [12], a compression corner model was used. It has a 30° compression corner with its dimension shown in the figure. A thermocouple was used for the temperature measurement, which was placed on the surface of the model. AATSP$_{ind04}$ was applied onto the model surface by the dipping deposition method discussed in section 2.1.

Fig. 11. Photograph of a compression corner model

3.2 Wind tunnel measurement setup

Fig. 12 shows a schematic description of The Hypersonic and High Enthalpy Wind Tunnel at The University of Tokyo, Kashiwa Campus. The flow conditions set in our TSP measurements are summarized in Table 3. Fig. 13 shows a schematic description of a model location and an optical setup. A compression corner model was placed in Mach 7.1 flow after the flow stabilization period. We received a signal when placing the model into the test section. This was used to trigger the image acquisition and reference temperature measurement. A model stabilization period is needed after placing the model, which was 2.5

s. The model was then released from the flow after 6.5 s. Three xenon lamp sources were used to illuminate the AATSP$_{ind04}$ coated model. Band-pass filters of 340 ± 50 nm were placed in front of the illumination to give UV excitation. A 12-bit high-speed CCD camera (Phantom v12.1) was used to acquire AATSP$_{ind04}$ images. An optical filter of 620 ± 50 nm was placed in front of the camera. Camera frame rate was set at 25 Hz.

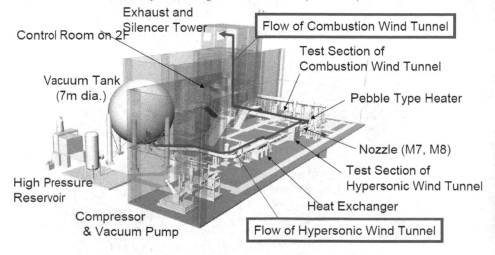

Fig. 12. Schematic description of Hypersonic and High Enthalpy Wind Tunnel at UT Kashiwa. (Figures obtained from http://daedalus.k.u-tokyo.ac.jp/wt/WTpamphE.pdf)

	conditions
mach number	7.1
stagnation pressure	0.95 MPa
static pressure	200 Pa
stagnation temperature	900 K
test duration	6.5 s

Table 3. Flow conditions of TSP measurement

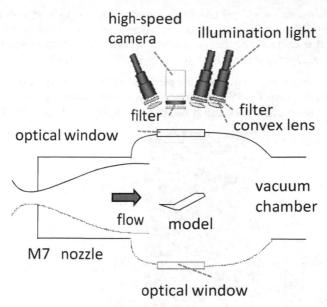

Fig. 13. Schematic of AA-TSP measurement setup

3.3 Global temperature measurement

Fig. 14 shows the *insitu* temperature calibration. Temperature data was monitored from the thermocouple, and the luminescent signal at the corresponding location was related. The

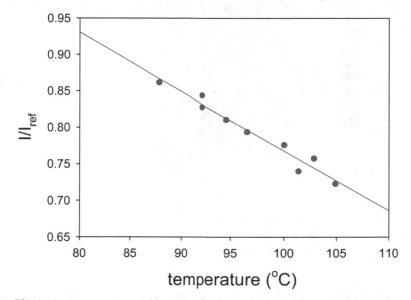

Fig. 14. The *insitu* temperature calibration. (Figures obtained from Kuriki et al. [10])

reference condition was a wind-off condition whose temperature and pressure were 20 °C and 0.18 kPa, respectively. The first order polynomial was used for fitting the calibration points. Based on this calibration, we can convert the luminescent images to the temperature distribution. The temperature sensitivity of the calibration was -0.82 %/°C.

Fig. 15 shows a temperature map and cross sectional distribution obtained from AATSP$_{ind04}$. Results are shown in every 1 s. We can see that the front edge of the model was heated the most. After 4 s of the measurement duration, this area was heated up to 413 K (140 °C).

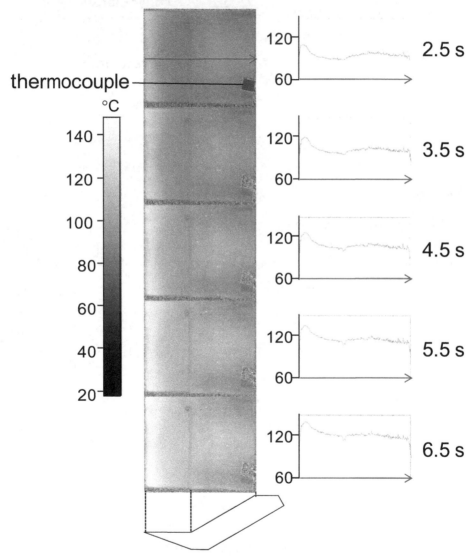

Fig. 15. Temperature map and cross sectional distribution in every 1 s. (Figures obtained from Kuriki et al. [10])

4. Conclusion

We have developed a quantum dot-based anodized aluminum temperature-sensitive paint for global temperature measurement. Based on the solvent study for applying quantum dots on an anodized aluminum coating, we found that chloroform as a dipping solvent gave the best signal level and temperature sensitivity. The resultant sensor provided temperature sensitivities from 100 K to 500 K. A total of thirteen temperature points were measured. By using a four-parameter sigmoidal fit to the temperature calibration, the best temperature sensitivity of -1.3 %/K was obtained at 245 K. By separating three temperature regions for engineering applications, three constant temperature sensitivities were obtained. The regions were separated into a low-temperature region of 100 K to 200 K, a middle- temperature region of 250 K to 350 K, and a high-temperature region of 400 K to 500 K. The temperature sensitivities of these regions were -0.9 %/K, -1.1 %/K, and -0.1 %/K, respectively.

The best AA-TSP was used in hypersonic wind tunnel application. It gave the global temperature measurement on a compression corner model at the hypersonic flow of Mach 7.1. The global temperature measurements related to the running time were obtained. The front edge of the model was most heated, and its temperature was raised up to 413 K for 4 s of measurement time.

5. Acknowledgment

Authors would like to acknowledge Prof. Osamu Imamura at Nihon University, Prof. Kojiro Suzuki, and Mr. Takeo Okunuki at The University of Tokyo for technical supports on The Hypersonic and High Enthalpy Wind Tunnel at The University of Tokyo, Kashiwa Campus.

6. References

Bernhard, W. & Umesh, G. (1981). Polymer Preprints. *American Chemical Society, Division of Polymer Chemistry*, Vol.22, pp. 308-309

Hines, M. A. & Guyot-Sionnest, P. (1996). Synthesis and characterization of strongly luminescing ZnS-capped CdSe nanocrystals. *J. Phys. Chem.*, Vol.100, pp. 468-471

Ishiguro, Y; Nagai, H.; Asai, K. & Nakakita, K. (2007). Visualization of Hypersonic Compression Corner Flows using Temperature- and Pressure-Sensitive Paints. *AIAA Paper 2007-118*, American Institute for Aeronautics and Astronautics

Kameda, M.; Tabei, T.; Nakakita, K.; Sakaue, H. & Asai, K. (2005). Image measurement of unsteady pressure fluctuation by a pressure-sensitive coating on porous anodized aluminum. *Meas. Sci. Technol.*, Vol.16, pp. 2517-2524

Kim, J. H; Min, B. R.; Won, J.; Kim, C. K. & Kang, Y. S. (2004). *Journal of Polymer Science, Part B: Polyer Physics*, Vol.42, pp. 621-628

Kuriki, K; Sakaue, H.; Imamura, O. & Suzuki, K. (2010). Temperature-Cancelled Anodized-Aluminum Pressure-Sensitive Paint for Hypersonic Compression Corner Flows. *AIAA Paper 2010-0673*, American Institute for Aeronautics and Astronautics

Liu, T. & Sullivan, J. P. (2004). *Pressure and Temperature Sensitive Paints*, Springer, ISBN 3-540-22241-3, Heidelberg, Germany (Chapter 1 and Chapter 3)

Nakakita, K; Yamazaki, T.; Asai, K.; Teduka, N.; Fuji, N. & Kameda, M. (2000). Pressure Sensitive Paint Measurement in a Hypersonic Shock Tunnel. *AIAA Paper 2000-2523*, American Institute for Aeronautics and Astronautics

Sakaue, H. (2005). Luminophore application method of anodized aluminum pressure sensitive paint as a fast responding global pressure sensor. *Rev. Sci. Instrum.*, Vol.76, 084101

Sakaue, H.; Aikawa, A. & Iijima, Y. (2010). Anodized Aluminum as Quantum-Dot Support for Global Temperature Sensing from 100 to 500 Kelvin. *Sensors and Actuators B: Chemical*, Vol.150, No.2, pp. 569-573

Somayajulu, D. R. S.; Murthy, C. N.; Awasthi, D. K.; Patel, N. V. & Sarkar, M. (2001). *Bulletin of Materials Science*, Vol.24, pp. 397-400

Walker, G. W.; Sundar, V. C.; Rudzinski, C. M. & Wun, A. W. (2003). Quantum-dot optical temperature probes. *Appl. Phys. Lett.*, Vol.83, pp. 3555-3557

8

Silicon Oxide Films Containing Amorphous or Crystalline Silicon Nanodots for Device Applications

Diana Nesheva[1], Nikola Nedev[2], Mario Curiel[3], Irina Bineva[1],
Benjamin Valdez[2] and Emil Manolov[1]

[1]*Institute of Solid State Physics, Bulgarian Academy of Sciences, Sofia*
[2]*Institute of Engineering, Autonomous University of Baja California, Mexicali, B. C.*
[3]*Centro de Nanociencias y Nanotecnología,*
Universidad Nacional Autónoma de México Ensenada, B. C.
[1]*Bulgaria*
[2]*Mexico*

1. Introduction

The impressive recent growth of the portable systems market (mobile PC, MP3 audio player, digital camera, mobile phones, hybrid hard disks, etc.), has increased the interest of the semiconductor industry on non-volatile memory (NVM) technologies. The demand for mobility applications is the main reason for the fast development of NVM technologies and products. Therefore lower power consumption, lower system costs, ever higher capacity and system performance are required. The conventional floating-gate memories satisfy these needs at present but the scaling of these memories is becoming increasingly difficult primarily because reliability problems limiting the bottom (tunnel) oxide to a thickness of around 10 nm.

The application of discrete storage nodes in conventional non-volatile memory has been considered as one of the key items for ensuring a better reliability of the non-volatile memory and to increase scaling. The basic idea is to replace the floating gate (FG) of NVMs by many discrete trapping centers. The most important effect of the localized trapping is the fact that a single leakage path due to a defect (intrinsic or created during the operation by the applied write/erase electric field) in the oxide can only discharge a few storage nodes. In the devices with discrete storage nodes, the storage medium consists either of natural traps, normally in a nitride layer (Eitan et al., 2000; Minami et al., 1994), or of semiconductor nanocrystals (Tiwari et al., 1995).

Two kinds of semiconductor structures containing nanocrystals are studied recently for non-volatile memory purposes. Ones of them are metal-insulator-semiconductor (MIS) structures containing semiconductor nanocrystals in the insulator layer. In such structures nanocrystals serve as charge storage media replacing the floating gate in conventional memory field effect transistors (FETs) (Horváth&Basa, 2009; Lombardo et al., 2004; Steimle et al., 2007; Tiwari et al., 1995). The other ones are the phase-change memory structures (which are not subject of consideration in this chapter), where the channel layer itself is switched between nanocrystalline and amorphous state with high and low conductivity, respectively (Hudgens&Johnson, 2004).

Since MIS structures containing semiconductor nanocrystals are less vulnerable to charge loss through isolated defects in the bottom oxide they provide opportunity for reduction of the size of the memory device and for operation at lower voltages compared to continuous floating gate flash devices. They use processes of direct charge tunnelling to the nanocrystals, Fowler-Nordheim tunnelling or channel hot electron (CHE) injection. The result is capturing of one or a few electrons in a potential well with well defined spatial position (Tsoukalas et al., 2005). The captured electron controls the current through a conducting channel situated close to the position of the captured electron and thus the nanocrystal plays the role of a floating gate. The electron stays in the nanocrystal for a certain time, which determines the duration of the "memory" effect in the device (i.e. the retention time).

At room temperature the memory effect, i.e. the charge capture in nanocrystals, manifests itself as a well defined shift of the capacitance-voltage (C-V) dependence of MIS structure (as a result of its charging or discharging) and as a change of the threshold voltage, called memory window, of the transistors fabricated with the same gate dielectric. The reduced density of states in the NCs restricts the states available for electrons and holes to tunnel to. Besides the Coulomb blockade effect arises from a larger electrostatic energy associated with placing a charged particle onto a rather small capacitance. Due to the both effects less charged carriers are used in the operation of the device that results in low-power operation without sacrificing speed. In addition, because the charge loss through lateral paths is suppressed the constraints on electrical contact isolation can be relaxed which permits thinner oxides and therefore lower voltage/power operation. An important advantage of the NVMs containing NCs is that they have shown a superior endurance at increased temperatures than the standard polysilicon floating gate NVMs and those using charge storage in natural traps. More detailed information on the principles of operation and non-volatile memories containing semiconductor or metal nanoparticles in various matrices can be found in a number of recent reviews [see for example (Horváth&Basa, 2009; Steimle et al., 2007; Tsoukalas et al., 2005).

The attention in this chapter is addressed to properties of memory structures in which the standard floating gate is replaced by crystalline or amorphous silicon nanoparticles (NPs), as storage nodes, separated by silicon oxide. The insulator of such MIS structure normally consists of three layers (Dimitrakis et al., 2005; Tiwari et al., 1996; Tsoi et al., 2005): (i) ultra thin tunnel SiO_2 layer grown on the crystalline Si wafer followed by (ii) composite layer of Si nanocrystals (NCs) in a SiO_2 matrix (nc-Si- SiO_2) and (iii) control SiO_2 layer, which insulates the NCs from the control gate. The middle nc-Si-SiO_2 layer has been mostly prepared by ion implantation of Si in thermal SiO_2 and subsequent annealing at high temperature (≥ 900 °C) (Normand et al., 2004 and references therein; Carreras et al., 2005; Ng et al., 2006a), by applying some chemical vapor deposition (CVD) technique for Si nanocrystals fabrication and subsequent CVD deposition of silicon dioxide (Lombardo et al., 2004; Oda et al., 2005; Rao et al., 2004), by deposition of a ultra thin amorphous Si layer and a subsequent oxidation of this layer at a high temperature (Kouvatsos et al., 2003; Tsoi et al., 2005) or by thermal evaporation of SiO powder under selected oxygen pressure (Lu et al., 2005, 2006).

The principal materials science aspects of the fabrication of two dimentional arrays of Si nanocrystals in thin SiO_2 layers and at tunable distances from the interface with the Si substrate have been considered in Ref. (Claverie et al., 2006). Below we mention only few interesting points in this respect. The effect of annealing temperature and gas ambient on the properties of nc-Si-SiO$_2$ has been studied in Ref. (Ioannou-Sougleridis et al., 2003). It has been ascertained that the samples annealed at 900 °C show significant memory effect which is a result of electron or hole capture in defect states with approximately same density for both types of carriers (~ 2x10^{12} cm^{-2}). The hydrogen annealing has shown that the main part of the defect states in these samples is due to unsaturated bonds at the nc-Si/SiO$_2$ interface. In the samples annealed at 1100 °C, the charge capturing does not depend substantially on the hydrogen annealing, which is explained assuming much lower concentration of defect states (~ 2.5x10^{11} cm^{-2}). In another study (Yu et al., 2003) dependence of the charge effect on the thickness of the nc-Si-SiO$_2$ layer is reported. Depth profiling of the charging effect of the nc-Si has revealed (Liu et al., 2006) that the charging effect decreases with the increase of nc-Si concentration and vanishes when a densely stacked nanocrystal layer is formed. The phenomenon is attributed to charge diffusion among the nanocrystals.

The impact of programming mechanisms Fowler-Nordheim/Fowler-Nordheim or channel hot electron/Fowler-Nordheim on the Performance and Reliability of NVM devices based on Si anocrystals has been explored at both room temperature and 85 °C in Ref. (Ng et al., 2006b). It has been shown that the channel hot electron programming has a larger memory window, a better endurance, and a longer retention time as compared to Fowler-Nordheim programming. Moreover, the channel hot electron programming yields less stress-induced leakage current than the Fowler-Nordheim programming, suggesting that it produces less damage to the gate oxide and the oxide/Si interface.

It has been demonstrated that the application of low-energy ion beam implantation has resulted in preparation of very promising NVMs with Si nanocrystals (Horváth&Basa, 2009; Tsoukalas et al., 2005) since it allows (Normand et al., 1998) to avoid wide distribution of NCs through the oxide layer and to obtain NCs at a specific location into the host matrix. It has been demonstrated that memory cells fabricated following a procedure including implantation with 1 keV silicon ions to a dose of 2×10^{16} cm^{-2} and subsequent annealing at 950 °C for 30 min in N$_2$ + 1.5% O$_2$ (Normand et al., 2003; Tsoukalas et al., 2005) displayed more than 10^6 write/erase ± 9 V/10 ms cycles without observation of gate degradation. The long time extrapolation of the data retention characteristics at two different temperatures, 25 and 85 °C, after + 9 V/− 9 V 10 ms pulses has shown that those Si NC memory device could achieve 10 years data retention at 85 °C. The memory transistors produced displayed about 2 V memory window for the write/erase pulses of ± 9 V, 10 ms.

Low-energy implantation (at 2 keV) has also been applied (Ng et al., 2006a) for preparation of MIS structures in which the tunneling oxide thickness was 3 nm or 7 nm for realization of direct or Fowler-Nordheim tunneling during the charging process. For direct tunneling a memory window width of about 1 V was obtained using charging pulses of ± 12 V, 1 μs. For Fowler-Nordheim tunneling the memory window width was about 0.5 V using charging pulses of ± 12 V, 1 ms. Thus, it has been shown that devices with a thicker tunnel layer required longer charging pulses, and they degraded faster, as well, but the extrapolated for 10 years memory window width at 85 °C was about 0.3 V for both type of structures.

Successful preparation of Si nanocrystals for memory applications has been also realized by chemical vapour deposition. Devices for non-volatile purposes were produced using 3.8-5.0 nm thick tunnel oxide (Rao et al., 2004) and a threshold voltage shift of ~ 1.5 V has been obtained by voltage pulses of 14 V, 100 μs or 12 V, 1 ms. Retention characteristics have also been very good, but the memory window shifted slightly with increasing number of write/erase cycles.

Recently tunnel barrier engineering has been suggested (Jung&Cho, 2008) as a promising way for tunnel oxide scaling. It uses multiple dielectric stacks to enhance field-sensitivity and thus to allow for shorter writing/erasing times and/or lower operating voltages than single SiO_2 tunnel oxide without altering the ten-year data retention constraint. Experiments on memory structures containing one-three Si NC layers prepared by evaporation of SiO powder and layer by layer growth (Lu et al., 2005, 2006) have shown that in structures with two or three NC layers two or three different saturation voltages were obtained with increasing bias related to charge injection to and capture at the first, second, or third NC layer (Lu et al., 2005). Charge captured in the first layer yielded a memory window width of about 2.5 V, charge captured in the second NC layer yielded an additional flat-band voltage shift of about 2.5 V, while charge captured in the third layer yielded an additional shift of about 1.5 V. It has been observed that the multilayer storage in memory structures prepared by different deposition methods yielded very good retention, as well (Han et al., 2007; Lu et al., 2005; Nassiopoulou et al., 2009).

For decades the radiation effects have been a serious problem for electronics used in defense and space systems and therefore the study of the radiation effects on metal-oxide-semiconductor (MOS) based devices, including MOS capacitors, has been an active research area over the past decades. An advantage of the nanocrystalline NVMs is their higher tolerance to radiation ionization which can result in memory applications in avionics, nuclear power stations, nuclear waste disposal sites, military, medicine etc. (Gerardi et al., 2011). Besides, MOS structures and transistors with nanocrystals are developed for dosimeter purposes.

Generally MOS structures pose high spatial resolution and their production is compatible with the technology used in the microelectronic industry. Two types MOS transistors are used for detecting of ionizing radiation: radiation-sensing field effect transistor (RADFET) (Buehler et al., 1993; Holmes-Siedle&Adams, 1986; Hughes et al., 1988; Price et al., 2004; Ramani et al., 1997; Stanic et al., 2005) and floating gate transistors (Edgecock et al., 2009; Kassabov et al., 1991; McNulty et al., 2006; Scheick et al., 1998, 1999; Tarr et al., 1998, 2004). Normally the RADFET is a p-channel MOS transistor whose SiO_2 gate is grown at specific conditions and its thickness is around 1 μm (Sarrabayrouse, 1991). A part of the charge carriers generated by the ionizing radiation is trapped in the oxide and the integrated irradiation dose is detected as a change in the transistor threshold voltage. To ensure good sensitivity of these dosimeters an electric field is normally applied on the control gated during irradiation. The radiation induces changes are irreversible and therefore RADFET dosimeters are for a single use only. At the floating gate dosimeter before irradiation the floating gate is charged by electron injection from the control gate (Kassabov et al., 1991; Tarr et al., 1998) or substrate (Martin et al., 2001) and this creates an electrical field in the oxide layer. Therefore no application of external electric field is necessary during irradiation which makes their application more convenient. In addition, the radiation induces changes which are reversible and multiple usage of those dosimeters is possible. Recently, experiments have been performed (Aktag et al., 2010) on introduction of Ge nanocrystals in the oxide layer of floating gate dosimeters. It has been observed that the presence of nanocrystals improves the radiation resistance of detectors but decreases their sensitivity.

In this chapter a review is made of the most important results obtained during the last five years by the authors and their collaborators in the field of development of metal-insulator-silicon structures with dielectric film containing amorphous or crystalline silicon nanoparticles, which are suitable for non-volatile memory and radiation applications. **In the introductory part** a brief review of the results of other research groups on MIS structures containing Si nanoparticles for non-volatile memory and detector applications has been made. **The next second part** gives information about the preparation of SiO_x films of different compositions and the annealing conditions used for the growth of amorphous (na-Si) and crystalline (nc-Si) silicon nanoparticles. Infrared absorption and Rutherford backscattering data give information on the oxygen content, while X-ray Diffraction and Reflectivity, X-ray Photoelectron Spectroscopy, Transmission Electron Microscopy, Atomic Force Microscopy, Infra Red Transmission and, Raman Scattering spectroscopy and Spectroscopic Ellipsometry data give information about the effect of furnace annealing on the properties of the SiO_x films. **In the third part** newly developed techniques for preparation of MIS structures containing amorphous and crystalline Si nanoparticles suitable for non-volatile memory application are described. The fabricated MIS structures are characterized by high frequency (100 kHz and 1 MHz) capacitance/conductance-voltage (C/G-V) measurements at applied dc voltage varying within a voltage range ± 15 V. **The last fourth part** presents recent data on preparation of MIS structures containing Si nanocrystals and their response to γ-radiation.

2. Preparation of SiO_x films and growth of amorphous and crystalline Si nanoparticles

It is well known that the film preparation and processing are very important for the performance of the MIS structures and devices with nanocrystals. They determine the film composition, which in the case of Si NP growth by SiO_x film annealing is of critical importance for the nanocrystal size, the filling of the oxides matrix with NCs (NC filling factor), the quality of the interface with the Si substrate, etc. Therefore this part of the chapter describes results from a thorough characterization of the SiO_x films. This information is important since these films are an essential part of the two(multi)layer insulator in the MIS structures described in Parts 3 and 4; they are the layers that contain Si NPs.

2.1 Preparation of SiO_x films

SiO_x layers with an initial composition of $x = 1.1$ and 1.3 and thickness of ~15 nm were prepared by thermal evaporation of SiO at a vacuum of 1×10^{-3} Pa on n- or p-type (100) c-Si (4 - 6 and 1 $\Omega \times cm$, respectively) substrates maintained at room temperature. The SiO evaporation was carried out from a tantalum crucible provided with a molybdenum cylindrical screen and thus evaporation in a quasi-closed volume takes place (Nesheva et al., 2003). The film thickness and deposition rate were monitored by a quartz microbalance system. Before the films deposition the silicon wafers were cleaned chemically using a standard procedure for the microelectronics industry. For Raman scattering, infra red and optical transmission measurements SiO_x films with same compositions but with thickness of

0.2 and 1 μm were deposited on c-Si and quartz substrates. The film composition has been determined by means of Rutherford Back Scattering (Nesheva et al., 2003).

All as-deposited layers were annealed at 250 °C for 30 min in an Ar atmosphere to keep them stable at room conditions. In order to grow Si nanoparticles an additional annealing at 700 °C in Ar or 1000 °C in N_2 atmosphere for 60 min was carried out.

2.2 Si nanoparticle growth, oxygen matrix densification, interface quality

The films annealed at high temperatures as well as control ones (annealed at 250 °C) were characterized by Transmission Electron Microscopy (TEM), Raman Spectroscopy, X-ray Diffraction (XRD) and Reflectivity (XRR), X-ray Photoelectron Spectroscopy (XPS) and Atomic Force Microscopy (AFM). Lattice-resolution TEM was carried out at 200 kV with a JEOL 2100 with a point-to-point resolution of 0.22 nm in the TEM mode. XRD/XRR measurements were performed using Philips X´pert Diffractometer (Cu Kα source), XPS analysis by Kratos Axis Ultra spectrometer (Al Kα (1486.6 eV) X-ray source), Raman spectra using Witec α-SNOM and AFM micrographs were taken by Asylum Research Microscope in tapping mode. Cross-sectional TEM (XTEM) samples were also prepared by gluing film-to-film two Si wafers and then cutting vertical sections which were first mechanically thinned to a thickness of 25 μm. Final thinning to electron transparency was accomplished by ion milling at very low angles (15° - 10°) from both cross-sectional sides. Optical and infra red transmission measurements were also carried out by means of Cary 5E, Perkin-Elmer Spectrum One FTIR and Bruker Vertex 70 spectrophotometers, respectively.

Raman scattering spectra of SiO_x films with thickness of 1 μm deposited on quartz substrates and annealed at 700 °C and 1000 °C for 60 min are shown in Fig.1. A broad band centered at ~ 470 cm^{-1} is observed in the spectrum of the film annealed at 700 °C. It is typical for amorphous silicon (Iqbal&Veprek, 1982; Nesheva et al., 2002) and its observation can be

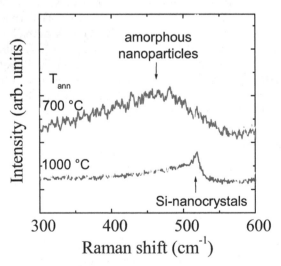

Fig. 1. Raman scattering spectra of SiO_x films (1 μm thick) deposited on quartz substrates and annealed at 700 °C or 1000 °C.

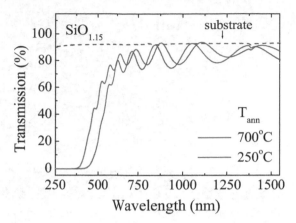

Fig. 2. Optical transmission spectra of two $SiO_{1.15}$ layers annealed at 250 and 700 °C. A red shift is observed upon annealing at 700 °C.

related to existence of pure amorphous silicon phase in these films. Such band is not seen in the spectrum of the film on quartz substrate annealed at 1000 °C, which confirms that the band at 470 cm-1 is related to a-Si phase rather than to light scattering from the SiO_x matrix or SiO_2 substrate. An asymmetric band peaked at ~ 518 cm-1 is observed in the spectrum of the film annealed at 1000 °C, which cannot originate from the quartz substrate and its appearance is an evidence for the existence of Si nanocrystals (Iqbal&Veprek, 1982).

The existence of pure silicon phase in the films annealed at 700 °C has also been confirmed by optical transmission measurements carried out on SiO_x layers on quartz substrates. Figure 2 shows optical transmission spectra of a sample with $x = 1.15$ measured before and after annealing at 700 °C. It is seen that the annealing causes a "red" shift of the spectrum. Since the increase of the oxygen content in the matrix should result in a "blue" shift, the observed "red" shift is an indication that absorption in a pure silicon phase takes place. A value of 2.64 eV has been obtained for the optical band gap of the amorphous Si particles grown upon annealing of samples with $x = 1.15$ at 700 °C and a value of 2.1 nm has been estimated for the average diameter of the largest amorphous Si NPs in our films (Nesheva et al., 2008).

Figures 3 (a) and (b) show XTEM micrographs of a control (250 °C) and an annealed at 1000 °C sample with $x = 1.3$ and thickness of 15 nm. Figure 3 (a) exhibits an amorphous structure and a nearly atomically flat interface, while Figure 3 (b) exhibits randomly oriented nanocrystals with a diameter of ~ 4-5 nm and an increase of the roughness of the c-Si wafer/dielectric interface. The NCs are positioned closer to the c-Si wafer than to the top SiO_x surface and are separated from the $c-Si/SiO_x$ interface by an amorphous region with thickness ≥ 3 nm. The thicknesses of the SiO_x layers determined are of 16.2 nm and 13.8 nm for the control and annealed sample, respectively. While the thickness of the control film is close to the one set during the deposition, film densification of about 15% has been concluded in result of the annealing at 1000 °C for 60 min.

The XRD results of control and 1000 °C annealed samples obtained in the $2\Theta/\Omega$ scans show (Fig. 4) only two peaks at 33° and 68° corresponding to diffraction from the crystalline Si

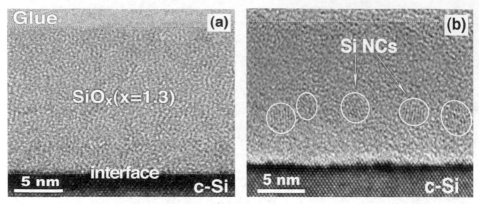

Fig. 3. Cross-sectional (XTEM) micrographs of a control sample (a) and an annealed at 1000 °C sample with nanocrystals in a dielectric matrix and rougher interface (b).

(100) substrate. No contribution due to the nanocrystals is observed because of the random NCs orientation, as revealed by TEM. The XRR spectra of a control and annealed at 700 and 1000 °C samples are shown in Fig. 5. The well defined interference fringes in the spectrum of the control sample (curve 1) imply an excellent interface between the dielectric and the c-Si wafer. With the increase of the annealing temperature from 700 °C to 1000 °C the amplitude of the interference fringes decreases (curves 2 and 3) indicating an increase of the interface roughness. The obtained results are in good agreement with the XTEM ones. The XRR technique was also applied to determine the thicknesses of the SiO_x layers. Values of 16.8 nm for the control and 16.9 and 14.8 nm for the annealed at 700 °C and 1000 °C layers, respectively, have been obtained. The errors in the thicknesses of the control and 1000 °C annealed samples are 3.7 and 7.2 %, respectively. Again, the sample annealed at 1000 °C show a small decrease of the initial thickness while the thickness of the film annealed at 700 °C is practically constant. This observation can be understood if assuming that at 700 °C the disorder of the substoichiometric SiO_x matrix is still rather high.

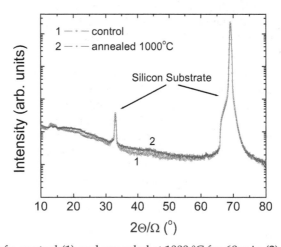

Fig. 4. $2\Theta/\Omega$ scans of a control (1) and annealed at 1000 °C for 60 min (2) samples.

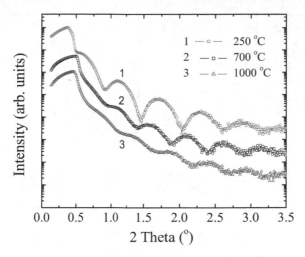

Fig. 5. XRR spectra of a control (1) and annealed at 700 (2) and 1000 ⁰C (3) samples.

XPS was used to obtain further information about the structure and chemical composition of the 15 nm films. In Figs 6 and 7 spectra of the Si 2p core level of control and annealed layers with expected $x = 1.3$ at incidence angles of 30⁰ and 90⁰, respectively, are presented. The Gaussian fitting curves are also shown. The main peak in the spectrum of the control sample at both angles (Figs. 6 (a) and 7 (a)) is positioned at 102.4 eV and corresponds to a SiO_x film with $x = 1.35$ (Alfonsetti et al., 1993), a value that is close to the composition set by the deposition conditions. After deconvolution three Gaussians peaked at 103.6, 102.4 and 100.1 eV were resolved. The peaks at 103.6 and 102.4 eV are attributed to Si-O bonds in pure SiO_2 and in Si_2O_3 compounds, respectively, while the shoulder at lower binding energy could be superposition of signals resulting from Si-O bonds in Si_2O suboxide and Si-Si bonds in silicon, having binding energies close to each other, 100.7 eV and 99.8 eV, respectively (Alfonsetti et al., 1993). In all spectra of the annealed films (Figs. 6 (b), (c), 7 (b), (c)) a well defined peak positioned at 103.8 eV corresponding to stoichiometric SiO_2 is present. This finding seems to be in disagreement with our previous results about the matrix composition after 700 ⁰C annealing obtained by IR spectroscopy (Donchev et al., Nesheva et al., 2003, 2008) but the IR measurements were carried out on thicker films (~ 0.2 μm) and give information about the volume properties of the films. The XPS signal originates from a region with thickness ≤ 10 nm and one can conclude that most likely the Si nanoscrystals close to the top surface get transformed into SiO_2 to a thickness of ~ 5-6 nm, due to native oxide formation after exposure to air, as confirmed by XTEM (Fig. 3 (b)). However, there is a substantial difference in the amplitude of the peak corresponding to pure Si phase (the doublet Si 2p 3/2, Si 2p 1/2 with binding energies of 99.4 and 99.9 eV) in the spectra of the layer annealed at 1000 ⁰C when measured at 30⁰ and 90⁰ (Figs. 6 (c), 7 (c)). At angle of incidence of 30⁰ the chemical composition of the first ~ 3 nm below the film surface is determined, while in the case of normal incidence the obtained information is for thicker region (~ 10 nm). The observed difference in the Si peak amplitude can be understood keeping in mind that the nanocrystals have a diameter of ~ 4-5 nm, they are positioned closer to the c-Si/dielectric interface and the top 6 - 7 nm of the SiO_x film are depleted of NCs (Fig. 3 (b)).

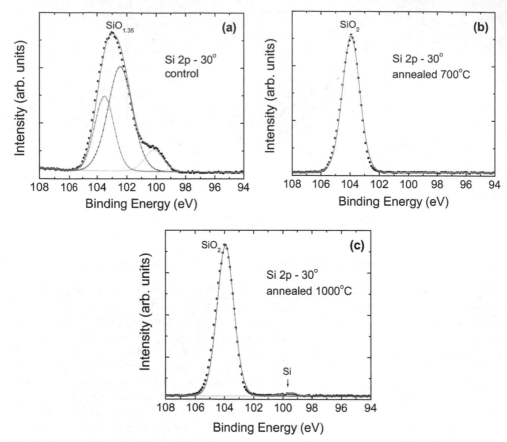

Fig. 6. Si 2p core level spectra of a control (a) and annealed at 700 °C (b) and 1000 °C (c) layers measured at 30° angle of incidence.

In the case of a-Si nanoparticles formed by annealing at 700 °C the same trend in the spectra measured at 30° and 90° is observed. Because of the smaller diameter of the a-Si NPs (~ 2 nm) and probably the same or similar to the nanocrystal spatial distribution the contribution of the Si phase in the XPS spectra is zero (Fig. 5 (b)) or very small at 90° angle of incidence (Fig. 6 (b)).

The AFM images in Fig. 8 (a)-(c) show the surface morphology of 15 nm SiO_x ($x = 1.3$) films annealed at 250 °C, 700 °C and 1000 °C, respectively. The scan area was set to 200 × 200 nm and the root mean square roughness (RMS) obtained has a value of 0.164 nm for the control sample (a), 0.258 nm for the 700 °C (b) and 0.129 nm for the 1000 °C (c) annealed samples, respectively. The AFM measurements revealed that the 700 °C annealing leads to an increase of the surface roughness which could be due to incomplete phase separation, while the increase of the annealing temperature to 1000 °C causes a decrease of the surface roughness and obtaining of smoother films. These results are consistent with the XRR data for the thickness variation after annealing at 700 °C and 1000 °C.

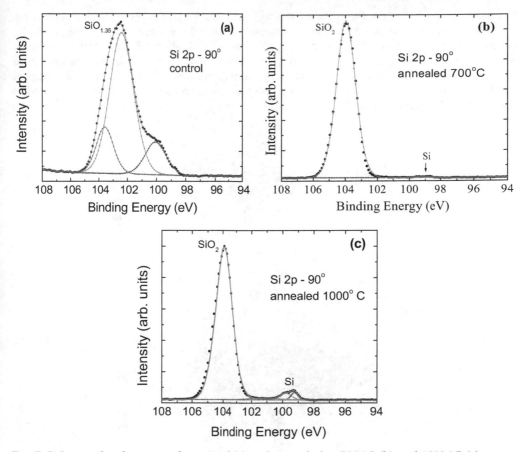

Fig. 7. Si 2p core level spectra of a control (a) and annealed at 700 °C (b) and 1000 °C (c) layers measured at 90° angle of incidence.

The np-Si-SiO$_x$ layers are composite materials in which the increase of the volume fraction f of the filler (amorphous or crystalline Si NPs) decreases the distance between the nanoparticles and at a certain concentration contact between some adjacent particles may occur and two- or three-dimensional networks of semiconductor nanoparticles may be formed. As mentioned above, network formation is undesired in the case of NVM devices with NPs and therefore the knowledge of f is important.

Assuming that there is no loss of oxygen atoms upon film annealing, the following relation between f and the atomic densities of the initial layer ρ_{SiOx} and pure silicon phase ρ_{Si} has been found (Nesheva et al., 2008):

$$f = (\rho_{SiOx} / \rho_{Si})\,(x + 1)^{-1}[1 - (x/y)] \qquad (1)$$

where x is the initial oxygen content and y is the oxygen content in the matrix of the annealed layers. Approximate values of f have been obtained using this relation in which the x and y values were determined from the IR transmission data (Donchev et al., in press). It has been obtained that the filling factor changes with the initial oxygen content and for films annealed at 700 °C and 1000 °C it varies between 0.05 and 0.2 and 0.07 and 0.28, respectively, when x decreases from 1.7 down to 1.15. The investigation of the carrier transport mechanism in 1 µm thick films have not shown network formation in high temperature annealed films. In the films annealed at 700 °C containing a-Si NPs, Poole–Frenkel transport mechanism has been observed while tunneling has been dominating in the films annealed at 1000 °C (Nesheva et al., 2008).

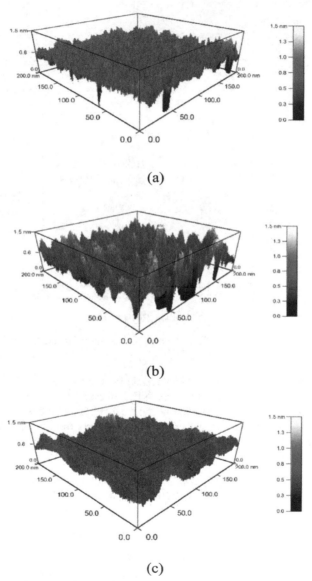

(a)

(b)

(c)

Fig. 8. AFM images of a control (a) and annealed at 700 °C (b) and 1000 °C (c) SiO$_x$ films.

3. MOS structures containing amorphous or crystalline Si nanoparticles for non-volatile memory applications

3.1 Experimental details

The experimental structures used in memory effect study (Nedev et al., 2008a, 2008b; Nesheva et al., 2007) were fabricated by deposition of a SiO_x ($x = 1.15$) layer with a thickness of ~ 15 nm on top of p- or n-type (100) crystalline silicon with resistivity of 1 and 4 - 6 $\Omega \times$cm, respectively, followed by radio frequency (r.f.) sputtering of a control silicon dioxide layer with a thickness of ~ 40 nm. Before the film deposition the silicon wafer was cleaned chemically following a standard for the microelectronics procedure. In order to form amorphous or crystalline silicon nanoparticles (with size ~ 5 nm), the structures were annealed for 60 min at 700 °C in Ar or 1000 °C in N_2, respectively. The annealing process was used not only for growing of silicon nanoparticles but also to form simultaneously a tunnel SiO_2 close to the interface with the silicon wafer free from nanoparticles (see Fig. 3 (b)). After the annealing Al metallization was carried out through a mask and MOS capacitors with area of $\sim 2 \times 10^{-3}$ cm^2 were formed. Aluminum was also used as a back contact to the crystalline silicon.

The MOS structures were characterized electrically by capacitance/conductance–voltage (C/G-V) measurements using Agilent E4980A Precision LCR Meter controlled by Agilent B1500A Semiconductor Device Analyzer. The polarity of the applied voltage given below concerns the top Al electrode.

3.2 Charge storage

The C-V characteristics of MOS structures with as-deposited SiO_x-SiO_2 gate dielectric have not displayed strong shift in the positive or negative direction, corresponding to charging from the c-Si substrate, when the gate voltage was swept in various ranges. Another important feature of the as-deposited structures is that they lose the trapped charge for several minutes.

Figures 9 (a) and (b) show high frequency C-V hysteresis curves of two annealed samples with amorphous and crystalline silicon nanoparticles, respectively, measured in the range ± 11 V (curves 2) and ± 15 V (curves 3). In all measurements ± scanning means that the gate bias was first swept from positive to negative voltage and then in the reverse direction. The first measurement on each MOS capacitor (initial curve) was carried out in a narrow enough range (see Fig. 9) in order to avoid charging of the structures. From the capacitance in accumulation a value of about 56 nm for the insulator effective thickness was obtained. This value is in good agreement with the total expected thickness of the deposited layers set by the deposition conditions, i.e. the sum of the SiO_x film thickness (~ 15 nm) and the thickness of the sputtered SiO_2 (~ 40 nm).

For both types of samples the application of a positive voltage shifts the C-V curves to the right, which corresponds to a negative charge trapped in the oxide close to the crystalline silicon wafer, and vice versa a negative bias shifts the curves in the opposite direction. As seen from Fig. 9 (a), (b) the shifted curves are parallel to the initial ones, which indicates uniform distribution of the trapped charge. Therefore one can expect that the charging is mainly due to capturing of electrons/holes in nanoparticles and not in traps in the dielectric, which could have varying spatial and energetic distributions. The value of the clockwise hysteresis is about 4.3 and 9 V (a) and 5.5 and 9.8 V (b) for ±11 V and ±15 V scanning ranges. The maximum average electric field across the gate dielectric, which corresponds to these scanning ranges, is about 2.0 MV/cm and 2.7 MV/cm, respectively. It is interesting that the C-V characteristics of the a-Si NP structure are much steeper, which indicates lower defect density at the c-Si wafer/silicon oxide interface.

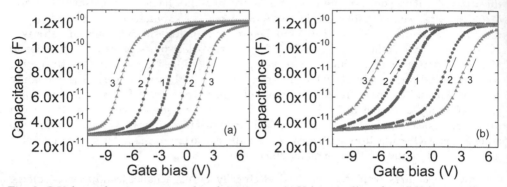

Fig. 9. C-V dependencies measured in the ranges ±11 V (curves 2) and ±15 V (curves 3) at 1 MHz of samples with (a) amorphous (b) crystalline silicon nanoparticles. The initial curve 1 was measured in the range +3 V, -7 V for both (a) and (b) samples.

Figures 10 (a) and (b) show the equivalent parallel conductance G vs. gate bias for the two types of samples. The measurements of G and C were carried out simultaneously in the same voltage scan. The observed shape of the equivalent parallel conductance, with a peak in weak inversion, corresponds to energy loss due to carrier generation and recombination through interface states (Nicollian&Brews, 2002). It is seen that the curves of structures with a-Si NPs are narrower than those of structures with Si NCs. Also the peak of the parallel

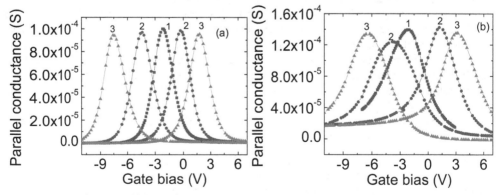

Fig. 10. Equivalent parallel conductance measured at 1 MHz of samples with (a) amorphous and (b) crystalline silicon nanoparticles. The scanning ranges are the same as for the C-V measurements in Fig. 9.

conductance has smaller value (with about 40 %) for the a-Si NP sample. These results confirm the above conclusion that the structures with a-Si NPs have a better SiO_2/c-Si interface than those with Si NCs and are in agreement with the XTEM data (Curiel et al., 2010a, 2010b) where epitaxial overgrowth and increased interface roughness have been found after annealing at 1000 °C (Fig. 3 (b)).

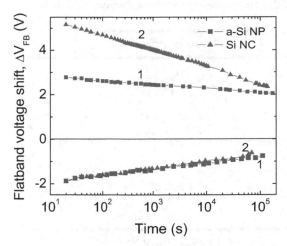

Fig. 11. Retention characteristics of structures with a-Si NPs (curves 1) and nanocrystals (curves 2) after charging structures negatively with voltage ramp sweep up to +12 V or positively with ramp sweep to -12 V.

The retention characteristics of structures with a-Si NPs and Si NCs were obtained by measuring the time-dependent variation of the flatband voltage (Fig. 11), which is proportional to the area density of the trapped charge (Sze, 1981). In both types of structures the charge loss follows approximately a logarithmic law. An essential advantage of the a-Si nanoparticle structures compared to the NC ones is the much slower discharging process observed, especially of trapped electrons. For example, 48 hours after charging with +12 V the MOS structures containing a-Si NPs still have 75% of the initial charge, while the structures with nanocrystals retain about 43%.

4. Radiation dosimeter based on metal-oxide-semiconductor structures with silicon nanocrystals

Figure 12 shows schematically a cross-section of the MOS structures used in dosimetry measurements. An essential difference between the dosimetry structures and the ones used in the memory studies is that an additional SiO_2 layer, 3.9 nm thick, was grown thermally before the deposition of the SiO_x film. The oxidation process was carried out in dry O_2 atmosphere at 850 °C. Except for the thermal oxidation and the greater thickness of the control oxide, which for dosimeters was 60 nm, the fabrication process was identical with the one used for fabrication of the memory study samples.

The principle of operation of the proposed dosimeter is based on generation of electron-hole pairs in the SiO_2 when the structure is exposed to ionizing radiation and separation of the generated carriers by local internal electric field created around each preliminary charged nanocrystal. For example if the NCs are negatively charged the holes generated in the SiO$_2$ are swept towards the nanocrystals, where they recombine with a part of the trapped electrons and reduce the net charge, while the generated electrons are swept towards the gate electrode. The discharge of the nanocrystals causes change of the MOS structure flatband voltage, which can be used to measure the absorbed dose.

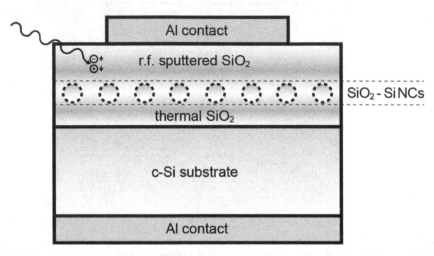

Fig. 12. Schematic cross-section of a MOS structure with three-layer gate dielectric: thermal SiO$_2$/SiO$_2$ – Si NC/sputtered SiO$_2$.

In order to set an initial flatband voltage ΔV_{FB0} prior to irradiation each dosimeter was charged by applying voltage pulses to the control gate with positive or negative polarity and with various amplitudes and durations. The positive pulses charge Si nanocrystals with electrons injected from the crystalline silicon wafer, while the negative ones charge NCs with holes. Figure 13 shows the initial as well as shifted C-V curves obtained after charging a structure with six sequential 10 V/5 s pulses. Each pulse causes parallel shift in the positive direction changing the flatband voltage with ~ 0.14 V; thus after the last pulse a shift of $\Delta V_{FB0} = 0.83$ V was obtained. The shifts in the positive direction correspond to a gradual increase of the negative charge in the gate dielectric. The characteristic after each charging pulse was measured in both directions in a narrow interval (0 – 2 V), in order to avoid change in the charge state of the nanocrystals; no hysteresis has been obtained.

Because of the thermal oxide, which makes the total SiO$_2$ thickness between the nanocrystals and the c-Si wafer greater than 7 nm and the thicker control oxide (60 nm) the structures used in dosimetry experiments showed much longer retention time than the memory ones (Nedev et al., 2011). Two weeks after charging the capacitors they showed an average relative change of the flatband voltage due to discharging at ambient conditions of ~ 2 %.

Charged samples were subjected to various integral gamma irradiation doses from 3 to 200 Gy which were accumulated in steps at a dose rate of 37 Gy/h. The γ-irradiation was carried out in air of 75 to 80% humidity by means of a 38 000 Ci ^{60}Co source with an average energy $E_\gamma = 1.25$ MeV.

Fig. 14 shows changes of the flatband voltage versus absorbed dose for two capacitors having an initial flatband voltage shift of ΔV_{FB0} = 0.8 and 0.67 V. The two curves have similar shape within an initial interval, 0 - 100 Gy, in which approximately linear dependence between ΔV_{FB} and the dose is observed. The obtained sensitivities for the linear region are $S \sim 2.1$ and 2.3 mV/Gy, respectively, and no correlation between ΔV_{FB0} and S was found (Nedev et al., 2011). The reduced sensitivity at doses higher than 100 Gy can be related to discharging of nanocrystals and decrease of the oxide internal electric field.

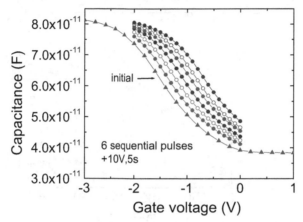

Fig. 13. C-V curves measured at 1 MHz of a MOS structure with Si nanocrystals charged with six sequential pulses, each of them with amplitude of +10 V and duration of 5 s. The initial curve is also presented.

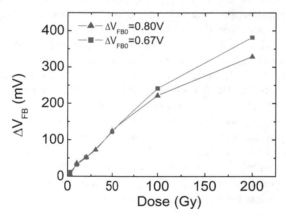

Fig. 14. Flatband voltage changes of p-Si MOS structures charged with electrons versus absorbed γ dose.

To test the possibility to reuse this type of dosimeters, structures irradiated to 200 Gy were recharged to the same ΔV_{FB0}, and subjected to a second γ-irradiation under the same conditions. The dependencies of the flatband voltage changes on the dose for the first and second irradiation for two dosimeters are shown in Fig. 15. As seen the curves have similar shapes but the second irradiation causes smaller response. Most likely the higher sensitivity observed in the first irradiation experiment is partially due to capturing of holes in existing or generated by the radiation deep traps in the sputtered SiO_2 and/or in defects at the sputtered SiO_2/SiO_2-Si NCs interface, a process which is irreversible at room temperature. Thus, although the structures have the same initial change of the flatband voltage ΔV_{FB0} before both irradiations, their charge states were different, and this could be the reason for the different characteristics measured after the first and second irradiation.

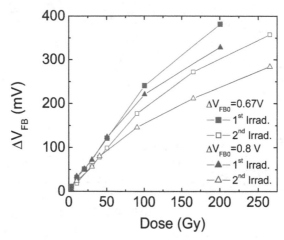

Fig. 15. Flatband voltage changes after the first and the second γ-irradiation.

To improve the quality of the control (top) oxide and at the same time to increase the sensitive volume of the sensor (the one which contains nanocrystals) a new approach for preparation of multilayer dielectric containing three-dimensional ensembles of silicon NCs has been proposed. The method is based on two-step annealing process at 1000 °C in different atmospheres. The first step is annealing in pure nitrogen to grow nanocrystals with a given size distribution; it is followed by a second annealing in nitrogen + oxygen atmosphere to complete the NC growth and at the same time to oxidize the already formed NCs, close to the top surface and thus to obtain a control SiO_2 region. By keeping the total annealing time 60 min it may be expected that the nanocrystals grown far from the top surface will have approximately the same size and spatial distribution as in the case of 60 min N_2 annealing. Figure 16 shows a XTEM micrograph of ~ 100 nm thick SiO_x ($x = 1.3$) film annealed for 50 min in pure N_2 and then for 10 min in 90 % N_2 + 10 % O_2 atmosphere. It proves that a two layer dielectric with ~ 75 nm SiO_2-Si NCs region and ~ 25 nm SiO_2 region free from nanocrystals has been successfully produced.

MOS structures with multilayer gate dielectric containing NCs obtained by the new two-step annealing process are currently tested as radiation dosimeters and NVMs.

5. Conclusions

In this paper, we have reviewed some interesting results of other research groups on MIS structures containing Si nanocrystals for non-volatile memory and detector applications. It has been shown that nanoparticle memories continue to develop new ideas and technologies for nanoparticle formation and MIS structure arrangement as self-assembling techniques have been applied and proven to be very useful for structure preparation. Furthermore, a brief description has been made of the most important results in this field obtained during the last five years by the authors and their collaborators. MOS structures for non-volatile memory applications have been prepared with dielectric film containing amorphous or crystalline silicon nanoparticles in which the tunnel oxide layer is formed during the structure annealing. It has been shown that the defect density at the c-Si wafer/silicon oxide interface is lower in the structures with a-Si NPs than in those containing Si NCs which is due to epitaxial overgrowth and increased interface roughness after annealing at 1000 ^{0}C. Another essential advantage of the a-Si nanoparticle structures compared to the NC ones is the much slower discharging process observed, especially of trapped electrons. Both advantages in combination with the lower growing temperature required for the a-Si NP formation make them rather perspective for non-volatile memory applications.

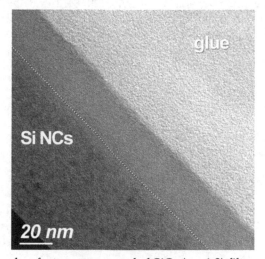

Fig. 16. XTEM micrographs of a two-step annealed SiO$_x$ (x = 1.3) film with thickness of ~ 100 nm. The contrast observed in the image confirms that the second annealing step (oxidation) forms a free from nanocrystals region with thickness of ~ 25 nm.

Recent authors' data on preparation of MOS structures containing Si nanocrystals and their response to γ-radiation have also been presented. The detector operation is based on generation of electron-hole pairs in the SiO$_2$ when the structure is exposed to ionizing radiation and separation of the generated carriers by the local internal electric field which is created around each nanocrystal by a preliminary charging of Si NCs. Upon a negative preliminary charging of the Si nanocrystals no spontaneous nanocrystal discharge occurs for more than 2 weeks. The γ-irradiation with doses in the range 0-100 Gy causes approximately linear variation of the flatband voltage. An advantage of this detector when compared with those using conventional floating gate devices is the expected improved radiation hardness because of the suppressed lateral currents. A new two-step annealing procedure has been applied on SiO$_x$ single layers deposited by thermal evaporation of SiO in vacuum on c-Si in order to prepare three layer MOS structures containing nanocrystals for device applications.

The above described authors' results indicate that the developed techniques for MOS structures preparation based on thermal evaporation of SiO in vacuum are well suited for preparation of Si nanoparticle memories and radiation detectors. All processing steps including the thermal annealing required for Si NP formation are compatible with the contemporary CMOS technology. Follow-up experiments should take place in a production environment and focus on the integration of the methods in the technology of commercial memory and detector devices.

6. Acknowledgements

The authors are very grateful to all their collaborators who have taken part in the development of the MIS structures for memory and detector applications. The measurements were carried out in part in the Frederick Seitz Materials Research Laboratory Central Facilities, University of Illinois, which are partially supported by the U.S. Department of Energy under grants DE-FG02-07ER46453 and DE-FG02-07ER46471. The collaboration with the TEM Laboratory, CNyN, UNAM, of Dr. Oscar Contreras Lopez and especially with Francisco Ruiz Medina is also gratefully acknowledged.

7. References

Aktag, A., Yilmaz, E., Mogaddam, N.A.P., Aygun, G., Cantas, A. & Turan, R. (2010). Ge Nanocrystals Embedded in SiO2 in MOS Based Radiation Sensors. *Nuclear Instruments and Methods in Physics Research, Section B-Beam Interactions with Materials and Atoms*, Vol. 268, No. 22, pp 3417-3420, ISSN 0168-583X

Alfonsetti, R., Lozzi, L., Passacantando, M., Picozzi, P. & Santucci, S. (1993). XPS studies on SiOx thin films. *Applied Surface Science*, Vol. 70-71, No. 1, (June 1993), pp 222-225, ISSN 0169-4332

Buehler, M.G., Blaes, B.R., Soli, G.A. & Tardio, G.R. (1993). On-chip p-MOSFET dosimetry. *IEEE Transactions on Nuclear Science*, Vol. 40, No. 6, pp 1442-1449, ISSN 0018-9499

Carreras, J., Garrido, B. & Morante, J.R. (2005). Improved charge injection in Si nanocrystal non-volatile memories. *Microelectronics Reliability*, Vol. 45, (May - June, 2005), pp 899-902, ISSN 00262714

Claverie, A., Bonafos, C., Assayag, G.B., Schamm, S., Cherkashin, N., Paillard, V., Dimitrakis, P., Kapetenakis, E., Tsoukalas, D., Muller, T., Schmidt, B., Heinig, K.H., Perego, M., Fanciulli, M., Mathiot, D., Carrada, M. & Normand, P. (2006). Materials Science Issues for the Fabrication of Nanocrystal Memory Devices by Ultra Low Energy Ion Implantation. *Journal Defect and Diffusion Forum*, Vol. 258 - 260, (October, 2006), pp 531-541 ISSN 1662-9507

Curiel M., I. Petrov, N. Nedev, D. Nesheva, M. Sardela, Y. Murata, B. Valdez, E. Manolov, and I. Bineva, Formation of Si nanocrystals in thin SiO2 films for memory device applications, Materials Science Forum Vol. 644 (2010b) pp. 101-104, ISSN:0255-5476.

Curiel, M., Nedev, N., Nesheva, D., Soares, J., Haasch, R., Sardela, M., Valdez, B., Sankaran, B., Manolov, E., Bineva, I. & Petrov, I. (2010a). Microstructural characterization of thin SiOx films obtained by physical vapor deposition. *Materials Science & Engineering B*, Vol. 174, pp 132-136, ISSN 0921-5107

Donchev, V., Nesheva, D., Todorova, D., Germanova, K. & Valcheva, E. (in press). Computer simulation study of infra-red transmission spectra of Si-SiOx nanocomposites. *Thin Solid Films*, ISSN 0040-6090

Edgecock, R., Matheson, J., Weber, M., Villani, E.G., Bose, R., Khan, A., Smith, D.R., Adil-Smith, I. & Gabrielli, A. (2009). Evaluation of commercial programmable floating gate devices as radiation dosimeters. *Journal of Instrumentation*, Vol. 4, (February 2009), pp P02002, ISSN 1748-0221

Eitan, B., Pavan, P., Bloom, I., Aloni, E., Frommer, A. & Finzi, D. (2000). NROM: a novel localized trapping, 2-bit nonvolatile memory cell. *IEEE Electron Device Letters*, Vol. 21, No. 11, pp 543-545, ISSN 0741-3106

Gerardi, C., Cester, A., Lombardo, S., Portoghese, R. & Wrachien, N. (2011). Nanocrystal Memories: An Evolutionary Approach to Flash Memory Scaling and a Class of Radiation-Tolerant Devices, In: *Radiation effects in semiconductors*, Iniewski, K., pp 103-150. CRC Press,Taylor&Francis Group. ISBN 9781439826942, Boca Raton

Han, K.I., Park, Y.M., Kim, S., Choi, S.-H., Kim, K.J., Park, I.H. & Park, B.-G. (2007). Enhancement of Memory Performance Using Doubly Stacked Si-Nanocrystal Floating Gates Prepared by Ion Beam Sputtering in UHV *IEEE Transactions on Electron Devices*, Vol. 54, No. 2, (Feb. 2007), pp 359 - 362, ISSN 0018-9383

Holmes-Siedle, A. & Adams, L. (1986). RADFETs: A Review of the Use of Metal Oxide Silicon Devices as Integrating Dosimeters. *Radiation Physics and Chemistry*, Vol. 28, No. 2, pp 235-244, ISSN 0969-806X

Horváth, Z.J. & Basa, P. (2009). Nanocrystal Non-Volatile Memory Devices. *Materials Science Forum:Thin Films and Porous Materials*, Vol. 609, (January, 2009), pp 1-9 ISSN 1662-9752

Hudgens, S. & Johnson, B. (2004). Overview of Phase-Change Chalcogenide Nonvolatile Memory Technology. *MRS Bulletin*, Vol. 29, pp 829-832, ISSN 0883-7694

Hughes, R.C., Huffman, D., Snelling, J.V., Zipperian, T.E., Ricco, A.J. & Kelsey, C.A. (1988). Miniature Radiation Dosimeter for in vivo Radiation Measurements. *International Journal of Radiation Oncology Biology Physics*, Vol. 14, pp 963-7, ISSN 0360-3016

Ioannou-Sougleridis, V., Nassiopoulou, A.G. & Travlos, A. (2003). Effect of high temperature annealing on the charge trapping characteristics of silicon nanocrystals embedded within SiO. *Nanotechnology*, Vol. 14, No. 11, pp 1174, ISSN 0957-4484

Iqbal, Z. & Veprek, S. (1982). Raman scattering from hydrogenated microcrystalline and amorphous silicon. *Journal of Physics C: Solid State Physics*, Vol. 15, pp 377-392, ISSN 0022-3719

Jung, J. & Cho, W.-J. (2008). Tunnel Barrier Engineering for Non-Volatile Memory. *Journal of Semiconductor Technology and Science*, Vol. 8, No. 1, pp 32-39, ISSN 1598-1657

Kassabov, J., Nedev, N. & Smirnov, N. (1991). Radiation dosimeter based on floating gate MOS transistor. *Radiation Effects and Defects in Solids*, Vol. 116, No. 1-2, pp 155-158, ISSN 1042-0150

Kouvatsos, D.N., Ioannou-Sougleridis, V. & Nassiopoulou, A.G. (2003). Charging effects in silicon nanocrystals within SiO2 layers, fabricated by chemical vapor deposition, oxidation, and annealing *Applied Physics Letters*, Vol. 82, No. 3, (15 January 2003), pp 397, ISSN 0003-6951

Liu, Y., Chen, T.P., Ng, C.Y., Ding, L., Zhang, S., Fu, Y.Q. & Fung, S. (2006). Depth Profiling of Charging Effect of Si Nanocrystals Embedded in SiO2: A Study of Charge

Diffusion among Si Nanocrystals. *Journal of Physical Chemistry B*, Vol. 110, No. 33, (August 2, 2006), pp 16499-16502, ISSN 1520-5207

Lombardo, S., Corso, D., Crupi, I., Gerardi, C., Ammendola, G., Melanotte, M., De Salvo, B. & Perniola, L. (2004). Multi-bit storage through Si nanocrystals embedded in SiO2. *Microelectronic Engineering*, Vol. 72, pp 411-414, ISSN 0167-9317

Lu, T.Z., Alexe, M., Scholz, R., Talelaev, V. & Zacharias, M. (2005). Multilevel charge storage in silicon nanocrystal multilayers. *Applied Physics Letters*, Vol. 87, No. 20, (November 2005), ISSN 0003-6951

Lu, T.Z., Alexe, M., Scholz, R., Talalaev, V. & Zacharias, M. (2006). Si nanocrystal based memories: Effect of the nanocrystals density. *Journal of Applied Physics*, Vol. 100, pp 014310, ISSN 0003-6951

Martin, M.N., Roth, D.R., Garrison-Darrin, A., McNulty, P.J. & Andreou, A.G. (2001). FGMOS dosimetry: design and implementation. *IEEE Transactions on Nuclear Science*, Vol. 48, No. 6, pp 2050 - 2055, ISSN 0018-9499

McNulty, P.J., Poole, K.F., Crissler, M., Reneau, J., Cellere, G., Paccagnella, A., Visconti, A., Bonanomi, M., Stroebel, D., Fennell, M. & Perez, R. (2006). Sensitivity and dynamic range of FGMOS dosemeters. *Radiation Protection Dosimetry*, Vol. 122, No. 1-4, (March 26 2007), pp 460-2, ISSN 1742-3406

Minami, S.-i., Ujiie, K., Terasawa, M., Komori, K., Furusawa, K. & Kamigaki, Y. (1994). A 3 Volt 1 Mbit Full-Featured EEPROM Using a Highly-Reliable MONOS Device Technology. *IEICE Transactions on Electronics*, Vol. Vol.E77-C, No. 8, (August 20 1994), pp 1260-1269, ISSN 0916-8516

Nassiopoulou, A.G., Olzierski, A., Tsoi, E., Salonidou, A., Kokonou, M., Stoica, T. & Vescan, L. (2009). Laterally ordered 2-D arrays of Si and Ge nanocrystals within SiO2 thin layers for application in non-volatile memories *International Journal of Nanotechnology*, Vol. 6, No. 1/2, pp 18 - 34, ISSN 1741-8151

Nedev, N., Nesheva, D., Manolov, E., Brüggemann, R., Meier, S. & Levi, Z. (2008a). Memory Effect in MOS Structures Containing Amorphous or Crystalline Silicon Nanoparticles. *Proceedings of 26th International Conference on Microelectronics (MIEL 2008)*, ISBN 978-1-4244-1881-7, Nis, Serbia, May 2008.

Nedev, N., Nesheva, D., Manolov, E., Brüggemann, R., Meier, S., Levi, Z. & Zlatev, R. (2008b). MOS structures containing silicon nanoparticles for memory device applications. *Journal of Physics: Conference Series*, Vol. 113, pp 012034, ISSN 1742-6596

Nedev, N., Manolov, E., Nesheva, D., Krezhov, K., Nedev, R., Curiel, M., Valdez, B., Mladenov, A. & Levi, Z. (2011). Radiation dosimeter based on Metal-Oxide-Semiconductor structures containing silicon nanocrystals. *Key Engineering Materials* (accepted), ISSN 1013-9826

Nesheva, D., Raptis, C., Perakis, A., Bineva, I., Aneva, Z., Levi, Z., Alexandrova, S. & Hofmeister, H. (2002). Raman scattering and photoluminescence from Si nanoparticles in annealed SiOx thin films. *Journal of Applied Physics*, Vol. 92, No. 8, pp 4678-4683, ISSN 0003-6951

Nesheva, D., Bineva, I., Levi, Z., Aneva, Z., Merdzhanova, T. & Pivin, J.C. (2003). Composition, structure and annealing-induced phase separation in SiOx films produced by thermal evaporation of SiOin vacuum. *Vacuum*, Vol. 68, pp 1-9, ISSN 0042-207X

Nesheva, D., Nedev, N., Manolov, E., Bineva, I. & Hofmeister, H. (2007). Memory effect in MIS structures with amorphous silicon nanoparticles embedded in ultra thin SiOx matrix. *Journal of Physics and Chemistry of Solids*, Vol. 68, pp 725-728, ISSN 0022-3697

Nesheva, D., Nedev, N., Levi, Z., Brüggemann, R., Manolov, E., Kirilov, K. & Meier, S. (2008). Absorption and transport properties of Si rich oxide layers annealed at various temperatures. *Semiconductor Science Technology*, Vol. 23, (March 2008), pp 045015-8, ISSN 1361-6641

Ng, C.Y., Chen, T.P., Sreeduth, D., Chen, Q., Ding, L. & Du, A. (2006a). Silicon nanocrystal-based non-volatile memory devices. *Thin Solid Films*, Vol. 504, No. 1-2, pp 25-27, ISSN 0040-6090

Ng, C.Y., Chen, T.P., Yang, M., Yang, J.B., Ding, L., Li, C.M., Du, A. & Trigg, A. (2006b). Impact of programming mechanisms on the performance and reliability of nonvolatile memory devices based on Si nanocrystals. *IEEE Transactions on Electron Devices*, Vol. 53, No. 4 (March 27), pp 663 - 667, ISSN 0018-9383

Nicollian, E.H. & Brews, J.R. (2002). *MOS (Metal Oxide Semiconductor) Physics and Technology*, Wiley-Interscience, ISBN ISBN: 978-0-471-43079-7, New York

Normand, P., Tsoukalas, D., Kapetanakis, E., Berg, J.A.V.D., Armour, D.G., Stoemenos, J. & Vieude, C. (1998). Formation of 2-D Arrays of Silicon Nanocrystals in Thin SiO2 Films by Very-Low Energy Si+ Ion Implantation. *Electrochemical and Solid-State Letters*, Vol. 1, No. 2, (June 11, 1998), pp 88-90, ISSN 1099-0062

Normand, P., Kapetanakis, E., Dimitrakis, P., Tsoukalas, D., Beltsios, K., Cherkashin, N., Bonafos, C., Benassayag, G., Coffin, H., Claverie, A., Soncini, V., Agarwal, A. & Ameen, M. (2003). Effect of annealing environment on the memory properties of thin oxides with embedded Si nanocrystals obtained by low-energy ion-beam synthesis *Applied Physics Letters*, Vol. 83, No. 1, ISSN 0003-6951

Normand, P., Dimitrakis, P., Kapetanakis, E., Skarlatos, D., Beltsios, K., Tsoukalas, D., Bonafos, C., Coffin, H., Benassayag, G., Claverie, A., Soncini, V., Agarwal, A., Sohl, C. & Ameen, M. (2004). Processing issues in silicon nanocrystal manufacturing by ultra-low-energy ion-beam-synthesis for non-volatile memory applications. *Microelectronic Engineering*, Vol. 73-74, pp 730-735, ISSN 0167-9317

Oda, S., Huang, S.-Y., Salem, M.A. & Mizuta, H. (2005). Charge storage in silicon nanocrystals and device application. *Proceedings of First Internantional Workshop on Semiconductor Nanocrystals, Seminano2005*, ISBN 963 7371 20 6, ISBN 963 7371 18 4, Budapest, Hungary, September 2005.

Price, R.A., Benson, C., Joyce, M.J. & Rodgers, K. (2004). Development of a RADFET linear array for intracavitary in vivo dosimetry during external beam radiotherapy and brachytherapy. *IEEE Transactions on Nuclear Science*, Vol. 51, No. 4, (Aug. 2004), pp 1420-1426, ISSN 0018-9499

Ramani, R., Russell, S. & O'Brien, P.F. (1997). Clinical Dosimetry Using MOSFETS. *International Journal of Radiation, Oncology, Biology, Physics*, Vol. 37, No. 4, pp 959, ISSN 0360-3016

Rao, R.A., Steimle, R.F., Sadd, M., Swift, C.T., Hradsky, B., Straub, S., Merchant, T., Stoker, M., Anderson, S.G.H., Rossow, M., Yater, J., Acred, B., Harber, K., Prinz, E.J., White Jr, B.E. & Muralidhar, R. (2004). Silicon nanocrystal based memory devices for NVM and DRAM applications. *Solid-State Electronics*, Vol. 48, No. 9, pp 1463-1473, ISSN 0038-1101

Sarrabayrouse, G. (1991). MOS radiation dosimeter: sensitivity and stability *Proceedings of First European Conference on Radiation and its Effects on Devices and Systems, RADECS 91*, ISBN 0-7803-0208-7, La Grande-Motte, France, 9-12 Sept. 1991

Scheick, L.Z., McNulty, P.J. & Roth, D.R. (1998). Dosimetry based on the erasure of floating gates in the natural radiation environments in space. *IEEE Transactions on Nuclear Science*, Vol. 45, No. 6, pp 2681 - 2688, ISSN 0018-9499

Scheick, L.Z., McNulty, P.J., Roth, D.R., Davis, M.G. & Mason, B.E. (1999). Measurements of dose with individual FAMOS transistors. *IEEE Transactions on Nuclear Science*, Vol. 46, No. 6, pp 1751 - 1756, ISSN 0018-9499

Stanic, S., Asano, Y., Ishino, H., Igarashi, A., Iwaida, S., Nakano, Y., Terazaki, H., Tsuboyama, T., Yoda, I. & ESontar, D. (2005). Radiation monitoring in Mrad range using radiation-sensing field-effect transistors. *Nuclear Instruments and Methods in Physics Research Section A: Accelerators, Spectrometers, Detectors and Associated Equipment*, Vol. 545, No. 1-2, pp 252-260, ISSN 0168-9002

Steimle, R.F., Muralidhar, R., Rao, R., Sadd, M., Swift, C.T., Yater, J., Hradsky, B., Straub, S., Gasquet, H., Vishnubhotla, L., Prinz, E.J., Merchant, T., Acred, B., Chang, K. & Jr., B.E.W. (2007). Silicon nanocrystal non-volatile memory for embedded memory scaling. *Microelectronics Reliability*, Vol. 47, No. 4-5, (April-May 2007), pp 585-592, ISSN 0026-2714

Sze, S.M. (1981). *Physics of Semiconductor Devices* (2nd edition), Wiley, ISBN 0-471-05661-8, New York

Tarr, N.G., Mackay, G.F., Shortt, K. & Thomson, I. (1998). A floating gate MOSFET dosimeter requiring no external bias supply. *IEEE Transactions on Nuclear Science*, Vol. 45, No. 3, pp 1470 - 1474 ISSN 0018-9499

Tarr, N.G., Shortt, K., Wang, Y. & Thomson, I. (2004). A sensitive, temperature-compensated, zero-bias floating gate MOSFET dosimeter *IEEE Transactions on Nuclear Science*, Vol. 51, No. 3, pp 1277 - 1282, ISSN 0018-9499

Tiwari, S., Rana, F., Chan, K., Hanafi, H., Chan, W. & Buchanan, D. (1995). Volatile and non-volatile memories in silicon with nano-crystal storage. *Proceedings of International Electron Devices Meeting, IEDM'95*, ISBN 01631918 Washington, DC, USA, 10 -13 December 1995

Tiwari, S., Rana, F., Hanafi, H., Hartstein, A., Crabbe, E.F. & Chan, K. (1996). A silicon nanocrystals based memory. *Applied Physics Letters*, Vol. 68, No. 10, (April 1996), pp 1377-80, ISSN 0003-6951

Tsoi, E., Normand, P., Nassiopoulou, A.G., Ioannou-Sougleridis, V., Salonidou, A. & Giannakopoulos, K. (2005). Silicon nanocrystal memories by LPCVD of amorphous silicon, followed by solid phase crystallization and thermal oxidation *Journal of Physics: Conference Series*, Vol. 10, pp 31, ISSN 1742-6588

Tsoukalas, D., Dimitrakis, P., Kolliopoulou, S. & Normand, P. (2005). Recent advances in nanoparticle memories. *Materials Science and Engineering: B*, Vol. 124-125, pp 93-101, ISSN 0921-5107

Yu, Z., Aceves, M., Carrillo, J. & Flores, F. (2003). Single electron charging in Si nanocrystals embedded in silicon-rich oxide *Nanotechnology*, Vol. 14, No. 9, pp 959, ISSN 0957-4484

9

Magnetic Mn$_x$Ge$_{1-x}$ Dots for Spintronics Applications

Faxian Xiu[1*], Yong Wang[2,3], Jin Zou[2] and Kang L. Wang[4]
[1]Electrical and Computer Engineering, Iowa State University, Ames, IA
[2]Division of Materials, The University of Queensland, Brisbane
[3]Materials Science and Engineering, Zhejiang University, Hangzhou
[4]Device Research Laboratory, Department of Electrical Engineering
University of California, Los Angeles, California
[1,4]USA
[2]Australia
[3]China

1. Introduction

Dilute magnetic semiconductors (DMSs) attract tremendous interest as emerging candidates for the microelectronics industry due to their uniqueness in exhibiting spin dependent magneto-electro-optical properties. Their distinctive material characteristics such as spin dependent coupling between semiconductor bands and the localized states promises magnetoelectric effect - the modulation of magnetic properties by an applied electric field - in semiconductors (Kulkarni et al., 2005, Ohno et al., 2000). Thus, a wide variety of semiconductor devices can be envisaged (Brunner, 2002a), such as spin polarized light-emitting diodes, lasers (Schulthess and Butler, 2001, Schilfgaarde and Mryasov, 2001) and spin transistor logic devices (Chen et al., 2005, Brunner, 2002b, Ohno et al., 2000). The development of these devices is considered as a possible route for extending the semiconductor scaling roadmap to spin-added electronics (Liu et al., 2005).

DMS materials can be developed by alloying semiconductors with several percentages of magnetic transition elements (with 3-d orbitals), such as Fe, Mn, Co, Ni, and V (Bolduc et al., 2005). The anticipated outcome due to this doping is the fact that the transition elements occupying substitutional sites hybridize with the semiconductor host via the sp-d exchange interaction and help to enhance the spin dependent transport, which in turn increase both the magnetization and the Curie temperature (T_c) of the semiconductor system (Jungwirth et al., 2006). This means that the transition metal doping can generate strong spin dependent coupling states in semiconductor systems, which may be further modulated by an electric field (Ohno et al., 2000). Since the band gap engineering and crystal structures of a DMS material are compatible with other semiconductors, they offer significant integration advantages as well (Brunner, 2002a).

---

* Corresponding Author

The dependence of the transition temperature on the density of mobile carriers is considered as a good evidence for a genuine DMS. However, the key challenge for the growth of DMS materials is the ability to control synthesis of transition metals in semiconductors, aimed at increasing the doping concentration in order to realize room-temperature ferromagnetism. Thus, achieving both semiconducting and ferromagnetic property at room temperature has become a great challenge. Unfortunately, this is hampered by low solubility of transition metal species with a strong tendency to form metallic clusters or heterogeneous regions within the matrix. So the primary research on DMS growth is to produce samples with uniform doping and being free of metallic precipitates.

Substantial theoretical and experimental work were carried out in the past two decades, either to predict or to measure the magnetic properties on a variety of semiconductor materials, including binary and ternary compounds such as III-V, II-VI, IV-VI, VI oxides, III nitrides and group IV systems. Among them, (Ga,Mn)As became the first and the mostly studied DMSs since 1996 (Lu and Lieber, 2006); but to our knowledge, some studies lacked the threshold doping profile to push its curie temperature to room temperature and above (currently around 190 K (Lyu and Moon, 2003, van der Meulen et al., 2008, Maekawa, 2006, Wang et al., 1995)). Since then, substantial research has been carried out in III-V and II-VI semiconductors to improve the T_c by increasing the solubility limit and meanwhile minimizing self-compensation effect of magnetic structures, using radical techniques such as delta doping of modulation doped quantum structures (Bhatt et al., 2002, Cho et al., 2008, Berciu and Bhatt, 2001, Lauhon et al., 2002). Another approach to achieve high-T_c DMS is to use transition metal doped wide bandgap nitrides and oxides, such as GaN, AlN, ZnO, TiO_2. In these materials, high magnetic impurity concentrations are needed due to the lack of long-range interaction among magnetic spins. However, there have been only limited reports on the carrier mediated effect in the wide band gap materials at room temperature (Nepal et al., 2009, Kanki et al., 2006, Philip et al., 2006). Though considerable amount of work has been done on III-V and II-VI DMSs (Ohno et al., 2000, Chiba et al., 2006a), the lack of high T_c and the difficulty in achieving electric field modulated ferromagnetism (carrier mediated ferromagnetism) at or above room temperature deters them from being considered for semiconductor integration.

Group IV semiconductors are of particular interest to the spintronics technology because of their enhanced spin lifetime and coherent length due to their low spin-orbit coupling and lattice inversion symmetry (Rheem et al., 2007, Choi et al., 2005a). There is also considerable interest for transition metal doped group IV semiconductors (Miyoshi et al., 1999, Park et al., 2002, Tsui et al., 2003, D'Orazio et al., 2004, Kazakova et al., 2005, Li et al., 2005, Demidov et al., 2006, Jamet et al., 2006, Collins et al., 2008, Ogawa et al., 2009, Tsuchida et al., 2009) to the semiconductor industry, owing in part to their excellent compatibility with silicon process technology. Several transition metals – including the use of single and co-doping of elements such as Mn, Cr, Co, Fe, Co-Mn, and Fe-Mn – have been used as magnetic dopants in both Si and Ge either through blanket implantation or through in-situ doping during epitaxial growth process. Both these materials are reported to be ferromagnetic (Cho et al., 2002, Jamet et al., 2006, Pramanik et al., 2003). The reported experimental results also clearly showed the presence of carrier mediated ferromagnetism in DMS Ge (Chen et al., 2007a, Xiu et al., 2010). A recent study predicted that the T_c could be increased by enhancing the substitutional doping of Mn in Ge and Si, via co-doping - by adding conventional electronic dopants such as As or P during the Mn doping process (Maekawa, 2004). Theories (Lou et al., 2007) and experiments (D'Orazio et al., 2004, Li et al., 2007) further suggest that the ferromagnetic transition temperature is related to the ratio of interstitial to substitutional Mn similar to that in the group III-V system (Jungwirth et al., 2006).

Considering the fact that, nanosystems such as nanowires and quantum dots (QDs) are the versatile building blocks for both fundamental studies in nanoscale and the assembly of present day functional devices, low dimensional Ge DMS systems such as Mn$_x$Ge$_{1-x}$ nanowires (van der Meulen et al., 2008, Kazakova et al., 2005, Majumdar et al., 2009b, Seong et al., 2009) and QDs (Guoqiang et al., 2008) have also been reported to have room temperature ferromagnetism. The above results are supported by modeling results (Wang and Qian, 2006) although carrier mediated exchange has not yet to be experimentally verified in nanowires. Similarly, high Curie temperatures have been observed in other transition metal doped semiconductor nanostructures such as II-VI (including ZnO (Chang et al., 2003, Cui and Gibson, 2005, Baik and Lee, 2005), ZnS (Brieler et al., 2004, Radovanovic et al., 2005), CdS (Radovanovic et al., 2005)) and III-V (including GaN (Radovanovic et al., 2005, Choi et al., 2005b), GaAs (Jeon et al., 2004)), but only inconclusive reports have been published on the magneto-electric transport in these nanostructures. It could be attributed to the complicated ferromagnetic properties caused by precipitates, second-phase alloys and nanometer-scale clusters. Therefore, it is important and indispensable to differentiate their contributions. Unfortunately, there is still a lack of fundamental understanding of the growth process and spin dependent states in these systems that gives an impetus to extensive investigations into these and related material systems.

In this chapter, we will first introduce the state-of-art theoretical understanding of ferromagnetism in group IV DMSs, particularlly pointing out the possible physics models underlying the complicated ferromagnetic behavior of Mn$_x$Ge$_{1-x}$. Then, we primarily focus on the magnetic characterizations of epitaxially grown Mn$_x$Ge$_{1-x}$ QDs. It is found that when the material dimension decreases (to zero), it evidences a change of material structures being free of precipitates and a significant increase of T_c over 400 K. The premilinary explanation on these behaviors can be attributed to the carrier confinement in the DMS nanostructures which strengthens hole localization and subsequently enhances the thermal stability of magnetic polarons, thus giving rise to a higher T_c than those of bulk films.

Another important aspect of DMS research lies in the electric field controlled ferromagnetism. We emphasize the existing experimental efforts and achievement in this direction, and provide a detailed description on the field controlled ferromagnetism of the Mn$_x$Ge$_{1-x}$ nanostructuers. We show that the ferromagnetism for the Mn$_x$Ge$_{1-x}$ nanostructures, such as QDs, can be manipulated via the control of gate voltage in a gate metal-oxide-semiconductor (MOS) structure (up to 100 K). These experimental data suggest a hole-mediated ferromagnetism as controlled by a gate bias and promise a potiental of using Mn$_x$Ge$_{1-x}$ nanostructures to achieve spintronic devices. Beyond the scope of DMSs, we will present the metallic Mn$_x$Ge$_{1-x}$ with x~20%, in which a constant Curie temperature was observed. The emphasis, however, is the manipulation of these metallic dots into self-assembled periodic arrays for potential device applications. Finally, we render general comments on the development of the Mn$_x$Ge$_{1-x}$ system and possible research schemes to develop spintronics devices based on this material system.

2. Theories of ferromagnetism in group IV DMS

As in other popular dilute magnetic semiconductors such as (Ga, Mn)As, the formation of magnetic order and the origin of ferromagnetism in Mn_xGe_{1-x} can be described (within the parametric limits) either using complementary mean field approximation models such as the Zener-Kinetic exchange (Ji et al., 2004), the Ruderman-Kittel-Kasuya-Yoshida (RKKY) (Zhao et al., 2003) interaction and percolation theory (Yuldashev et al., 2001) or using density functional theory such as *ab initio* full-potential augmented plane wave (FLAPW) (Stroppa et al., 2003) electronic calculations. Interestingly, the Zener-kinetic exchange model proposed in 1950 could be used for interpreting many of the experimental results in transition metal doped semiconductors (Ji et al., 2007, Akai, 1998, Dietl et al., 2001) besides the transition metals themselves. All the above theoretical approaches have tried to address and to a certain extent successfully answered chemical, electronic and magnetic properties of the transition metal doped group IV semiconductors with in the context of (1) the nature of transition metal impurities, (2) the origin of ferromagnetism, (3) the influence of dopants on the T_c, (4) the solubility limits of carrier density on the chemical nature, (5) the influence of carrier density on the electronic structure, and (6) the influence of different hosts on the electronic structure. Within the above theoretical approaches, these models assume that the ferromagnetism in transition metal doped Ge system is mediated through exchange coupling between the carriers and the interacting lattice. This means that, the interaction is through p-d orbital hybridization between the d levels and the valence band p states of Ge. According to these models, as the number of dopant carriers increases both the electronic mobility and the ferromagnetic ordering of the system also increases along with the T_c. Thus, the assumption is that the strong exchange coupling between the p and d orbitals leads to strong kinetic exchange coupling between the spin polarized holes and spin polarized TM substitutional vacancies. Among all the 3-d TM impurities, [TM=V, Cr, Mn, Fe, Co, Ni], Mn is the most favored dopant. This is because compared to other dopants, Mn favors less clustering, low non-uniform doping distribution and higher substitutional doping over interstitials. The *Ab initio* electronic structure calculations within density functional theory show that, such doping behaviors could result in relatively higher carrier hole concentrations and increased magnetic moments (Park et al., 2002, Stroppa et al., 2003). Complementary theoretical work using the frozen-magnon scheme also shows that the ferromagnetism is produced only through holes and Mn is a good source for generating holes in Ge system (Picozzi and Lezaic, 2008).

In addition, some of the other factors that determine the ferromagnetic strengths in Ge are (1) disorder of Mn site locations, (2) distance between Mn-Mn atoms, (3) solid solubility limits and preferential surface orientation for interstitial and substitution site formation (Luo et al., 2004, Erwin and Petukhov, 2002, Prinz, 1998), and (4) co-doping of Co or Cr with Mn (They were found to reduce cluster formation but not to significantly contribute to the ferromagnetism in the system (Jedema et al., 2002, Chen et al., 2009)). According to full potential linearized augmented plane wave (FLAPW) calculations made on the RKKY model, the coupling between the Mn atoms could be ferromagnetic (FM) or anti-ferromagnetic (AFM) in nature but the interaction is very localized to Mn sites and depends on the distance between Mn atoms: the density of states decreases rapidly as the Ge atom moves away from Mn (Stroppa et al., 2003). However, disorder of Mn site locations is found to influence the ferromagnetic strength and the transition temperature. In addition, the impact of inherent disorder is also confirmed both by experiments and by the mean field approximation (Valenzuela and Tinkham, 2004, Garzon et al., 2005, Yuldashev et al., 2001). Some of the above models found themselves hard to explain the long range exchange interaction and coupling strength when there is varying dopant concentration and disorder ranging from high resistive state to half metallic state. On the other hand, there are attempts to find solutions for disorder dependent magnetic variation using Percolation theory based mean field approximation calculation (Yuldashev et al., 2001, Schlovskii and L., 1984). Percolation theory provides possible explanations for the experimentally observed low values of saturation magnetization in low carrier density Mn$_x$Ge$_{1-x}$ because the infinite cluster of percolating bound magnetic polarons triggering the long-range ferromagnetic order leaves out a large number of Mn moments at temperatures less than T_c. However, the role of disorder with respect to insulating and conductive metallic phases and their effects on mobility is still not well understood. Conversely, the FLAPW calculations using the density functional theory shows that Mn favors long-range ferromagnetic alignment over short-range anti-ferromagnetic alignment and this increases with the increase of Mn content which is in agreement with many of the experimental results (Stroppa et al., 2003). The estimated Curie temperature is within the range of 134-400 K and is in agreement with experimental and theoretical results (Wang and Qian, 2006).

However, there are also conflicting reports that long range FM order could only happen at low temperature (12 K) and that at low temperature MnGe could show only spin glass behavior due to the inter-cluster formation between the FM Mn-rich clusters (Poli et al., 2006, Lou et al., 2007). With these reports, it is understood that the magnetic phases are more favored over paramagnetic phases by an energy difference of 200 meV/Mn atom (Stroppa et al., 2003). The report also states that the Mn$_x$Ge$_{1-x}$ system is very close to half metallicity irrespective of the amount of doping concentration (Stroppa et al., 2003, Peressi et al., 2005). Electronic band structure calculations using density of states (DOS) show that the binding energy of the Ge 4s states are basically unaffected by the exchange splitting of Mn d states and the valence band maximum of the minority spin channel in Mn$_x$Ge$_{1-x}$ reaches E_F at Gamma (T) such that the spin gap becomes zero (Stroppa et al., 2003, Peressi et al., 2005). In other words, the DOS shows a valley around E_F and is strictly zero at E_F in the minority spin channel. On the other hand, the energy gap is indirect and the bands around E_F in the majority spin channel arise from Mn d and Ge p hybridizations so as to give rise to hole pockets closer to the gamma point with a localized magnetic moment of about 3μ_B. This could also be explained using chemical and structural configurations of Mn in Ge. The different chemical species determine the relative positions of the anion and cation atomic energy levels in the Mn$_x$Ge$_{1-x}$ while the different anion size dictates essentially the equilibrium lattice constants (Stroppa et al., 2003). This essentially explains the carrier mediated effect in Mn$_x$Ge$_{1-x}$ even though the system is closer to half metallic.

The aim of all these theoretical work is to build foundation for achieving carrier mediated ferromagnetism at temperatures above room temperature. It is understood that this can be achieved only by growing cluster free and uniformly doped Mn$_x$Ge$_{1-x}$ samples. Experimental reports show that T_c as high as 400 K is obtained for cluster free Mn$_x$Ge$_{1-x}$ samples grown (1) using sub-surfactant epitaxial method (Kimura and Otani, 2008) (2) epitaxial QDs (Xiu et al., 2010) and (3) nanowire structures (Kazakova et al., 2005, Majumdar et al., 2009b, van der Meulen et al., 2008). However, there are only very few theoretical studies depicting the relation between the quantum confinement and its influence on the origin of ferromagnetism in low dimensional DMS structures (Johnson and Silsbee, 1985, Patibandla et al., 2006),(Kelly et al., 2003). It apparently lacks of systematic theoretical study particularly on Mn_xGe_{1-x} QDs. Therefore, it will be interesting to understand the ability of the quantum confinement phenomenon to retain spin polarization in Mn_xGe_{1-x} QDs structure. Studies on III-V QDs using LSDA approximation show that, unlike in the bulk structures, adding a single or multiple carriers in a magnetic QD can both strongly change the total carrier spin and the temperature of the onset of magnetization and this can be modulated by modifying the quantum confinement and the strength of coulomb interactions. These theoretical results must have ramification in the high Curie temperature Mn_xGe_{1-x} QDs as well since our studies also reveal similar characteristics. But, all of the effect including high T_c that we observe cannot be explained without invoking tiny defects either at the interface or on the surface although the conventional transmission electron microscopy (TEM) cannot reveal their existence due to the limited resolution. All of the above material and electronic properties has been studied in order to understand how feasible a DMS material is for carrier mediated ferromagnetism.

3. Growth and characterizations of group IV DMS and nanostructures

3.1 Overview of the Mn doped Ge

As mentioned above, the Ge-based DMS has attracted extensive attention due to its possibility to be integrated with the mainstream Si microelectronics, in which they may be used to enhance the functionality of Si integrated circuits (Park et al., 2002). In particular, the hole mediated effect discovered in Mn_xGe_{1-x} DMS opens up tremendous possibilities to realize spintronic devices with advantages in reducing power dissipation and increasing new functionalities, leading to perhaps normally off computers. To date, there are many reports on the Mn_xGe_{1-x} growth and characterizations by molecular beam epitaxy (MBE)(Park et al., 2002, Li et al., 2005, Jaeger et al., 2006, Jamet et al., 2006, Tsui et al., 2003, Ahlers et al., 2006a, Ahlers et al., 2006b, Bihler et al., 2006, Bougeard et al., 2006, Devillers et al., 2007, Li et al., 2007, Park et al., 2001, Wang et al., 2008a, Pinto et al., 2005b), ion implantation (Park et al., 2005, Lin et al., 2008, Verna et al., 2006, D'Orazio et al., 2002, Liu et al., 2004, Lifeng et al., 2004, Passacantando et al., 2006, Ottaviano et al., 2007), and bulk crystal growth(Cho et al., 2002, Biegger et al., 2007). We can divide them here, for an easier orientation, into two groups: those that cover fundermantal studies of phase formation, ferromagnetism, and transport properties(Park et al., 2002, Li et al., 2005, Majumdar et al., 2009a, Wang et al., 2008a, Gunnella et al., 2005, Liu et al., 2006, Jaeger et al., 2006, Jamet et al., 2006, Li et al., 2006b, Zhu et al., 2004, Miyoshi et al., 1999, Li et al., 2006a, D'Orazio et al., 2003, Zeng et al., 2006, Cho et al., 2002, Liu and Reinke, 2008, Gambardella et al., 2007, Cho et al., 2006, Morresi et al., 2006, Sugahara et al., 2005, Wang et al., 2008b, Ogawa et al., 2009, Ahlers et al., 2006a, Ahlers et al., 2006b, Bihler et al., 2006, Bougeard et al., 2006, Devillers et al., 2007, D'Orazio et al., 2004, Erwin and Petukhov, 2002, Li et al., 2007, Majumdar et al., 2009b, Park et al., 2001, Passacantando et al., 2007, Seong et al., 2009, Chen et al., 2007b, Biegger et al., 2007, Pinto et al., 2005b, Verna et al., 2007, Pinto et al., 2003, Yu et al., 2006, Yada et al., 2008, Tsuchida et al., 2009, Gambardella et al., 2005), and those that aim to minimize cluster formation and enhance Curie temperatures via co-doping methods(Tsui et al., 2007, Paul and Sanyal, 2009, Gareev et al., 2006, Demidov et al., 2006, Collins et al., 2008, Tsui et al., 2003).

Experimental results show that the Mn doping process in Mn$_x$Ge$_{1-x}$ is complex. Both the T_c and the saturation magnetization depend on the interplay of a variety of factors, which are ultimately determined by growth conditions and post-annealing process (Jamet et al., 2006). The concentration and distribution of Mn dopants, the carrier density, the presence of common defects such as Mn interstitials and Mn clusters significantly influence the magnitude and interactions of the magnetic coupling (Biegger et al., 2007). In early 2002, Park *et al.* (Park et al., 2002) pinoeerly demonstrated ferromagnetic Mn$_x$Ge$_{1-x}$ thin films grown by low-temperature MBE. It was found that the Mn$_x$Ge$_{1-x}$ films were p-type semiconductors in character with hole concentration of 10^{19}~10^{20} cm^{-3}, and exhibited pronounced extraordinary Hall effect. The T_c of the Mn$_x$Ge$_{1-x}$ thin films increased linearly with Mn concentration from 25 to 116 K (Mn≤3.5 %). Field controlled ferromagnetism was also observed in a simple gated structure through application of a gate voltage (± 0.5 V), showing a clear hole mediated effect at 50 K. Such a small gate voltage presents an excellent comparability with conventional low voltage circuitry. The origin of ferromagnetic order was understood in the frame of local-spin density approximation (LSDA), where strong hybridization between Mn d states of T$_2$ symmetry with Ge p states leads to the configuration $e(\uparrow)^2 T_2(\uparrow)^2 T_2(\downarrow)^1$, with a magnetic moment of 3μ$_B$. However, the LSDA calculations gave overestimated T_c which was attributed to the incomplete activation of Mn in experiments and the absence of hole compensation in the simulation (Park et al., 2002).

Since then, various preparation techniques were employed in order to produce Mn-doped Ge DMS, including aforementioned MBE (Li et al., 2005, Jaeger et al., 2006, Jamet et al., 2006, Tsui et al., 2003, Ahlers et al., 2006a, Ahlers et al., 2006b, Bihler et al., 2006, Bougeard et al., 2006, Devillers et al., 2007, Li et al., 2007, Park et al., 2001, Wang et al., 2008a, Pinto et al., 2005b), single-crystal growth(Cho et al., 2002, Biegger et al., 2007), and ion implantation(Park et al., 2005, Lin et al., 2008, Verna et al., 2006, D'Orazio et al., 2002, Liu et al., 2004, Lifeng et al., 2004, Passacantando et al., 2006, Ottaviano et al., 2007), aiming to further increase T_c and to obtain the electric field controllability at room temperature (Biegger et al., 2007). Among these efforts, Cho *et al.* (Cho et al., 2002) reported the synthesis of Mn doped bulk Ge single crystals with 6 % of Mn and a high ferromagnetic order at about 285 K via a vertical gradient solidification method. The origin of the high T_c was found to be complex because of the presence of dilute and dense Mn doped regions. Jaeger *et al.* (Jaeger et al., 2006), however, pointed out that the magnetic properties of these samples were clearly dominated by the presence of the intermetallic compound Mn$_{11}$Ge$_8$. This was indeed observed by Biegger *et al.* (Biegger et al., 2007), where similar Mn$_x$Ge$_{1-x}$ crystals were produced via Bridgman's crystal growth technique and intermetallic compounds were found in both Mn-rich and Mn-poor regions. These experiments suggest that the bulk Mn$_x$Ge$_{1-x}$ crystals produced at over 1000 °C may not be an appropriate method for the Mn$_x$Ge$_{1-x}$ DMS preparation (Biegger et al., 2007).

Apart from the bulk cystal growth, MBE has been widely recognized as a power tool for DMS growth since it can provide non-equilibrium conditions to enhance the incorporation of Mn dopant (Park et al., 2002). Via this technique, Mn doping was found to be extremely sensitive to growth conditions, particularlly to the growth temperature. Excellent work in this aspect can be found in references (Jamet et al., 2006, Ahlers et al., 2006b, Park et al., 2001, Bougeard et al., 2006, Li et al., 2007, Demidov et al., 2006, Pinto et al., 2005b). In general, at low growth temperatures ($T_b < 120$ °C), the diffusion of Mn atoms leads to the formation of

Mn-rich nanostructures, such as nanocolumns (Devillers et al., 2007) and nanodots (Bougeard et al., 2006) with irregular shapes. These nanostructures could contain a high Mn concentration up to 38 % (Jamet et al., 2006) while the surrounding matrix remains a low Mn concentration less than 1 %. A perfect lattice coherence was found in the proximity of the nanostructures, indicative of a strong compression for coherent nanostructure and a large tension for the surrounding matrix (Devillers et al., 2007). These observations are in good agreement with therorectial predictions of a spinodal decomposition (Fukushima et al., 2006), which occurs under a layer-by-layer growth mode with strong pairing attraction between Mn atoms and a tendency of surface diffusion of Mn atoms. These theoretical calculations successfully explained the formation of nanostructures in (Ga, Mn)N and (Zn, Cr)Te) (Fukushima et al., 2006, Devillers et al., 2007), which can be also readily applied to the Mn_xGe_{1-x} system (Devillers et al., 2007). The magnetic properties, however, exhibit various characteristics depending on the nanostructure size and the Mn concentration. In most cases, zero-field cooled (ZFC) and field cooled (FC) magnetizations show superferromagnetic properties with a blocking temperature of ~15 K. It is now clear that such a low blocking temperature is originated from the coherent nanostructures with a diameter below several nanometers. Curie temperatures of these nanostructures were also identified to be below 170 K and varied with different Mn concentration. No Mn_5Ge_3 metallic phases were observed at low temperature regime ($T_b < 120$ °C).

Dougeard et al. (Bougeard et al., 2006) also found that when the dimension of the nanostructures further reduced, the films showed no overall spontaneous magnetization down to 2 K. TEM results were interpreted in terms of an assembly of superparamagnetic moments developing in the dense distribution of nanometer-siezed clusters. In addition to the single crystalline Mn_xGe_{1-x}, homogeneous ampohous nanoclusters were also observed under certain growth conditions and contributed to ferromagnic order below 100 K (Devillers et al., 2007, Sugahara et al., 2005). To render a general understanding of the Mn_xGe_{1-x} magnetic properties, Jaeger et al. (Jaeger et al., 2006) presented a detailed study on $Mn_{0.04}Ge_{0.96}$ and $Mn_{0.20}Ge_{0.80}$ thin films grown at low temperatures and related their magnetic behavior to spin-glass. A frozen magnetic state at low temperatures was observed and attributed to the formation of nanostructures. This is in a good agreement with Devillers et al. (Devillers et al., 2007). Therefore, it can be concluded that the Mn_xGe_{1-x} magetic semiconductor is not a conventional ferromagnet since its magnetic properties are significantly complicated by the formation of nanostructures and metallic clusters.

When the growth temperature falls into 120~145 °C, the Mn_5Ge_3 nanoclusters start to develop and dominate the magnetic property with a Curie temperature of 296 K (Devillers et al., 2007, Jamet et al., 2006). This phase, frequently observed at high temperature growth, is the most stable (Mn, Ge) alloy. The other stable compound $Mn_{11}Ge_8$ was also observed in nanocrysllites surrounded with pure Ge (Park et al., 2001). Although they are ferromagnetic, the metallic character considerably jeopardizes their potential for spintronic applications. Under a narrow growth window around 130 °C, another phase $MnGe_2$ nanocolumns could be developed with a ~33 % of Mn and a T_c beyond 400 K (Jamet et al., 2006). A remarkable feature of these clusters is the large magneto-resistance due to geometrical effects. Further increasing the growth temperature leads to the dominating magnetic behavior from the Mn_5Ge_3 clusters. As the coherent nanostructures grow larger at high temperature growth, higher blocking temperatures (~30 K) were also obtained (Devillers et al., 2007).

To minimize the phase separation, Mn codoping with Co was attempted to stabilize structures at high Mn doping concentration (Tsui et al., 2003, Tsui et al., 2007, Collins et al., 2008). It showed that codoping with Co can dramatically reduce phase separation and diffusion of Mn within the Ge lattice while it magnetically complements Mn. The measured strain states indicate the critical role played by substitutional Co with its strong tendency to dimerize with interstitial Mn. (Collins et al., 2008) The highest T_c achieved so far is about 270 K (Tsui et al., 2003). Similiarly, Gareev *et al.* (Gareev et al., 2006) reported on the codoping of Fe to achieve a T_c of 209 K. Recent poazrized neutron reflectivity measurements provided evidence to show no segregation and lower clustering tendencies for higher Fe doping (Paul and Sanyal, 2009), which further supported co-doping approach in Mn_xGe_{1-x}.

Ion implantation was also attempted due to its efficiency of Mn incorporation (Passacantando et al., 2006). However, extensive experiments revealed that Mn-rich precipitates were developed and buried in a crystalline Ge matrix (Park et al., 2005, Lin et al., 2008, Verna et al., 2006, D'Orazio et al., 2002, Liu et al., 2004, Lifeng et al., 2004, Passacantando et al., 2006, Ottaviano et al., 2007) in a much similar manner to that of the thin films grown by MBE (Park et al., 2001, Bougeard et al., 2006, Li et al., 2007, Demidov et al., 2006).

Although there has been much progress in producing high quality Mn_xGe_{1-x} thin films, the formation of uncontrollable metallic clusters yet remains a challenge. High T_c DMS thin films without any precipitations seem to be a critical obstacle for the fabrications of practical spintronic devices functioning at room temperature, not to mention the room-temperature controlled ferromagnetism.

3.2 Mn_xGe_{1-x} by molecular beam epitaxy

To gain further insight into the complex growth of this material system, we have performed detailed structural and magnetic characterizations of Mn_xGe_{1-x} grown by MBE with different thickness. Our experiments showed that the Mn doping behavior and magnetic properties are dramatically different when the material dimension becomes smaller. (Chen et al., 2007a, Xiu et al., 2010) Bulk Mn_xGe_{1-x} films always tend to form clusters no matter what growth conditions are. When it comes to one and zero dimensions, however, the cluster formation may be possibly limited because the Mn concentration cannot accumulate enough and the strain can be easily accommodated to minimize the metallic clusters (Wang et al., 2008a, Ottaviano et al., 2007). We further show that self-assembled Mn_xGe_{1-x} DMS QDs can be successfully grown on Si substrate with a high T_c in excess of 400 K. Finally, the electrical field controlled ferromagnetism was also demonstrated in MOS structures using these QDs as the channel layers, as elaborated in section 4.

3.2.1 Self-assembled DMS Mn_xGe_{1-x} QDs

$Mn_{0.05}Ge_{0.95}$ QDs were grown on *p*-type Si substrates. Cross-section TEM was carried out to determine the structural characteristics and the Mn composition. HR-TEM image reveals that a QD has a dome shape with a base diameter of about 30 nm and a height of about 8 nm (Fig. 1(a)) The interface between the dot and the Si substrate has excellent lattice coherence. A careful inspection reveals that the dot is single-crystalline without evidence of pronounced dislocations or stacking faults. However, because of the heavy Mn doping it is possible that some amount of point defects (such as Mn interstitials) may be present inside the dot, which is beyond the detect capability of conventional TEM.

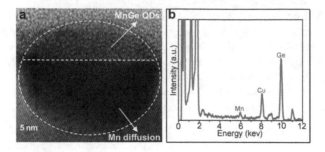

Fig. 1. Structural properties of $Mn_{0.05}Ge_{0.95}$ quantum dots grown on a p-type Si substrate. (a) A high-resolution TEM cross-section image of typical $Mn_{0.05}Ge_{0.95}$ quantum dot showing the detailed lattice structure. (b) Typical EDS spectrum showing that both Mn and Ge are present in the $Mn_{0.05}Ge_{0.95}$ quantum dot. The average Mn concentration in the quantum dots was estimated to be 4.8±0.5 %.

Directly underneath the $Mn_{0.05}Ge_{0.95}$ QD, Mn diffuses into Si substrate and forms a strained MnSi area, which has the same diameter as the top $Mn_{0.05}Ge_{0.95}$ QD, but a height of about 16 nm (Fig. 1(a)). This diffusion behavior is not unusual as it was also observed in (In, Mn)As and (In, Cr)As QDs systems. (Holub et al., 2004, Zheng et al., 2007) However, the migrations of Mn into the substrate make it difficult to accurately determine Mn concentration inside the dot. To address this challenge, we have performed energy dispersive x-ray spectroscopy (EDS) experiments (in a scanning TEM mode) to analyze the Mn composition at nanoscale (Fig. 1(b)). When the EDS was performed, electron probes with several nanometers in diameter were illustrated on the $Mn_{0.05}Ge_{0.95}$ QDs and their underlying Si substrate. Consequently Si peaks are constantly observed in the EDS spectra. To estimate the Mn concentrations, we performed a quantitative analysis of atomic percentages of Mn and Ge and artificially discounted the Si peak. The EDS analysis over many QDs reveals a Mn/Ge atomic ratio of about 0.144:1. Since approximately 1/3 volume fraction of Mn is distributed in the Mn_xGe_{1-x} QD (Xiu et al., 2010), the average Mn concentration can be estimated to be 4.8 ± 0.5 %. Note that the deviation was determined by a thorough study of many $Mn_{0.05}Ge_{0.95}$ QDs. Both the HR-TEM investigations and the composition analysis suggest that each individual QD is a single-crystalline DMS system.

Magnetic properties were studied using a superconducting quantum interference device (SQUID) magnetometer at various temperatures. Figure 2(a) and two corresponding insets show the temperature-dependent hysteresis loops when the external magnetic field is parallel to the sample surface (in-plane). The field-dependent magnetization indicates a strong ferromagnetism above 400 K. The saturation magnetic moment per Mn atom is roughly estimated to be 1.8 μ_B at 5 K. A fraction of roughly 60 % of Mn is estimated to be activated assuming that each Mn has a moment of 3 μ_B (Park et al., 2002, Stroppa et al., 2003). The Arrott plots were also made in order to evaluate the T_c (Ohno et al., 2000) (Fig. 2(b)). By neglecting the high order terms, the magnetic field can be expressed in the following equation (Arrott, 1957),

$$H = \frac{1}{\chi}M + \beta M^3 \quad \text{or} \quad \frac{H}{M} = \frac{1}{\chi} + \beta M^2 \qquad (1)$$

where H is the external magnetic field, M is the magnetic moment from the sample, χ is the susceptibility, and β is a material dependent constant.

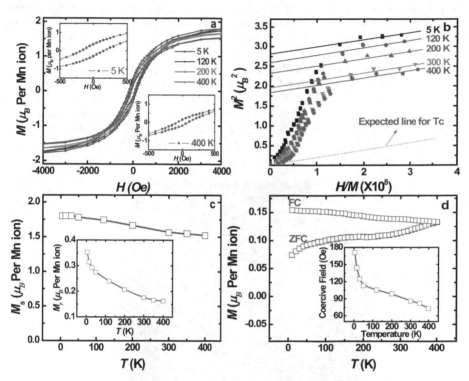

Fig. 2. Magnetic properties of the Mn$_{0.05}$Ge$_{0.05}$ quantum dots grown on a p-type Si substrate. (a) Hysteresis loops measured at different temperatures from 5 to 400 K. (b) The Arrott plots were made to obtain the Curie temperature. Consistent with (a), the Curie temperature is projected to be above 400 K; (c) The temperature dependence of saturation moments. The inset gives the remnant moments with respect to temperature; (d) Zero-field cooled and field cooled magnetizations of quantum dots with a magnetic field of 100 Oe; the inset shows the coercivity values at different temperatures. The external magnetic field is in parallel with the sample surface.

According to Equation (1), M^2 can be plotted as a function of H/M. When H/M is extrapolated to $M^2=0$, the intercept on the H/M axis gives $1/\chi$. The T_c for this material can be obtained when $1/\chi$ vanishes. Figure 2(b) shows the Arrott plots at several temperatures. It can be seen that even at 400 K the intercept ($1/\chi$) on the H/M axis does not vanish, which means that the susceptibility still has a finite value and the T_c has not been reached yet. By using the slope obtained at 400 K, a dashed line can be drawn as shown in Fig. 2(b), in which a T_c is projected to be beyond 400 K. This is in a good agreement with the data from the hysteresis loops showing the magnetic order above 400 K. Figure 2(c) and corresponding inset show the temperature-dependent saturation and remnant moments per Mn ion, respectively. Both of them demonstrate weak temperature dependences and substantial amount of magnetization moments remain even at 400 K.

ZFC and FC magnetizations were measured with a magnetic field of 100 Oe as shown in Fig. 2(d). The magnetic moments do not drop to zero, suggesting a high T_c beyond 400 K, which is in a good agreement with the Arrott plots in Fig. 2(b). From these two curves, one can also infer the formation of a single phase in this material system, *i.e.*, DMS QDs. The wide separation of the ZFC and FC curves in the temperature range of 5 to 400 K shows the irreversibility of susceptibilities, possibly arising from strain-induced anisotropy as a large lattice mismatch exists between Si and Ge (Dietl et al., 2001). The temperature-dependent coercivity is shown in Fig. 2(d) inset. The coercivity decreases from 170 Oe (at 5 K) to 73 Oe (at 400 K). The small coercivity in the entire temperature range measured features a soft ferromagnetism which originates from Mn ions diluted in the Ge matrix (Kazakova et al., 2005). The above magnetic properties support the fact that the $Mn_{0.05}Ge_{0.95}$ QDs exhibit a DMS type ferromagnetic order.

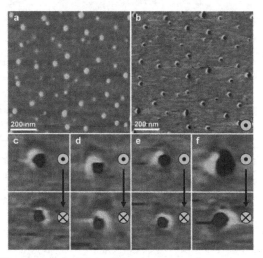

Fig. 3. AFM and MFM images of the $Mn_{0.05}Ge_{0.95}$ quantum dots measured at 320 K. (a) Typical AFM image of $Mn_{0.05}Ge_{0.95}$ quantum dots; (b) Corresponding MFM image with the tip magnetization pointing toward the sample; (c)~(f), Enlarged MFM images of individual quantum dots taken from (b). From these MFM measurements, opposite contrasts were observed when applying an opposite magnetization to the tip.

Atomic force microscopy (AFM) and magnetic force microscopy (MFM) measurements were carried out to investigate the morphology and ferromagnetism of $Mn_{0.05}Ge_{0.95}$ QDs at 320 K, respectively. The average dot size is 50 nm in base diameter and 6 nm in height. The dot density is about 6×10^9 cm^{-2} (Fig. 3(a)). The corresponding MFM image was taken by lifting up the MFM probe 25 nm above the topographic height of the sample in a phase detection mode (Fig. 3(b)). The appearance of bright-and-dark areas in the MFM image clearly shows the formation of magnetic domains in the Mn$_{0.05}$Ge$_{0.95}$ QDs, which is similar to (In, Mn)As DMS QDs (Jeon et al., 2002). Figures 3(c)~3(f) show enlarged MFM images of several individual Mn$_{0.05}$Ge$_{0.95}$ QDs. By reversing the tip magnetization, opposite contrast was observed for each dot, indicating that the magnetic signals originated from the top Mn$_{0.05}$Ge$_{0.95}$ QDs. Note that each QD is a single domain "particle". During the magnetization process, the domain would rotate preferentially along the magnetic field to produce net magnetization moments. Since the experiments were performed at 320 K, the formation of metallic phases such as Mn$_5$Ge$_3$ and Mn$_{11}$Ge$_8$ can be easily ruled out because they have low Curie temperatures of 296~300 K. Overall, the above MFM results agree well with the TEM observations and the ferromagnetic order at high temperature obtained in the SQUID measurements.

3.3 Discussions

We have reviewed the growth of Mn$_{0.05}$Ge$_{0.95}$ by MBE, ion implantation, and bulk sintering, emphasizing the growth temperature effect on the phase formation and magnetic properties. While the current motivation is driven by the prospect of enhancing Curie temperatures, the experiments show a strong tendency of metallic precipitates developed in bulk and relatively thick Mn$_x$Ge$_{1-x}$ crystals. The fundermental understanding here points to a spin-glass-like feature in Mn$_x$Ge$_{1-x}$ due to the formation of various lattice coherent nanostructures and metallic precipitates, which exhibit magnetic blocking behaviors under different temperature regimes. The challenge to eliminate these clusters seems to be overwhelming and nearly impossible. However, when the material dimension decreases, it evidences a change of material structures being free of precipitates and a significant increase of Curie temperatures over 400 K. The premilinary explanation on these behaviors can be attributed to the carrier confinement in a DMS QD which strengthens hole localization and subsequently enhances the thermal stability of magnetic polarons giving rise to a higher T_c than those of bulk films. Nevertheless, the experimental data suggest that with nanoscale structures, the quantum confinement effect comes into being which significantly influences the exchange coupling between the confined holes and the localized Mn^{2+}. Future progress in this direction will involve a detailed theoretical treatise to understand the quantum confinement effect on magnetic properties in a quantitive picture.

4. Electric field controlled ferromagnetism

4.1 Introduction

Electric field control of ferromagnetism has a potential to realize spin field-effect transistors (spin FETs) and nonvolatile spin logic devices via carrier-mediated effect (Awschalom et al., 2002, Chiba et al., 2006b). With the manipulation of carrier spins, a new generation of nonvolatile (green) computing systems could be eventually developed for many low-power-dissipation applications in all fields, including sensor network, health monitoring, information, sustainable wireless system, etc. Since Datta and Das (Datta and Das, 1990) first introduced the concept of spin FETs in 1990, enormous efforts were dedicated to creating a device wherein the carrier transport is modulated by electrostatic control of carrier spins (Philip et al., 2006, Ohno et al., 2000, Appelbaum and Monsma, 2007, Chiba et al., 2006a, Nazmul et al., 2004, Nepal et al., 2009, Koo et al., 2009, Chiba et al., 2008, Boukari et al., 2002). To understand and exploit this controllability, several theoretical models were proposed to explain the ferromagnetic coupling in DMS on a microscopic scale (also elaborated in the theoretical section): the Zener Kinetic-exchange (Jungwirth et al., 2006), double-exchange (Dietl et al., 2000), the RKKY interaction (Matsukura et al., 1998, Yagi et al., 2001). These models share a common feature that a spontaneous ferromagnetic order is carrier-mediated through the increase of carrier concentrations. One of major challenges, however, is to search an ideal material with room-temperature controllable spin states (Philip et al., 2006, Kanki et al., 2006, Dietl and Ohno, 2006). In recent years, emerged DMSs became one of the promising candidates since they could possibly offer high T_c in excess of 300 K (Dietl et al., 2000). The demonstration of the carrier medicated ferromagnetism involving correlated electron/hole systems leads to a para- to ferro-magnetism phase transition (Park et al., 2002, Dietl et al., 2000, Chen et al., 2007a, Chiba et al., 2006b). In principle, the collective alignment of spin states in these DMSs can be manipulated by the modulation of carrier concentrations through gate biasing in a FET structure (Ohno et al., 2000, Dietl and Ohno, 2006). For this kind of spin FETs, the "source" and "drain" may be completed through "nanomagnet", which are in turn controlled by the gate; and no carrier transport is needed. Clearly, one may also involve the control of source-drain conductance by gate-voltage-induced precession of injected spins (from the source). Since early 2000s, a significant progress was achieved (Ohno et al., 2000, Chiba et al., 2006a, Chiba et al., 2008, Chiba et al., 2006b), in which the ferromagnetism of a (In, Mn)As channel layer could be effectively turned on and off via electric fields in a gated FET. Such extraordinary field modulated ferromagnetism immediately rendered the development of future spintronic devices. However, the manipulation of ferromagnetism was limited because of low T_c of the Mn-doped III-V materials (Weisheit et al., 2007). Therefore, a search for new DMS materials with $T_c > 300$ K and carrier mediated ferromagnetism becomes a current global challenge (Jungwirth et al., 2006, Dietl and Ohno, 2006).

4.2 Electric field controlled ferromagnetism in self-assembled DMS $Mn_{0.05}Ge_{0.95}$ QDs

Since the self-assembled $Mn_{0.05}Ge_{0.95}$ QDs have a high T_c above 400 K and are free of metallic phases (Mn_5Ge_3 and $Mn_{11}Ge_8$), it can be potentially used as a channel layer in a MOS device to study the modulation of ferromagnetism by electric field. The device structure consists of a metal gate (Au, 200 nm), Al_2O_3 (40 nm), $Mn_{0.05}Ge_{0.95}$ QDs (6 nm), a wetting layer (<0.6 nm), a p-type Si substrate (1×10^{18} cm^{-3}), and a back metal contact (Au, 200 nm). A relative thick dielectric of Al_2O_3 (40 nm) is employed to ensure a small leakage current to a level below 10^{-6} A/cm^2. Figure 4 shows the electric field controlled ferromagnetism performed at 50, 77, and 100 K, corresponding to (a)-(c), (d)-(f), and (g)-(i), respectively. Due to the similarity of the data, we take 77 K as an example to describe the device operation in Figs. 4(d)-4(f). Figures 4(d) and 4e) show the hysteresis loops by SQUID with negative and positive biases on the MOS gate at 77 K, respectively. Under a negative bias, the holes are attracted into the channel of the device (accumulation). In this circumstance, however, the hysteresis loop does not show a remarkable change (Fig. 4(d)). This could be explained by the fact that even at 0 V, the QDs device is already accumulated with enough holes to induce ferromagnetism. In other words, the hole mediated effect is sufficient to align a majority of the activated Mn ions along one direction in each individual QD. Further increasing negative bias does not change much on the hole concentrations. On the contrary, with the positive bias, a large amount of holes are depleted into the p-type Si so that hole mediated effect is notably reduced. In fact, the surface of the device can have a high concentration of electrons if the leakage is limited (Inversion). As a result, the Mn ions start to misalign because of the lack of holes. The saturation moment per Mn ion decreases more than 10 times as the gate bias increases from 0 to + 40 V (Fig. 4(e)). It should be noted that, at +40 V, the saturation and remnant moments of the Mn$_{0.05}$Ge$_{0.95}$ QDs become fairly weak, resembling a "paramagnetic-like" state. Figure 4(f) summarizes the change of remnant moments as a function of gate voltage. A similar trend in the carrier density as a function of the gate bias was observed from CV curves (not shown here), suggesting a strong correlation between hole concentrations and the ferromagnetism, *i.e.* the hole-mediated effect. The inset in Fig. 4(f) displays an enlarged picture to clearly show the change of remnant moments with respect to the gate bias. By increasing the measurement temperature to 100 K (Figs. 4(g)-4(i)), the modulation of the ferromagnetism becomes less pronounced compared to those at 50 and 77 K due to the increased leakage current in our MOS devices. The above results evidently demonstrate that the hole mediated effect does exist in this material system.

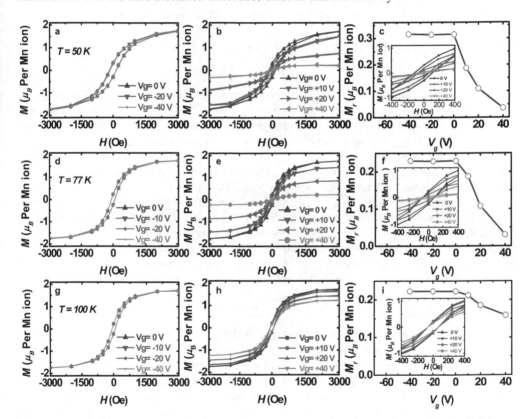

Fig. 4. Control of ferromagnetism of Mn$_{0.05}$Ge$_{0.95}$ quantum dots by applying electric field at 50 K ((a)~(c)), 77 K ((d)~(f)), and 100 K ((g)~(i)). (a), (d), and (g): Hysteresis loops with zero and negative bias of -10, -20 and -40 V on the gate; (b), (e), and (h): The hysteresis loops with zero and positive bias of +10, +20, and +40 V; (c), (f), and (i): Remnant moments with respect to the gate bias. Insets of (c), (f) and (i) are enlarged figures from the central part of (b), (e) and (h) to clearly show the change of remnant moments, respectively.

4.3 Discussions

Electrical field controlled ferromagnetism has been successfully demonstrated in Mn-doped III-V DMSs (Chiba et al., 2008, Chiba et al., 2006b, Chiba et al., 2006a, Dietl and Ohno, 2006, Ohno et al., 2000). It is well established that the variation of hole concentrations renders such a controllability. (Ohno et al., 2000) The low T_c of III-V DMS, however, presents a challenge to further realize room-temperature controlled ferromagnetism (Weisheit et al., 2007). Alternatively, the recent experiments on $Mn_{0.05}Ge_{0.95}$ QDs show a high T_c above room temperature and gate modulated ferromagnetism over 100 K (Xiu et al., 2010). While the leakage current suppressed the gate modulation, room-temperature controlled ferromagnetism would not be impossible because of the high T_c of this system. Research in this direction may open up a pathway for achieving room-temperature Ge-based (and other) Spin-FETs and spin logic devices. These spintronic devices could potentially replace the conventional FETs with lower power consumptions and provide additional new device functionalities, which is beyond today's mainstream CMOS technology of microelectronics.

5. Metallic Mn_xGe_{1-x} (x~20%) nanodot arrays

5.1 Introduction

There is also a need to develop ferromagnet/semiconductor hybrid structures for semiconductor spintronics since they have magnetic and spin-related functions and excellent compatibility with semiconductor device structures. (Tanaka, 2002, Wang et al., 2011) By embedding magnetic nanocrystals into conventional semiconductors, a unique hybrid system can be developed, allowing not only utilizing the charge properties but also the spin of carriers, which immediately promises next-generation non-volatile magnetic memories and sensors. (Dietl, 2006, Kuroda et al., 2007) On the other hand, spin-injections into the semiconductor can be dramatically enhanced via coherent nanostructures, which considerably reduce undesired spin scatterings. (Kioseoglou et al., 2004) Although magnetic hybrid systems, such as MnAs/GaAs, have been extensively studied over several decades, the control (over the spatial location, shape and geometrical configuration) of the magnetic nanostructures (for instance MnAs) still remains a major challenge to further improve the performance of the related magnetic tunnel junctions (MTJs) and spin valves. (Dietl, 2008)

5.2 Growth methods

We employed a concept of stacked Mn_xGe_{1-x} (x~20%) (Hereafter, we denote it as MnGe) nanodots by alternatively growing MnGe and Ge layers with designated thicknesses (nominal 3 nm thick MnGe and 11 nm thick Ge). It is well known that Mn doping in Ge induces compressive strain because of its larger atomic size, (Slater, 1964) assuming that no lattice defects are generated during the doping process, i.e., lattice coherence. Mn-rich MnGe nanodots induced by the spinodal decomposition should be strained if the lattice coherence between the nanodots and the matrix remains. Once the strained nanodots are developed, a thin Ge spacer layer, subsequently deposited with an optimized thickness, will retain the perfect lattice coherence with the underneath nanodots. This enables the existing nanodots to exert strain on the Ge spacer layer and produce "strained spots", which, in turn, become preferred nucleation sites for successive nanodots. Eventually, multilayered and vertically aligned MnGe nanodot arrays can be produced, similar to the scenarios of stacked InAs/GaAs (Xie et al., 1995, Solomon et al., 1996) and Ge/Si (Liao et al., 2001) quantum dots. Indeed, by employing this innovative approach, we achieved the growth of coherent self-assembled MnGe nanodot arrays with an estimated density of $10^{11}\sim10^{12}$ cm^{-2} within each MnGe layer.

Ten periods of MnGe nanodots were epitaxially grown on Ge (100) and GaAs (100) substrates by the MBE system. TEM and EDS in the scanning TEM (STEM) mode were performed to understand the nanostructures and compositional variations of the resulting thin films. Figure 5a and c are typical plane-view and cross-sectional TEM images and show the general morphology of the MnGe nanostructures, viewed along the <100> and <011> directions, respectively. A high-density of dark nanodots can be clearly seen in both cases.

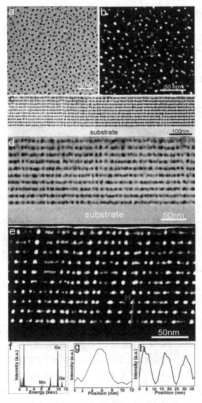

Fig. 5. Transmission electron microscopy (TEM), scanning TEM and energy dispersive X-ray spectroscopy (EDS) results of the multilayer MnGe nanodots. (a), a typical low magnification plane-view TEM image, in which the dark spots are MnGe nanodots. (b), the STEM plane view image. The white spots indicate the MnGe nanodots. (c), a low magnification cross-sectional view. Ordered MnGe nanodots (dark) are clearly seen. (d), a higher magnification TEM image. (e), the STEM cross-section image. The white spots denote the MnGe nanodots. (f), the EDS profile, showing the Mn and Ge peaks. (g, h), line scan profiles of the marked line in (b), (e) using Mn K peak, respectively, confirming that the white spots are Mn-rich nanodots.

Based on the magnified cross-sectional image shown in Figure 5d, the nanodot arrays are clearly observed with 10 stacks along the growth direction although not perfectly vertical. In order to determine the composition of the dark dots, EDS analyses in the STEM mode were carried out and typical plane-view and cross-sectional STEM images are shown in Figure 12b and e, respectively. Figure 5f is the EDS result taken from a typical dot and shows clearly the Mn and Ge peaks. Figure 5g and h present EDS line scans using the Mn K peak for the dots marked by G and H in Figure 5b and e, respectively, indicating high concentrations of Mn inside the dots. Taking all these comprehensive TEM results into account, it is concluded that the nanodots are Mn-rich when compared with the surrounding matrix.

5.3 Magnetotransport properties of nanodot arrays

The resistivity measurements were carried out to probe the carrier transport under different temperatures. It was found that the temperature-dependent resistivities rapidly increase with decreasing temperature due to the carrier freeze-out effect at low temperatures, which is typically observed in doped semiconductors. (Sze, 1981) Considering the embedded MnGe nanodots, the rise in resistivities at low temperatures also suggests a strong localization of carriers, which takes place at the Mn sites and/or at the MnGe/Ge interfaces, similar to the scenario of MnSb clusters in InMnSb crystals. (Ohno et al., 1999) The temperature dependent resistivity can be generally described by (Ma et al., 2005)

$$\rho(T) = \rho_0 \exp[(\frac{T_0}{T})^{1/\alpha}] \qquad (2)$$

where $\rho(T)$ is the temperature-dependent resistivity; ρ_0 and T_0 denote material parameters, a is a dimensionality parameter: $\alpha = 2$ for one-dimensional (1D), $\alpha = 3$ for 2D, and $\alpha = 4$ for 3D systems. In order to reveal the carrier transport mechanisms at different temperature regions, fittings were performed in the plots of lnρ as a function of T^{-a} (Figure 6a). The best fittings were found when a equals to 1 and 4 in the high-temperature and low-temperature regions, respectively, corresponding to the carrier transport via the band conduction (Han et al., 2003) (thermal activation of acceptors) and the 3D Mott's variable range hopping processes. (Ma et al., 2005) According to the fitting results to Equation (2), the obtained nanodot arrays show a dominated hopping process below 10 K. At such a low temperature, the majority of free holes are recaptured by the acceptors. As a result, the free-hole band conduction becomes less important and hole hopping directly between acceptors in the impurity band contributes mostly to the conductivity. (Han et al., 2003) Above 100 K, the conduction is dominated by the thermal activation of the holes (the band conduction). A thermal activation energy (E_a) of 15 meV can be obtained from Equation (2) with $\alpha = 1$ and $E_a = T_0 K_B$, where K_B is the Boltzmann constant. This activation energy does not correspond to any known acceptor energy levels due to Mn doping in Ge, consistent with results shown in Ref. (Pinto et al., 2005a).

To explore practical applications for our extraordinary nanodot arrays, the MR measurements were performed from 2 to 300 K with an external magnetic field up to 10 Tesla. Figure 6b shows the plots of temperature-dependent MR at given magnetic fields (5 and 10 Tesla) for the nanodot arrays. Under a strong magnetic field, the MR in the region of variable range-hopping conduction can be described by (Schlovskii, 1984, Bottger, 1985)

$$MR(H) = \exp[\frac{C}{(\lambda^2 T)^{1/3}}] - 1 \qquad (3)$$

where the magnetic length λ equals to $(c\hbar / eH)^{1/2}$ and C is a field and temperature independent constant. Note that the Equation (3) is only valid in a strong-field limit. (Schlovskii, 1984, Ganesan and Bhat, 2008, Bottger, 1985) The inset in Figure $_6$b shows the best fitting results, in which a linear behavior of MR versus $T^{-1/3}$ is obtained, further confirming the hopping conduction mechanisms ($T \le 8$ K). Note that the absolute values of MRs were used for the fitting purpose. These fitting results are reasonably close to the obtained hopping regions determined from the zero-magnetic-field resistivity measurements ($T \le 10$ K, Figure 6a).

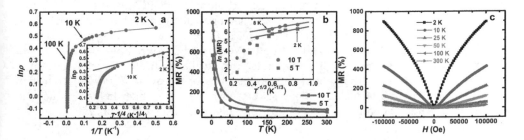

Fig. 6. Magneto-transport measurements for the MnGe nanodot arrays ((A)~(C)) and the nanocolumns ((D)~(F)). For the MnGe nanodot arrays, (A) shows the temperature-dependent resistivity (lnρ vs T^{-1}) and the inset displays the plot of lnρ vs $T^{-1/4}$; (B) is temperature-dependent magneto-resistance (MR) under fixed magnetic fields of 5 and 10 Tesla and inset shows the plot of ln(MR) vs $T^{-1/3}$; (C) demonstrates positive MRs at different temperatures and different magnetic fields.

It is striking to observe that the coherent MnGe nanodot arrays present a large and positive MR up to 900 % at 2 K (Figure 6c). Traditionally, the positive MR is attributed to the Lorentz force in the semiconductor matrix, which deflects the carriers during the transport process. (Ganesan and Bhat, 2008) The resulting MR is positive and proportional to $(\mu H)^2$ under low magnetic fields ($H \le 1$ Tesla in our case) where μ is the semiconductor mobility (units $m^2 V^{-1} S^{-1}$ or T^{-1}) and H is the magnetic field. However, with a simple calculation, the estimated orbital MR is too small to explain the large MR observed from the nanodot arrays. Instead, we anticipate that, besides the effect of orbital MR, the high-density magnetic nanodots could significantly contribute to the large MR ratios due to an enhanced geometric MR effect, from which the current path may be significantly deflected when external magnetic fields were applied to the magnetic nanostructures. (Yuldashev et al., 2001, Solin et al., 2000) To elucidate the underlying physics of the geometrical effect, we consider a thin Hall bar geometry with a measurement current applied in the x-direction, a Hall voltage in the y direction, z direction normal to the sample surface, and an external magnetic field H parallel to z. For semiconductors, the current density and the total electric field can be described by $j = \bar{\sigma} E$, where the magneto-conductivity tensor is given by (Yuldashev et al., 2001, Solin et al., 2000)

$$\bar{\sigma}(H) = \begin{pmatrix} \dfrac{\sigma}{1+\beta^2} & \dfrac{\sigma\beta^2}{1+\beta^2} & 0 \\[3mm] \dfrac{-\sigma\beta^2}{1+\beta^2} & \dfrac{\sigma}{1+\beta^2} & 0 \\[3mm] 0 & 0 & \sigma \end{pmatrix}. \tag{4}$$

Here, $\beta = \mu H$. At zero magnetic field, β vanishes. The conductivity tensor is diagonal when lacking of the magnetic field; and the current density can be simply described by $j = \sigma E$. Since the electric field is normal to the surface of a metallic inclusion and $j \parallel \sigma E$, the current flowing through the material is concentrated into the metallic region which behaves like a "short circuit". As a result, the inclusion of metallic clusters can lead to a higher conduction than that of a homogeneous semiconductor. (Yuldashev et al., 2001, Solin et al., 2000) However, at high magnetic fields ($\beta \gg 1$), the off-diagonal terms of $\bar{\sigma}(H)$ dominate. Equivalently, the Hall angle between j and E approaches 90° ($j \perp E$); and the current becomes tangent to the nanodots. This further indicates that the current is deflected to flow around the nanodots, resembling an "open circuit" state. The transition from the "short circuit" at the zero field to the "open circuit" at high fields produces an increase of resistance, i.e., a positive geometrically-enhanced MR. (Solin et al., 2000)

5.4 Discussions

The introduction of high percentage Mn (x~20%) generates metallic behavior in Mn doped Ge. Such heavily doped samples possess nanoscale MnGe dots. With a superlattice approach, we have successfully fabricated extraordinarily coherent and self-organized MnGe nanodot arrays embedded in the Ge and GaAs matrixes by low temperature MBE. A high yield of such aligned nanodot arrays was confirmed on different substrates, showing an ideal controllability and reproducibility. More importantly, giant positive magneto-resistances were obtained due to the geometrically-enhanced effect. We anticipate that these studies will advance the development of MnGe magnetic semiconductors and/or other similar systems. The obtained coherent and self-assembled nanostructures could be potentially used as the building blocks in the high-density magnetic memories, sensors, MTJs, and other spintronic devices, enabling a new generation of low dissipation magnetoelectronic devices.

6. Summary and prospectives

This chapter is a review of theoretical and experimental progress that has been achieved in understanding ferromagnetism and related electronic properties in the Mn_xGe_{1-x} DMSs and MnGe metallic nanodots. Interest in DMS ferromagnetism is motivated by the possibility to engineer systems that combine many of the technologically useful features of ferromagnetic and semiconducting materials. This goal has been achieved to an impressive degree in (III, Mn)V DMSs, and further progress can be anticipated in the future. However, due to the low T_c of (III, Mn)V DMSs, the spintronics research seemly reaches a critical bottbleneck, where achieving a high T_c DMS becomes an intriguing and challenging task. Fortunately, the Mn_xGe_{1-x} material system offers a possible route towards higher T_c. The structural and magnetic properties can be adjusted simply by modifying the system dimensions from 3-D thin films to 0-D quantum dots. A high T_c in excess of 400 K can be obtained and is presumably attributed to the quantum confinement effect, which strengthens the hybridation between the localized Mn impurities and iterant holes. Bound magnetic polarons may also exist since this system falls into a regime where Mn concentrations are much larger than that of holes. Nevertheless, the high T_c in the low-dimentional system suggests that the quantum structure exibit extraordinary properties that significantly differ from these of bulk films. The future progress would rely on the precise theoretical understanding of the quantum confinement effect on ferromagnetism.

It is well known that in order to achieve functioning spintronic devices working at ambient temperatures, it requires the following criteria: (i) the ferromagnetic transition temperature should safely exceed room temperature, (ii) the mobile charge carriers should respond strongly to changes in the ordered magnetic state, and (iii) the material should retain fundamental semiconductor characteristics, including sensitivity to doping and light, and electric fields produced by gate charges. For more than a decade, these three key issues have been the focus of intense experimental and theoretical research. Progress has been also made in achieving field controlled ferromagnetism in (III, Mn)As system, even though the controllability remains at low temperatures because of low T_c. Therefore, the critical challenge now is either to continue increasing T_c in (III, Mn)As, or to look for a new DMS system with both high Curie temperature (Tc>>300 K) and the field controlled ferromagnetism to satisfy all these three criteria. The experimental results of field controlled ferromagnetism in the Mn$_{0.05}$Ge$_{0.95}$ QDs suggest that the ferromagnetism in this system sensitively responds to the electrical field via the hole mediated effect, similar to that in (III, Mn)As system. Therefore, with a much higher T_c compared with III-V DMS, the Mn$_x$Ge$_{1-x}$ nanostructures could become one of the most promising candidates to achieve room-temperature operation.

On the other hand, the metallic MnGe nanodots with a high percentage of Mn (~20%) show perfect lattice coherence with the Ge matrix. Such a ferrromagnet/semiconudcotr hybrid system is extremely useful owning to their high-quality defect-free interface, where a high efficient spin injection can be realized without a need of oxides such as MgO (spin filter). We also hope to encourage more research efforts in this direction because of the excellent compatibility of MnGe nanodots with the current CMOS technology.

7. References

Ahlers, S., Bougeard, D., Riedl, H., Abstreiter, G., Trampert, A., Kipferl, W., Sperl, M., Bergmaier, A. & Dollinger, G. 2006a. Ferromagnetic Ge(Mn) nanostructures. *Physica E: Low-dimensional Systems and Nanostructures*, 32, 422-425.

Ahlers, S., Bougeard, D., Sircar, N., Abstreiter, G., Trampert, A., Opel, M. & Gross, R. 2006b. Magnetic and structural properties of Ge$_x$Mn$_{1-x}$ films: Precipitation of intermetallic nanomagnets. *Physical Review B (Condensed Matter and Materials Physics)*, 74, 214411.

Akai, H. 1998. Ferromagnetism and Its Stability in the Diluted Magnetic Semiconductor (In, Mn)As. *Physical Review Letters*, 81, 3002-3005.

Appelbaum, I. & Monsma, D. J. 2007. Transit-time spin field-effect transistor. *Applied Physics Letters*, 90, 262501.

Arrott, A. 1957. Criterion for Ferromagnetism from Observations of Magnetic Isotherms. *Physical Review*, 108, 1394-1396.

Awschalom, D. D., Loss, D. & Samarth, N. (eds.) 2002. *Semiconductor spintronics and quantum computation*: Springer-Verlag Berlin Heidelberg New York.

Baik, J. M. & Lee, J. L. 2005. Fabrication of vertically well-aligned (Zn,Mn)O nanorods with room temperature ferromagnetism. *Advanced Materials*, 17, 2745-+.

Berciu, M. & Bhatt, R. N. 2001. Effects of Disorder on Ferromagnetism in Diluted Magnetic Semiconductors. *Physical Review Letters*, 87, 107203.

Bhatt, R. N., Berciu, M., Kennett, M. P. & WAN, X. 2002. Diluted Magnetic Semiconductors in the Low Carrier Density Regime. *Journal of Superconductivity*, 15, 71-83.

Biegger, E., Staheli, L., Fonin, M., Rudiger, U. & Dedkov, Y. S. 2007. Intrinsic ferromagnetism versus phase segregation in Mn-doped Ge. *Journal of Applied Physics*, 101, 103912-5.

Bihler, C., Jaeger, C., Vallaitis, T., Gjukic, M., Brandt, M. S., Pippel, E., Woltersdorf, J. & Gosele, U. 2006. Structural and magnetic properties of Mn_5Ge_3 clusters in a dilute magnetic germanium matrix. *Applied Physics Letters*, 88, 112506.

Bolduc, M., Awo-Affouda, C., Stollenwerk, A., Huang, M. B., Ramos, F. G., Agnello, G. & Labella, V. P. 2005. Above room temperature ferromagnetism in Mn-ion implanted Si. *Physical Review B*, 71, 033302.

Bottger, H., Bryksin, V. 1985. *Hopping conduction in solids.*, Berlin, Akademie-Verlag.

Bougeard, D., Ahlers, S., Trampert, A., Sircar, N. & Abstreiter, G. 2006. Clustering in a Precipitate-Free GeMn Magnetic Semiconductor. *Physical Review Letters*, 97, 237202.

Boukari, H., Kossacki, P., Bertolini, M., Ferrand, D., Cibert, J., Tatarenko, S., Wasiela, A., Gaj, J. A. & Dietl, T. 2002. Light and Electric Field Control of Ferromagnetism in Magnetic Quantum Structures. *Physical Review Letters*, 88, 207204.

Brieler, F. J., Grundmann, P., Froba, M., Chen, L. M., Klar, P. J., Heimbrodt, W., Von Nidda, H. A. K., Kurz, T. & Loidl, A. 2004. Formation of Zn1-xMnxS nanowires within mesoporous silica of different pore sizes. *Journal Of The American Chemical Society*, 126, 797-807.

Brunner, K. 2002b. Si/Ge nanostructures. *Reports on Progress in Physics*, 65, 27-72.

Chang, Y. Q., Wang, D. B., Luo, X. H., Xu, X. Y., Chen, X. H., LI, L., Chen, C. P., Wang, R. M., Xu, J. & Yu, D. P. 2003. Synthesis, optical, and magnetic properties of diluted magnetic semiconductor Zn_{1-x} Mn_xO nanowires via vapor phase growth. *Applied Physics Letters*, 83, 4020-4022.

Chen, H., Zhu, W., Kaxiras, E. & Zhang, Z. 2009. Optimization of Mn doping in group-IV-based dilute magnetic semiconductors by electronic codopants. *Physical Review B (Condensed Matter and Materials Physics)*, 79, 235202.

Chen, J., Wang, K. L. & Galatsis, K. 2007a. Electrical field control magnetic phase transition in nanostructured Mn_xGe_{1-x}. *Applied Physics Letters*, 90, 012501.

Chen, Y. F., Lee, W. N., Huang, J. H., Chin, T. S., Huang, R. T., Chen, F. R., Kai, J. J., Aravind, K., Lin, I. N. & Ku, H. C. 2005. Growth and magnetic properties of self-assembled (In, Mn)As quantum dots. *Journal of Vacuum Science & Technology B: Microelectronics and Nanometer Structures*, 23, 1376-1378.

Chen, Y. X., Yan, S.-S., Fang, Y., Tian, Y. F., Xiao, S. Q., Liu, G. L., Liu, Y. H. & Mei, L. M. 2007b. Magnetic and transport properties of homogeneous Mn_xGe_{1-x} ferromagnetic semiconductor with high Mn concentration. *Applied Physics Letters*, 90, 052508.

Chiba, D., Matsukura, F. & Ohno, H. 2006a. Electric-field control of ferromagnetism in (Ga,Mn)As. *Applied Physics Letters,* 89, 162505.

Chiba, D., Matsukura, F. & Ohno, H. 2006b. Electrical magnetization reversal in ferromagnetic III-V semiconductors. *Journal of Physics D: Applied Physics,* 39, R215-R225.

Chiba, D., Sawicki, M., Nishitani, Y., Nakatani, Y., Matsukura, F. & Ohno, H. 2008. Magnetization vector manipulation by electric fields. *Nature,* 455, 515-518.

Cho, S., Choi, S., Hong, S. C., Kim, Y., Ketterson, J. B., Kim, B.-J., Kim, Y. C. & Jung, J.-H. 2002. Ferromagnetism in Mn-doped Ge. *Physical Review B,* 66, 033303.

Cho, Y. J., Kim, C. H., Kim, H. S., Lee, W. S., Park, S.-H., Park, J., Bae, S. Y., Kim, B., Lee, H. & Kim, J.-Y. 2008. Ferromagnetic Ge$_{1-x}$M$_x$ (M = Mn, Fe, and Co) Nanowires. *Chemistry of Materials,* 20, 4694-4702.

Cho, Y. M., Yu, S. S., Ihm, Y. E., Lee, S. W., Kim, D., Kim, H., Sohn, J. M., Kim, B. G., Kang, Y. H., Oh, S., Kim, C. S. & Lee, H. J. 2006. Neutron irradiation effects on polycrystalline Ge1-xMnx thin films grown by MBE. *Current Applied Physics,* 6, 482-485.

Choi, H.-J., Seong, H.-K., Chang, J., Lee, K.-I., Park, Y.-J., Kim, J.-J., Lee, S.-K., He, R., Kuykendall, T. & Yang, P. 2005a. Single-Crystalline Diluted Magnetic Semiconductor GaN:Mn Nanowires. *Advanced Materials,* 17, 1351-1356.

Choi, H. J., Seong, H. K., Chang, J., Lee, K. I., Park, Y. J., Kim, J. J., Lee, S. K., He, R. R., Kuykendall, T. & Yang, P. D. 2005b. Single-crystalline diluted magnetic semiconductor GaN : Mn nanowires. *Advanced Materials,* 17, 1351-+.

Collins, B. A., Chu, Y. S., He, L., Zhong, Y. & Tsui, F. 2008. Dopant stability and strain states in Co and Mn-doped Ge (001) epitaxial films. *Physical Review B (Condensed Matter and Materials Physics),* 77, 193301.

Cui, J. B. & Gibson, U. J. 2005. Electrodeposition and room temperature ferromagnetic anisotropy of Co and Ni-doped ZnO nanowire arrays. *Applied Physics Letters,* 87, 133108.

D'orazio, F., Lucari, F., Passacantando, M., Santucci, P. P. S. & Verna, A. Magneto-optical study of Mn ions implanted in Ge. Magnetics Conference, 2002. INTERMAG Europe 2002. Digest of Technical Papers. 2002 IEEE International, 2002. BB9.

D'orazio, F., Lucari, F., Pinto, N., Morresi, L. & Murri, R. 2004. Toward room temperature ferromagnetism of Ge:Mn systems. *Journal of Magnetism and Magnetic Materials,* 272-276, 2006-2007.

D'orazio, F., Lucari, F., Santucci, S., Picozzi, P., Verna, A., Passacantando, M., Pinto, N., Morresi, L., Gunnella, R. & Murri, R. 2003. Magneto-optical properties of epitaxial MnxGe1-x films. *Journal of Magnetism and Magnetic Materials,* 262, 158-161.

Datta, S. & Das, B. 1990. Electronic analog of the electro-optic modulator. *Applied Physics Letters,* 56, 665-667.

Demidov, E., Danilov, Y., Podol'skiĭ, V., Lesnikov, V., Sapozhnikov, M. & Suchkov, A. 2006. Ferromagnetism in epitaxial germanium and silicon layers supersaturated with managanese and iron impurities. *JETP Letters,* 83, 568-571.

Devillers, T., Jamet, M., Barski, A., Poydenot, V., Bayle-Guillemaud, P., Bellet-Amalric, E., Cherifi, S. & Cibert, J. 2007. Structure and magnetism of self-organized Ge$_{1-x}$Mn$_x$ nanocolumns on Ge(001). *Physical Review B (Condensed Matter and Materials Physics),* 76, 205306.

Dietl, T. 2006. Self-organized growth controlled by charge states of magnetic impurities. *Nature Materials*, 5, 673-673.

Dietl, T., Awschalom, D.D., Kaminska, M., & Ohno, H. 2008. *Spintronics*, Elsevier

Dietl, T. & Ohno, H. 2006. Engineering magnetism in semiconductors. *Materials Today*, 9, 18-26.

Dietl, T., Ohno, H. & Matsukura, F. 2001. Hole-mediated ferromagnetism in tetrahedrally coordinated semiconductors. *Physical Review B*, 63, 195205.

Dietl, T., Ohno, H., Matsukura, F., Cibert, J. & Ferrand, D. 2000. Zener Model Description of Ferromagnetism in Zinc-Blende Magnetic Semiconductors. *Science*, 287, 1019-1022.

Erwin, S. C. & Petukhov, A. G. 2002. Self-Compensation in Manganese-Doped Ferromagnetic Semiconductors. *Physical Review Letters*, 89, 227201.

Fukushima, T., Sato, K., Katayama-Yoshida, H. & Dederichs, P. H. 2006. Spinodal decomposition under layer by layer growth condition and high curie temperature quasi-one-dimensional nano-structure in dilute magnetic semiconductors. *Japanese Journal of Applied Physics Part 2-Letters & Express Letters*, 45, L416-L418.

Gambardella, P., Brune, H., Dhesi, S. S., Bencok, P., Krishnakumar, S. R., Gardonio, S., Veronese, M., Grazioli, C. & Carbone, C. 2005. Paramagnetic Mn impurities on Ge and GaAs surfaces. *Physical Review B*, 72, 045337.

Gambardella, P., Claude, L., Rusponi, S., Franke, K. J., Brune, H., Raabe, J., Nolting, F., Bencok, P., Hanbicki, A. T., Jonker, B. T., Grazioli, C., Veronese, M. & Carbone, C. 2007. Surface characterization of Mn_xGe_{1-x} and $Cr_yMn_xGe_{1-x-y}$ dilute magnetic semiconductors. *Physical Review B (Condensed Matter and Materials Physics)*, 75, 125211.

Ganesan, K. & Bhat, H. L. 2008. Growth, magnetotransport, and magnetic properties of ferromagnetic (In,Mn)Sb crystals. *Journal of Applied Physics*, 103, 6.

Gareev, R. R., Bugoslavsky, Y. V., Schreiber, R., Paul, A., Sperl, M. & Doppe, M. 2006. Carrier-induced ferromagnetism in Ge(Mn,Fe) magnetic semiconductor thin-film structures. *Applied Physics Letters*, 88, 222508.

Garzon, S., Zutic, I. & Webb, R. A. 2005. Temperature-Dependent Asymmetry of the Nonlocal Spin-Injection Resistance: Evidence for Spin Nonconserving Interface Scattering. *Physical Review Letters*, 94, 176601.

Gunnella, R., Morresi, L., Pinto, N., Murri, R., Ottaviano, L., Passacantando, M., D'orazio, F. & Lucari, F. 2005. Magnetization of epitaxial MnGe alloys on Ge(1??) substrates. *Surface Science*, 577, 22-30.

Guoqiang, Z., Tateno, K., Sogawa, T. & Nakano, H. 2008. Vertically aligned GaP/GaAs core-multishell nanowires epitaxially grown on Si substrate. *Applied Physics Express*, 064003 (3 pp.).

Han, J. P., Shen, M. R., Cao, W. W., Senos, A. M. R. & Mantas, P. Q. 2003. Hopping conduction in Mn-doped ZnO. *Applied Physics Letters*, 82, 67-69.

Holub, M., Chakrabarti, S., Fathpour, S., Bhattacharya, P., Lei, Y. & Ghosh, S. 2004. Mn-doped InAs self-organized diluted magnetic quantum-dot layers with Curie temperatures above 300 K. *Applied Physics Letters*, 85, 973-975.

Jaeger, C., Bihler, C., Vallaitis, T., Goennenwein, S. T. B., Opel, M., Gross, R. & Brandt, M. S. 2006. Spin-glass-like behavior of Ge:Mn. *Physical Review B (Condensed Matter and Materials Physics)*, 74, 045330.

Jamet, M., Barski, A., Devillers, T., Poydenot, V., Dujardin, R., Bayle-Guillemaud, P., Rothman, J., Bellet-Amalric, E., Marty, A., Cibert, J., Mattana, R. & Tatarenko, S.

2006. High-Curie-temperature ferromagnetism in self-organized Ge$_{1-x}$Mn$_x$ nanocolumns. *Nat Mater,* 5, 653-659.

Jedema, F. J., Heersche, H. B., Filip, A. T., Baselmans, J. J. A. & Van Wees, B. J. 2002. Electrical detection of spin precession in a metallic mesoscopic spin valve. *Nature,* 416, 713-716.

Jeon, H. C., Chung, K. J., Chung, K. J., Kang, T. W. & Kim, T. W. 2004. Enhancement of the ferromagnetic transition temperature in self-assembled (Ga$_{1-x}$Mn$_x$)As quantum wires. *Japanese Journal Of Applied Physics Part 2-Letters & Express Letters,* 43, L963-L965.

Jeon, H. C., Jeong, Y. S., Kang, T. W., Kim, T. W., Chung, K. J., Jhe, W. & Song, S. A. 2002. (In$_{1-x}$Mn$_x$)As Diluted Magnetic Semiconductor Quantum Dots with Above Room Temperature Ferromagnetic Transition. *Advanced Materials,* 14, 1725-1728.

Ji, Y., Hoffmann, A., Jiang, J. S. & Bader, S. D. 2004. Spin injection, diffusion, and detection in lateral spin-valves. *Applied Physics Letters,* 85, 6218-6220.

Ji, Y., Hoffmann, A., Jiang, J. S., Pearson, J. E. & Bader, S. D. 2007. Non-local spin injection in lateral spin valves. *Journal of Physics D: Applied Physics,* 5.

Johnson, M. & Silsbee, R. H. 1985. Interfacial charge-spin coupling: Injection and detection of spin magnetization in metals. *Physical Review Letters,* 55, 1790.

Jungwirth, T., Sinova, J., Masek, J., Kucera, J. & Macdonald, A. H. 2006. Theory of ferromagnetic (III,Mn)V semiconductors. *Reviews of Modern Physics,* 78, 809-864.

Kanki, T., Tanaka, H. & Kawai, T. 2006. Electric control of room temperature ferromagnetism in a Pb(Zr$_{0.2}$Ti$_{0.8}$)O$_3$/La$_{0.85}$Ba$_{0.15}$MnO$_3$ field-effect transistor. *Applied Physics Letters,* 89, 242506.

Kazakova, O., Kulkarni, J. S., Holmes, J. D. & Demokritov, S. O. 2005. Room-temperature ferromagnetism in Ge$_{1-x}$Mn$_x$ nanowires. *Physical Review B,* 72, 094415.

Kelly, D., Wegrowe, J.-E., Truong, T.-K., Hoffer, X. & Ansermet, J.-P. 2003. Spin-polarized current-induced magnetization reversal in single nanowires. *Physical Review B,* 68, 134425.

Kimura, T. & Otani, Y. 2008. Local domain structure of exchange-coupled NiFe/CoO nanowire probed by nonlocal spin valve measurement. *Journal of Applied Physics,* 103, 083915.

Kioseoglou, G., Hanbicki, A. T., Sullivan, J. M., Van 'T Erve, O. M. J., Li, C. H., Erwin, S. C., Mallory, R., Yasar, M., Petrou, A. & Jonker, B. T. 2004. Electrical spin injection from an n-type ferromagnetic semiconductor into a III-V device heterostructure. *Nature Materials,* 3, 799-803.

Koo, H. C., Kwon, J. H., Eom, J., Chang, J., Han, S. H. & Johnson, M. 2009. Control of Spin Precession in a Spin-Injected Field Effect Transistor. *Science,* 325, 1515-1518.

Kulkarni, J. S., Kazakova, O., Erts, D., Morris, M. A., Shaw, M. T. & Holmes, J. D. 2005. Structural and Magnetic Characterization of Ge0.99Mn0.01 Nanowire Arrays. *Chemistry of Materials,* 17, 3615-3619.

Kuroda, S., Nishizawa, N., Takita, K., Mitome, M., Bando, Y., Osuch, K. & Dietl, T. 2007. Origin and control of high-temperature ferromagnetism in semiconductors. *Nature Materials,* 6, 440-446.

Lauhon, L. J., Gudiksen, M. S., Wang, D. & Lieber, C. M. 2002. Epitaxial core-shell and core-multishell nanowire heterostructures. *Nature,* 420, 57-61.

Li, A. P., Wendelken, J. F., Shen, J., Feldman, L. C., Thompson, J. R. & Weitering, H. H. 2005. Magnetism in Mn_xGe_{1-x} semiconductors mediated by impurity band carriers. *Physical Review B,* 72, 195205.

Li, A. P., Zeng, C., Benthem, K. V., Chisholm, M. F., Shen, J., Rao, S. V. S. N., Dixit, S. K., Feldman, L. C., Petukhov, A. G., Foygel, M. & Weitering, H. H. 2007. Dopant segregation and giant magnetoresistance in manganese-doped germanium. *Physical Review B (Condensed Matter and Materials Physics),* 75, 201201.

Li, H., Wu, Y., Guo, Z., Luo, P. & Wang, S. 2006a. Magnetic and electrical transport properties of $Ge_{1-x}Mn_x$ thin films. *Journal of Applied Physics,* 100, 103908.

Li, H., Wu, Y., Liu, T., Wang, S., Guo, Z. & Osipowicz, T. 2006b. Magnetic and transport properties of Ge:Mn granular system. *Thin Solid Films,* 505, 54-56.

Liao, X. Z., Zou, J., Cockayne, D. J. H., Wan, J., Jiang, Z. M., Jin, G. & Wang, K. L. 2001. Annealing effects on the microstructure of Ge/Si(001) quantum dots. *Applied Physics Letters,* 79, 1258-1260.

Lifeng, L., Nuofu, C., Chenlong, C., Yanli, L., Zhigang, Y. & Fei, Y. 2004. Magnetic properties of Mn-implanted n-type Ge. *Journal of Crystal Growth,* 273, 106-110.

Lin, H.-T., Huang, W.-J., Wang, S.-H., Lin, H.-H. & Chin, T.-S. 2008. Carrier-mediated ferromagnetism in p-Si(100) by sequential ion-implantation of B and Mn. *Journal of Physics: Condensed Matter,* 9, 095004.

Liu, C., Yun, F. & Morko , H. 2005. Ferromagnetism of ZnO and GaN: A Review. *Journal of Materials Science: Materials in Electronics,* 16, 555-597.

Liu, H. & Reinke, P. 2008. Formation of manganese nanostructures on the Si(100)-(2X1) surface. *Surface Science,* 602, 986-992.

Liu, L., Chen, N., Wang, Y., Zhang, X., Yin, Z., Yang, F. & Chai, C. 2006. Growth and properties of magnetron cosputtering grown Mn_xGe_{1-x} on Si (001). *Solid State Communications,* 137, 126-128.

Liu, L., Chen, N., Yin, Z., Yang, F., Zhou, J. & Zhang, F. 2004. Investigation of Mn-implanted n-type Ge. *Journal of Crystal Growth,* 265, 466-470.

Lou, X., Adelmann, C., Crooker, S. A., Garlid, E. S., Zhang, J., Reddy, K. S. M., Flexner, S. D., Palmstrom, C. J. & Crowell, P. A. 2007. Electrical detection of spin transport in lateral ferromagnet-semiconductor devices. *Nat Phys,* 3, 197-202.

Lu, W. & Lieber, C. M. 2006. Semiconductor nanowires. *Journal of Physics D: Applied Physics,* 21.

Luo, X., Zhang, S. B. & Wei, S.-H. 2004. Theory of Mn supersaturation in Si and Ge. *Physical Review B,* 70, 033308.

Lyu, P. & Moon, K. 2003. Ferromagnetism in diluted magnetic semiconductor quantum dot arrays embedded in semiconductors. *The European Physical Journal B - Condensed Matter and Complex Systems,* 36, 593-598.

Ma, Y. J., Zhang, Z., Zhou, F., Lu, L., Jin, A. Z. & Gu, C. Z. 2005. Hopping conduction in single ZnO nanowires. *Nanotechnology,* 16, 746-749.

Maekawa, S. 2004. Spin-dependent transport in magnetic nanostructures. *Journal of Magnetism and Magnetic Materials,* 272-276, E1459-E1463.

Maekawa, S. 2006. *Concepts in Spin Electronics,* Oxford University Press.

Majumdar, S., Das, A. K. & Ray, S. K. 2009a. Magnetic semiconducting diode of p-Ge_{1-x} Mn_x/n-Ge layers on silicon substrate. *Applied Physics Letters,* 94, 122505.

Majumdar, S., Mandal, S., Das, A. K. & Ray, S. K. 2009b. Synthesis and temperature dependent photoluminescence properties of Mn doped Ge nanowires. *Journal of Applied Physics*, 105, 024302.

Matsukura, F., Ohno, H., Shen, A. & Sugawara, Y. 1998. Transport properties and origin of ferromagnetism in (Ga,Mn)As. *Physical Review B*, 57, R2037-R2040.

Miyoshi, T., Matsui, T., Tsuda, H., Mabuchi, H. & Morii, K. 1999. Magnetic and electric properties of Mn$_5$Ge$_3$/Ge nanostructured films. 85, 5372-5374.

Morresi, L., Pinto, N., Ficcadenti, M., Murri, R., D'orazio, F. & Lucari, F. 2006. Magnetic and transport polaron percolation in diluted GeMn films. *Materials Science and Engineering: B*, 126, 197-201.

Nazmul, A. M., Kobayashi, S. & Tanaka, S. S. M. 2004. Electrical and Optical Control of Ferromagnetism in III-V Semiconductor Heterostructures at High Temperature (~100 K). *Jpn. J. Appl. Phys.*, 43.

Nepal, N., Luen, M. O., Zavada, J. M., Bedair, S. M., Frajtag, P. & El-Masry, N. A. 2009. Electric field control of room temperature ferromagnetism in III-N dilute magnetic semiconductor films. *Applied Physics Letters*, 94, 132505.

Ogawa, M., Han, X., Zhao, Z., Wang, Y., Wang, K. L. & Zou, J. 2009. Mn distribution behaviors and magnetic properties of GeMn films grown on Si (001) substrates. *Journal of Crystal Growth*, 311, 2147-2150.

Ohno, H., Chiba, D., Matsukura, F., Omiya, T., Abe, E., Dietl, T., Ohno, Y. & Ohtani, K. 2000. Electric-field control of ferromagnetism. *Nature*, 408, 944-946.

Ohno, Y., Young, D. K., Beschoten, B., Matsukura, F., Ohno, H. & Awschalom, D. D. 1999. Electrical spin injection in a ferromagnetic semiconductor heterostructure. *Nature*, 402, 790-792.

Ottaviano, L., Passacantando, M., Verna, A., Parisse, P., Picozzi, S., Impellizzeri, G. & PRIOLO, F. 2007. Microscopic investigation of the structural and electronic properties of ion implanted Mn-Ge alloys. *physica status solidi (a)*, 204, 136-144.

Park, E. S., Kim, D. H. & Kim, W. T. 2005. Parameter for glass forming ability of ternary alloy systems. *Applied Physics Letters*, 86, 061907.

Park, Y. D., Hanbicki, A. T., Erwin, S. C., Hellberg, C. S., Sullivan, J. M., Mattson, J. E., Ambrose, T. F., Wilson, A., Spanos, G. & Jonker, B. T. 2002. A Group-IV Ferromagnetic Semiconductor: Mn$_x$Ge$_{1-x}$. *Science*, 295, 651-654.

Park, Y. D., Wilson, A., Hanbicki, A. T., Mattson, J. E., Ambrose, T., Spanos, G. & Jonker, B. T. 2001. Magnetoresistance of Mn:Ge ferromagnetic nanoclusters in a diluted magnetic semiconductor matrix. *Applied Physics Letters*, 78, 2739-2741.

Passacantando, M., Ottaviano, L., D'orazio, F., Lucari, F., Biase, M. D., Impellizzeri, G. & Priolo, F. 2006. Growth of ferromagnetic nanoparticles in a diluted magnetic semiconductor obtained by Mn$^+$ implantation on Ge single crystals. *Physical Review B (Condensed Matter and Materials Physics)*, 73, 195207.

Passacantando, M., Ottaviano, L., Grossi, V., Verna, A., D'orazio, F., Lucari, F., Impellizzeri, G. & Priolo, F. 2007. Magnetic response of Mn-doped amorphous porous Ge fabricated by ion-implantation. *Nuclear Instruments and Methods in Physics Research Section B: Beam Interactions with Materials and Atoms*, 257, 365-368.

Patibandla, S., Pramanik, S., Bandyopadhyay, S. & Tepper, G. C. 2006. Spin relaxation in a germanium nanowire. *Journal of Applied Physics*, 100, 044303.

Paul, A. & Sanyal, B. 2009. Chemical and magnetic interactions in Mn- and Fe-codoped Ge diluted magnetic semiconductors. *Physical Review B (Condensed Matter and Materials Physics)*, 79, 214438.

Peressi, M., Debernardi, A., Picozzi, S., Antoniella, F. & Continenza, A. 2005. Half-metallic Mn-doped Si_xGe_{1-x} alloys: a first principles study. *Computational Materials Science*, 33, 125-131.

Philip, J., Punnoose, A., Kim, B. I., Reddy, K. M., Layne, S., Holmes, J. O., Satpati, B., Leclair, P. R., Santos, T. S. & Moodera, J. S. 2006. Carrier-controlled ferromagnetism in transparent oxide semiconductors. *Nat Mater*, 5, 298-304.

Picozzi, S. & Lezaic, M. 2008. Ab-initio study of exchange constants and electronic structure in diluted magnetic group-IV semiconductors. *New Journal of Physics*, 055017.

Pinto, N., Morresi, L., Ficcadenti, M., Murri, R., D'orazio, F., Lucari, F., Boarino, L. & Amato, G. 2005a. Magnetic and electronic transport percolation in epitaxial $Ge_{1-x}Mn_x$ films. *Physical Review B*, 72, 165203

Pinto, N., Morresi, L., Ficcadenti, M., Murri, R., D.Orazio, F., Lucari, F., Boarino, L. & AMATO, G. 2005b. Magnetic and electronic transport percolation in epitaxial $Ge_{1-x}Mn_x$ films. *Physical Review B*, 72, 165203.

Pinto, N., Morresi, L., Gunnella, R., Murri, R., D'orazio, F., Lucari, F., Santucci, S., PICOZZI, P., Passacantando, M. & Verna, A. 2003. Growth and magnetic properties of MnGe films for spintronic application. *Journal of Materials Science: Materials in Electronics*, 14, 337-340.

Poli, N., Urech, M., Korenivski, V. & Haviland, D. B. Spin-flip scattering at Al surfaces. 2006. AIP, 08H701.

Pramanik, S., Bandyopadhyay, S. & Cahay, M. Spin transport in nanowires. Nanotechnology, 2003. IEEE-NANO 2003. 2003 Third IEEE Conference on, 2003. 87-90 vol.2.

Prinz, G. A. 1998. Magnetoelectronics. *Science*, 282, 1660-1663.

Radovanovic, P. V., Barrelet, C. J., Gradecak, S., Qian, F. & Lieber, C. M. 2005. General synthesis of manganese-doped II-VI and III-V semiconductor nanowires. *Nano Letters*, 5, 1407-1411.

Rheem, Y., Yoo, B.-Y., Beyermann, W. P. & Myung, N. V. 2007. Magneto-transport studies of single ferromagnetic nanowire. *Physica status solidi (a)*, 204, 4004-4008.

Schilfgaarde, M. V. & Mryasov, O. N. 2001. Anomalous exchange interactions in III-V dilute magnetic semiconductors. *Physical Review B*, 63, 233205.

Schlovskii, B. I., Efros, A. L. 1984. *Electronics properties of doped semiconductors* Berlin, Spinger-Verlag.

Schlovskii, B. I. & L., E. A. 1984. *Electronics properties of doped semiconductors*, Spinger-Verlag.

Schulthess, T. C. & Butler, W. H. 2001. Electronic structure and magnetic interactions in Mn doped semiconductors. *Jounal of Applied Physics*, 89, 7021-7023.

Seong, H.-K., Kim, U., Jeon, E.-K., Park, T.-E., Oh, H., Lee, T.-H., Kim, J.-J., Choi, H.-J. & Kim, J.-Y. 2009. Magnetic and Electrical Properties of Single-Crystalline Mn-Doped Ge Nanowires. *The Journal of Physical Chemistry C*, 113, 10847-10852.

Slater, J. C. 1964. Atomic Radii In Crystals. *Journal of Chemical Physics*, 41, 3199-&.

Solin, S. A., Thio, T., Hines, D. R. & Heremans, J. J. 2000. Enhanced room-temperature geometric magnetoresistance in inhomogeneous narrow-gap semiconductors. *Science*, 289, 1530-1532.

Solomon, G. S., Trezza, J. A., Marshall, A. F. & Harris, J. S. 1996. Vertically aligned and electronically coupled growth induced InAs islands in GaAs. *Physical Review Letters*, 76, 952-955.

Stroppa, A., Picozzi, S., Continenza, A. & Freeman, A. J. 2003. Electronic structure and ferromagnetism of Mn-doped group-IV semiconductors. *Physical Review B*, 68, 155203.

Sugahara, S., Lee, K. L., Yada, S. & Tanaka, M. 2005. Precipitation of amorphous ferromagnetic semiconductor phase in epitaxially grown Mn-doped Ge thin films. *Japanese Journal of Applied Physics Part 2-Letters & Express Letters*, 44, L1426-L1429.

Sze, S. M. 1981. *Physics of Semiconductor Devices*, New York, Wiley.

Tanaka, M. 2002. Ferromagnet (MnAs)/III-V semiconductor hybrid structures. *Semiconductor Science and Technology*, 17, 327-341.

Tsuchida, R., Asubar, J. T., Jinbo, Y. & Uchitomi, N. 2009. MBE growth and properties of GeMn thin films on (001) GaAs. *Journal of Crystal Growth*, 311, 937-940.

Tsui, F., Collins, B. A., He, L., Mellnik, A., Zhong, Y., Vogt, S. & Chu, Y. S. 2007. Combinatorial synthesis and characterization of a ternary epitaxial film of Co and Mn doped Ge (001). *Applied Surface Science*, 254, 709-713.

Tsui, F., He, L., Ma, L., Tkachuk, A., Chu, Y. S., Nakajima, K. & Chikyow, T. 2003. Novel Germanium-Based Magnetic Semiconductors. *Physical Review Letters*, 91, 177203.

Valenzuela, S. O. & Tinkham, M. 2004. Spin-polarized tunneling in room-temperature mesoscopic spin valves. *Applied Physics Letters*, 85, 5914-5916.

Van Der Meulen, M. I., Petkov, N., Morris, M. A., Kazakova, O., Han, X., Wang, K. L., Jacob, A. P. & Holmes, J. D. 2008. Single Crystalline $Ge_{1-x}Mn_x$ Nanowires as Building Blocks for Nanoelectronics. *Nano Letters*, 9, 50-56.

Verna, A., D'orazio, F., Ottaviano, L., Passacantando, M., Lucari, F., Impellizzeri, G. & Priolo, F. 2007. Magneto-optical investigation of high temperature ion implanted Mn_xGe_{1-x} alloy: evidence for multiple contributions to the magnetic response. *physica status solidi (a)*, 204, 145-151.

Verna, A., Ottaviano, L., Passacantando, M., Santucci, S., Picozzi, P., D'orazio, F., Lucari, F., Biase, M. D., Gunnella, R., Berti, M., Gasparotto, A., Impellizzeri, G. & Priolo, F. 2006. Ferromagnetism in ion implanted amorphous and nanocrystalline Mn_xGe_{1-x}. *Physical Review B (Condensed Matter and Materials Physics)*, 74, 085204.

Wang, H.-Y. & Qian, M. C. Electronic and magnetic properties of Mn/Ge digital ferromagnetic heterostructures: An ab initio investigation. 2006. Journal of Applied Physics, 99, 08D705.

Wang, K. L., Thomas, S. G. & Tanner, M. O. 1995. SiGe band engineering for MOS, CMOS and quantum effect devices. *Journal of Materials Science: Materials in Electronics*, 6, 311-324.

Wang, Y., Xiu, F., Wang, Y., Zou, J., Beyermann, W., Zhou, Y. & Wang, K. 2011. Coherent magnetic semiconductor nanodot arrays. *Nanoscale Research Letters*, 6, 134.

Wang, Y., Zou, J., Zhao, Z., Han, X., Zhou, X. & Wang, K. L. 2008a. Direct structural evidences of $Mn_{11}Ge_8$ and Mn_5Ge_2 clusters in $Ge_{0.96}Mn_{0.04}$ thin films. *Applied Physics Letters*, 92, 101913.

Wang, Y., Zou, J., Zhao, Z., Han, X., Zhou, X. & Wang, K. L. 2008b. Mn behavior in $Ge_{0.96}Mn_{0.04}$ magnetic thin films grown on Si. *Journal of Applied Physics*, 103, 066104.

Weisheit, M., Fahler, S., Marty, A., Souche, Y., Poinsignon, C. & Givord, D. 2007. Electric Field-Induced Modification of Magnetism in Thin-Film Ferromagnets. *Science*, 315, 349-351.

Xie, Q. H., Madhukar, A., Chen, P. & Kobayashi, N. P. 1995. Vertically Self-Organized INAS Quantum Box Islands on Gaas (100). *Physical Review Letters*, 75, 2542-2545.

Xiu, F., Wang, Y., Kim, J., Hong, A., Tang, J., Jacob, A. P., Zou, J. & Wang, K. L. 2010. Electric-field-controlled ferromagnetism in high-Curie-temperature $Mn_{0.05}Ge_{0.95}$ quantum dots. *Nature Materials*, 9, 337-344.

Yada, S., Sugahara, S. & Tanaka, M. 2008. Magneto-optical and magnetotransport properties of amorphous ferromagnetic semiconductor $Ge_{1-x}Mn_x$ thin films. *Applied Physics Letters*, 93, 193108.

Yagi, M., Noba, K.-I. & Kayanuma, Y. 2001. Self-consistent theory for ferromagnetism induced by photo-excited carriers. *Journal of Luminescence*, 94-95, 523-527.

Yu, S. S., Anh, T. T. L., Ihm, Y. E., Kim, D., Kim, H., Hong, S. K., Oh, S., Kim, C. S., Lee, H. J. & Woo, B. C. 2006. Magneto-transport properties of amorphous $Ge_{1-x}Mn_x$ thin films. *Current Applied Physics*, 6, 545-548.

Yuldashev, S. U., Shon, Y., Kwon, Y. H., Fu, D. J., Kim, D. Y., Kim, H. J., Kang, T. W. & Fan, X. 2001. Enhanced positive magnetoresistance effect in GaAs with nanoscale magnetic clusters. *Journal of Applied Physics*, 90, 3004-3006.

Zeng, C., Yao, Y., Niu, Q. & Weitering, H. H. 2006. Linear Magnetization Dependence of the Intrinsic Anomalous Hall Effect. *Physical Review Letters*, 96, 037204.

Zhao, Y.-J., Shishidou, T. & Freeman, A. J. 2003. Ruderman-Kittel-Kasuya-Yosida–like Ferromagnetism in MnxGe1-x. *Physical Review Letters*, 90, 047204.

Zheng, Y. H., Zhao, J. H., Bi, J. F., Wang, W. Z., JI, Y., Wu, X. G. & Xia, J. B. 2007. Cr-doped InAs self-organized diluted magnetic quantum dots with room-temperature ferromagnetism. *Chinese Physics Letters*, 24, 2118-2121.

Zhu, W., Weitering, H. H., Wang, E. G., Kaxiras, E. & Zhang, Z. 2004. Contrasting Growth Modes of Mn on Ge(100) and Ge(111) Surfaces: Subsurface Segregation versus Intermixing. *Physical Review Letters*, 93, 126102.

10

Spin-Based Quantum Dot Qubits

V. N. Stavrou[1] and G. P. Veropoulos[2]
[1]Department of Physics and Astronomy, University of Iowa, Iowa City
[2]Division of Physics, Hellenic Naval Academy, Hadjikyriakou,
and Division of Academic Studies, Hellenic Navy Petty Officers Academy, Skaramagkas,
[1]USA
[2]Greece

1. Introduction

Spin-based electronics in low dimensional structure (LDS) e.g. quantum wells (QWs) and quantum dots (QDs) (especially in QDs) have recently attracted a lot of interest for applications in quantum information technology (Burkard et al., 1999; Imamoglu et al., 1999; Nielsen & Chuang, 2000). Two-dimensional (2-d) QDs (Baruffa et al., 2010; Bertoni et al., 2005; Fessatidis et al., 1999; Foulkes et al., 2001; Hayashi et al., 2004; Hu & Sarma, 2000; 2001; Räsänen et al., 2003; Simserides et al., 2000; 2006; 2007; Stano & Fabian, 2005; 2006; Thorwart et al., 2005; Witzel et al., 2007) and self-assembled QDs (SAQDs) (Artús et al., 2000; Bányai & Koch, 1993; Fujisawa et al., 1998; 2006; Lee et al., 1998; 2004; Pryor et al., 1997; Pryor, 1998; Pryor & Flatté, 2003) among other heterostructures have been suggested for constructing quantum bits (qubits). Quantum computing architecture mainly uses the properties of QDs related to charge and spin of the confined carriers, among others. Thus, the carrier relaxation via the emission of phonons (Khaetskii & Nazarov, 2001; Mohanty, 2000; Stavrou & Hu, 2005; 2006; Stavrou, 2007; Woods et al., 2002) and the light polarization interplay with the spin-polarized states within QDs (Chye et al., 2002; Pryor & Flatté, 2003; Stavrou, 2008; 2009), which are electrically injected into QDs, are of special importance in quantum computing technology.

During the last decade, a few theoretical and experimental reports have been announced related to light polarization within self-assembled quantum dots (Cantele et al., 2001; 2002; Chye et al., 2002; Pryor & Flatté, 2003; Stavrou, 2008; 2009). More specifically, they have studied circular polarization dependence of dipole recombination of spin-polarized states within a self-assembled quantum dots. The comparison between the theoretical estimation (Pryor & Flatté, 2003) and the experimental report (Chye et al., 2002) shows that the theoretical model can successfully describe the experiment.

In this chapter, we review the circularly polarized light along the orientation of spin-polarized carriers in a system of two coupled SAQDs made with InAs/GaAs. The dependence of circular light polarization on the geomerty parameters, the absence of an external magnetic field and the presence of a magnetic field is the main part of our investigation We can highlight the most important results along the direction [110]: a) for large interdot distances (uncoupled single QDs), our results are consistent with the polarization along to the direction [110], of the emitted light for the case of a single QD ($\sim$ 5%, Ref. (Chye et al., 2002; Pryor & Flatté,

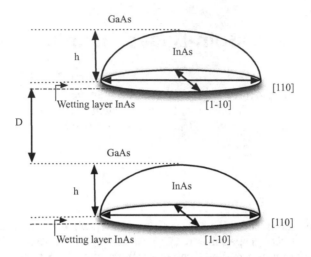

Fig. 1. The Geometry of the coupled ellipsoidal-shaped quantum dot heterostructure made with InAs/GaAs and a fixed wetting layer thickness to $0.3nm$. The two quantum dots are identical and the dot height is fixed to $h = 2.1nm$. The thickness of the wetting layer must be included in the height of the QDs in order to achieve the comparison to the experimental results.

2003)), b) for smaller distances the polarization increases receiving maximum value for our typical geometry $\sim 40\%$, c) in the case of large QDs the polarization decreases by increasing the magnetic field and for small dots the polarization increases, d) circular polarization along other polarization directions shows that light polarization does not depend on the above mentioned parameters.

This chapter is arranged as follows: In section 2, we briefly review the most recent SAQS fabrication methods. The eight-band $\mathbf{k} \cdot \mathbf{p}$ theory and light polarization theory are described in section 3. In section 4, we present and comment the numerical results. Lastly, the conclusions are placed in section 5.

2. Self-assembled QDs

SAQDs are nanostructures which have been worldwide used in nanotechnology e.g. quantum computing devices. The fabrication techniques of the SACQDs are not a simple formation mechanism but appears to be difficult and not easy to be controlled (Barabási & Stanley, 1995; Barabási, 1997; Darula & Barabási, 1997; Daudin et al., 1997; Dawson et al., 2007; Eaglesham & Cerullo, 1990; Lee et al., 1998; 2004; Rastelli et al., 2005; Tersoff, 1995; 1998; Tu & Tersoff, 2007). The various parameters (e.g. thermodynamic parameters) which influence the SACQDs growth are not well understood and as a result the shape of the SACQDs, which are very important in determining confining energies of electrons/holes, can not be easily achieved with high quality. Semiconductor QDs growth techniques which have been often used are the colloidal methods (Blackburn et al., 2003; Chestnoy et al., 1986; Wang & Herron, 1991)

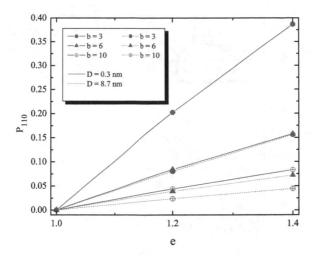

Fig. 2. The circular light polarization along the direction [110] dependence on the elongation for different separation distances, width-to-height ratio and fixed dot height $h = 2.1nm$. There is not an applied external magnetic field to the structure.

and the high-resolution electron lithography. (Lebens et al., 1990; Owen, 1985; Temkin et al., 1987) The main problem of these growth techniques is the incorporation of the defects. In fabricating SAQDs, these difficulties were overcome by using the Stranski-Krastanov (SK) coherent island growth mode. SK growth is one of the three primary modes by which thin films grow epitaxially at crystal surfaces/interfaces.

Theoretical and computational studies in epitaxial crystal growth is highly attractive in QDs sample fabrication. Several research groups have studied several growth parameters which play a crucial role during the self-organized growth of semiconductor quantum dot structures in the Stranski-Krastanov growth mode. Among others, the strain field of the lattice-mismatched systems, the growth temperature V/III ratio and the QD growth time are a few parameters which are of crucial importance in theoretical studies and fabrication technology. Kinetic Monte-Carlo simulations (Biehl & Much, 2003; 2005; Ghaisas & Madhukar, 1986; Kew et al., 1993; Khor & Sarma, 2000; Lam et al., 2002; Madhukar, 1983; Much et al, 2001; Ohr et al, 1992; Sitter, 1992) have been used to describe the wetting-layer and the island formation in heteroepitaxial growth.

3. Theory

Computational algorithms which have been used to solve Schrodinger equation in nanostructures are density functional theory (DFT) (Gross et al., 1991; Hohenberg & Kohn, 1964; Kohn & Sham, 1965; Sham & Kohn, 1966), Monte Carlo techniques (Foulkes et al., 2001; Räsänen et al., 2003; Thijssen, 1998), direct diagonalization techniques (Sadiku, 2001; Thijssen, 1998; Varga et al., 2011) and $\mathbf{k} \cdot \mathbf{p}$ theory (Bahder, 1990; 1992; Bimberg et al., 1999;

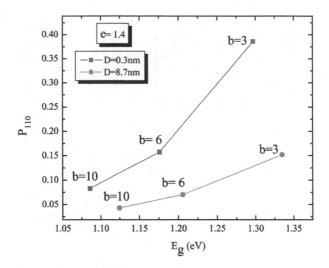

Fig. 3. The circular light polarization along the direction [110] as a function of the energy gap for different interdot distances and width-to-height ratio with fixed dot height $h = 2.1nm$ and elongation $e = 1.4$ in the absence of an external magnetic field.

Gershoni et al., 1993; Pryor et al., 1997; Pryor, 1998), among others. Although, $\mathbf{k} \cdot \mathbf{p}$ is used to describe single electron/hole wavefunctions in heterostructures, all the other above mentioned algorithms can also solve Schrodinger equation in heterostructures with a large number of electrons/holes.

In this section, we theoretically describe the $\mathbf{k} \cdot \mathbf{p}$ model for the case of single electron/hole quantum dots. The evaluation of wavefunctions will be used in our present research concerning the circular light polarization. The 8-band $\mathbf{k} \cdot \mathbf{p}$ model requires a basis of eight Bloch functions. The wavefunctions related to the Bloch functions are the following

$$\{\psi_{s,\uparrow}, \psi_{x,\uparrow}, \psi_{y,\uparrow}, \psi_{z,\uparrow}, \psi_{s,\downarrow}, \psi_{x,\downarrow}, \psi_{y,\downarrow}, \psi_{z,\downarrow}\} \tag{1}$$

where x, y, y are related to three directions of the space and the arrows denote the spin.

The 8-band $\mathbf{k} \cdot \mathbf{p}$ Hamiltonian (Bahder, 1990; 1992; Bimberg et al., 1999) is described by the following 2x2 block matrix form

$$\hat{\mathcal{H}} = \begin{pmatrix} G(\mathbf{k}) & \Gamma \\ -\overline{\Gamma} & \overline{G}(-\mathbf{k}) \end{pmatrix} \tag{2}$$

where Γ and G are 4x4 matrices. The matrix Γ couples the spin projections $\uparrow$ and $\downarrow$ due to spin-orbit (SO) interaction and matrix G consists of the potential energy, the kinetic energy, a SO interactions and a strain dependent part.

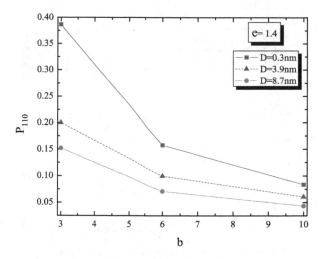

Fig. 4. The circular light polarization along the direction [110] as a function of width-to-height ratio with fixed elongation and different interdot distances with fixed dot height $h = 2.1nm$ in the absence of an external magnetic field.

$$\Gamma = \begin{pmatrix} 0 & 0 & 0 & 0 \\ 0 & 0 & 0 & \Delta_0/3 \\ 0 & 0 & 0 & -i\Delta_0/3 \\ 0 & -\Delta_0/3 & i\Delta_0/3 & 0 \end{pmatrix} \tag{3}$$

where Δ_0 is the splitting between the Γ_8 and Γ_7 valence bands (Bahder, 1990; 1992; Bimberg et al., 1999).

The matrix G is given by the summation of a potential energy part (G_1), a kinetic energy part (G_2), a SO interaction part (G_{SO}) and a strain dependent part (G_{st})

$$G = G_1 + G_2 + G_{SO} + G_{st} \tag{4}$$

where the matrices G_1, G_2, G_{SO} and G_{st} are respectively given by

$$G_1 = \begin{pmatrix} E_c & iPk_x & iPk_y & iPk_z \\ -iPk_x & E'_v & 0 & 0 \\ -iPk_y & 0 & E'_v & 0 \\ -iPk_z & 0 & 0 & E'_v \end{pmatrix} \tag{5}$$

$$G_2 = \begin{pmatrix} A'k^2 & Bk_yk_z & Bk_xk_z & Bk_xk_y \\ Bk_yk_z & L'k_x^2 + M(k_y^2 + k_z^2) & N'k_xk_y & N'k_xk_z \\ Bk_zk_x & N'k_xk_y & L'k_y^2 + M(k_x^2 + k_z^2) & N'k_yk_z \\ Bk_xk_y & N'k_xk_z & N'k_yk_z & L'k_z^2 + M(k_x^2 + k_y^2) \end{pmatrix} \tag{6}$$

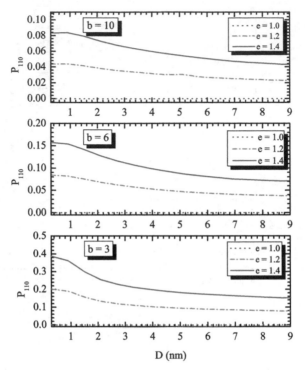

Fig. 5. The circular light polarization along the direction [110] as a function of the interdot distance for different elongation and width-to-height ratio with fixed dot height $h = 2.1nm$. There is not an applied external magnetic field to the structure.

$$
G_{SO} = \begin{pmatrix} 0 & 0 & 0 & 0 \\ 0 & 0 & -i\Delta_0/3 & 0 \\ 0 & -i\Delta_0/3 & 0 & 0 \\ 0 & 0 & 0 & 0 \end{pmatrix} \tag{7}
$$

and

$$
G_{st} = \begin{pmatrix} a_c(\epsilon_{xx} + \epsilon_{yy} + \epsilon_{zz}) & b'\epsilon_{yz} - iP\epsilon_{xj}k^j & b'\epsilon_{zx} - iP\epsilon_{yj}k^j & b'\epsilon_{xy} - iP\epsilon_{zj}k^j \\ b'\epsilon_{yz} + iP\epsilon_{xj}k^j & l\epsilon_{xx} + m(\epsilon_{yy} + \epsilon_{zz}) & n\epsilon_{xy} & n\epsilon_{xz} \\ b'\epsilon_{zx} + iP\epsilon_{yj}k^j & n\epsilon_{xy} & l\epsilon_{yy} + m(\epsilon_{xx} + \epsilon_{zz}) & n\epsilon_{yz} \\ b'\epsilon_{xy} + iP\epsilon_{zj}k^j & n\epsilon_{xz} & n\epsilon_{yz} & l\epsilon_{xx} + m(\epsilon_{yy} + \epsilon_{zz}) \end{pmatrix} \tag{8}
$$

where all the parameters used within the matrices are given by Refs (Bimberg et al., 1999; Gershoni et al., 1993; Saada, 1989).

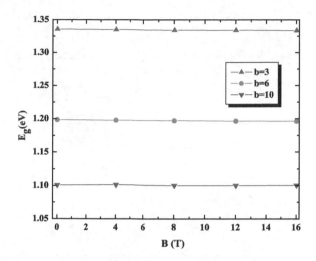

Fig. 6. The energy gap as a function of the applied magnetic field for different width-to-height ratio and fixed elongation (e=1.4).

In our study, we have considered two ellipsoidal caps QDs made with InAs which are separated by a distance D as illustrated in Fig.1. The QDs are embedded on a wetting layer of InAs and are surrounded by GaAs. Single electron and hole wavefunctions are calculated using the strain dependent $\mathbf{k} \cdot \mathbf{p}$ theory. In our calculations, we have fixed the height of the dots to $h = 2.1nm$ and varied the width-to-height ratio

$$b = (d_{[110]} + d_{[1-10]})/h \qquad (9)$$

and elongation

$$e = d_{[110]}/d_{[1-10]} \qquad (10)$$

The electron/hole wavefunctions were numerically computed on a real space grid with spacing equal to the wetting layer thickness $0.3nm$. Strain and carrier confinement split the heavy hole (HH) and light hole (LH) degeneracy and the states are doubly degenerate which are denoted by $|\psi >$ and $T|\psi >$ (time reverses of each other). The energy gap (E_g) of the coupled SAQD structure strongly depends on the interdot distance as it is shown in our investigation. All material parameters that have been used in the $\mathbf{k} \cdot \mathbf{p}$ simulations are taken by Ref. (Pryor1, Vurgaftman). Spin polarized ground states are constructed by taking a linear combination of the states comprising the doublet and adjusting the coefficient in order to maximize the expectation value of the pseudospin operator projected onto a direction l (Pryor & Flatté, 2003). The requested complex number α maximizes the following

$$\frac{[< \psi| + \alpha^* < \psi|T]\hat{l} \cdot \mathbf{S}[|\psi > +\alpha T|\psi >]}{1 + |\alpha|^2} \qquad (11)$$

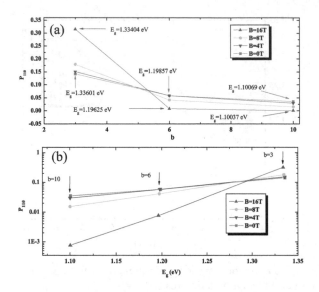

Fig. 7. (a) The circular light polarization along the direction [110] as a function of the width-to-height ratio for different applied magnetic fields with fixed dot height $h = 2.1nm$ and elongation $e = 1.4$. (b) The circular light polarization along the direction [110] as a function of the energy gap for different applied magnetic fields with fixed dot height $h = 2.1nm$ and elongation $e = 1.4$.

where **S** is the pseudospin operator (Pryor et al., 1997; Pryor, 1998; Pryor & Flatté, 2003) in the 8-band **k** · **p** theory

$$\vec{S} = \begin{pmatrix} \vec{\sigma}\Gamma_6 & 0 & 0 \\ 0 & \vec{J}\Gamma_8 & 0 \\ 0 & 0 & \vec{\sigma}\Gamma_7 \end{pmatrix} \tag{12}$$

and l the spin orientation.

Let us consider the situation in which the electron spin is polarized along the same direction l as the observed emitted light. The polarization which characterizes the emitted light is given by (Jackson, 1998)

$$P_l = \frac{I_l^{(+)} - I_l^{(-)}}{I_l^{(+)} + I_l^{(-)}} \tag{13}$$

where $I_l^{(\pm)}$ is the light intensity with $\pm$ helicity. The intensity of emission of circularly polarized light for the case of spin-polarized electron and unpolarized holes is given by

$$I_l^{(\pm)} = | < \psi_h |\hat{e}_l^{(\pm)} \cdot \mathbf{p}|\psi_e > |^2 + | < \psi_h |T\hat{e}_l^{(\pm)} \cdot \mathbf{p}|\psi_e > |^2 \tag{14}$$

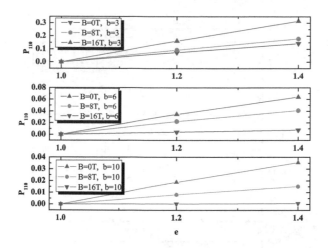

Fig. 8. The circular light polarization along the direction [110] as a function of the elongation for different applied magnetic fields and width-to-height ratio with fixed dot height $h = 2.1nm$.

and for the case of spin-polarized holes and unpolarized electron the intensity of circularly polarized light is given by

$$I_l^{(\pm)} = |<\psi_h|\hat{e}_l^{(\pm)} \cdot \mathbf{p}|\psi_e>|^2 + |<\psi_h|\hat{e}_l^{(\pm)} \cdot \mathbf{p}T|\psi_e>|^2 \tag{15}$$

where the indices e and h correspond to electron and holes respectively, $\mathbf{p}$ is the momentum operator and $\hat{e}_l^{(\pm)}$ is the circular polarization vector with helicity $\pm$ (which denotes circularly polarized light that propagates along direction l). It is worth mentioning that the last two equations give identical results as a result of the anticommutation relations between $\mathbf{p}$ and T. As a result the emitted light polarization is independent, no matter if the injected spin-polarized carriers are electrons or holes.

4. Results

In this work, we research the circular light polarization along the plane [110] (as illustrated in Fig.1) for 100% spin-polarized carriers along [110]. The dependence of light polarization on the elongation is shown at Fig.2 for different interdot distances and QD sizes (width-to-height ratio). For axially symmetric QDs the polarization is zero because the intensities with different helicity are equal and in the case of more elongated QDs the light polarization increases due to the azimuthal symmetry breaking. Furthermore, circular light polarization increases as QDs come close to each other and as the size of the dots decreases. This happens as a result of the wavefunction dependence on the QDs geometry. Increasing the interdot distance the electron energies in two lowest conduction band converge towards that of an electron in a single QD. The energy difference between the ground state and first excited state decreases as the distance increases. On the other hand, in the valence band the energy splitting decreases as

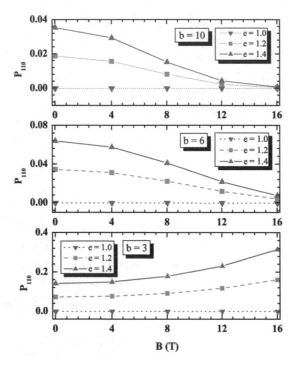

Fig. 9. The circular light polarization along the direction [110] as a function of the applied magnetic field for different elongation and width-to-height ratio with fixed dot height $h = 2.1nm$.

the separation distance increases. Both conduction and valence band ground energies increase by increasing the separation distance (Bimberg et al., 1999; Stavrou, 2008). As a result the energy gap increases by increasing the interdot distance. Fig. 3, presents the polarization as a function of energy gap for different geometric parameters. It is clear that increasing the width-to-height ratio light polarization decays and increases as the separation decreases because of the wavefunction dependence on the geometric parameters. In the case of large elongation ($e = 1.4$), small dot size ($b = 3$) and interdot distance ($D = 0.3nm$) the polarization has the largest value ($P_{110} = 0.38$, the emitted light is 38% polarized). Polarization has the smallest values for large dots for all the interdot distances as it is illustrated in Fig.3. The dependence of circular light polarization on the width-to-height ratio is presented in Fig.4. For small separation distance the polarization becomes large and as the separation distance increases the light polarization decreases.

A complete picture of the dependence of circular light polarization on all the above mentioned parameters is illustrated in Fig.5. The light polarization efficiency along the direction [110] strongly depends on the interdot distance, the size and the elongation of the QD. In the limit of large interdot distance (uncoupled QDs) and elongation $e = 1.4$ the results are consistent with the numerical calculations previously reported (Pryor & Flatté, 2003) and the experimental results (Chye et al., 2002) ($\sim$ 5% circularly polarized light for 100% polarized carriers). For other polarization directions like [001] the polarization efficiency it founds to be very close to 1 ($P_{001} = 0.99$) and almost independent on the interdot distance. So ($P_{001} = 0.99$) is the best choice if one is interested in the light polarization along the growth direction. Along the directions [100] and [010] the circular polarization vanishes $P_{100} = P_{010} = 0$ for any of our QD geometry.

The next part of our investigation is related to circular light polarization dependence on an applied external magnetic field. We have considered the case of a single QD in the presence of a magnetic field. The variation of the magnetic field releases the dependence on the Zeeman energy. As the magnetic field increases the energy splitting increases due to the Zeeman energy splitting. The dependence of the energy gap on the magnetic field is presented in Fig.6 for different sizes of QDs. The current investigation shows that the energy gap strongly depends on the size of the dot and less on the external magnetic field (Fig.6). In the case of small dots ($b = 3$) and $B = 0T$ (or $B = 16T$) the energy gap gets the value $E_g = 1.336eV$ (or $E_g = 1.334eV$). Therefore, for the above case, the energy gap difference is $0.002eV$ which is a tiny difference. The existence of the magnetic field does not change the energy difference (E_g) between the ground states in conduction band and valence band. Thus, the size of the dots is the dominant parameter that changes drastically the energy gap due to the fact that for the case of small QDs the electrons (holes) have larger energy than in the case of larger QDs which results in larger energy difference between the carrier in the bottom of the conduction and the top valence band state.

The dependence of circular polarization on the size and on the energy gap is given in Fig.7. In the case of fixed width-to-height ratio b=3, the smallest value of P_{110} is achieved for the case of B=0T and the largest for B=16T, although the energy gap slightly differs. Fig.7(b) presents the dependence of circular light polarization on the QD energy gap. As we have earlier shown (absence of magnetic field), in the case of the presence of magnetic field and $e = 1.0$ (axial symmetric QD), light polarization vanishes because of unchanging value of light intensity for $\pm$ helicity (Fig.8). Increasing the elongation, circular light polarization increases because of the azimuthal symmetry breaking. Light polarization increases as the magnetic field increases and decreases as the size of the quantum dots increases , as earlier explained.

The light polarization as a function of the external magnetic field, the separation distance, the size of the dots and the elongation is presented in Fig.9. As it is obvious, in small (large dots), polarizarion increases (decreases) by increasing the magnetic field. Responsible for this kind of behavior is the dependence of the energy gap on the size of QD and the Zeeman splitting due to the external magnetic field (Stavrou, 2009). As earlier mentioned, the energy gap of small dots is larger than that of large dots. Furthermore, the carrier wavefunctions involved in Eq. (13) depend on the size of the QDs and result in larger polarization for the case of small dots (large E_g). Increasing the magnetic field in small QDs, the light polarization gets larger values as a consequence of carrier wavefunctions dependence on the magnetic field. Although the carriers are 100% polarized, the emitted light is less than 100% polarized. In the case of $B = 0T$ and elongation $e = 1.4$ the results are consistent with the numerical calculations (Pryor & Flatté, 2003) and the experimental results (Chye et al., 2002) ($\sim$ 5% circularly polarized light for 100% polarized carriers) previously reported. Increasing the E_g, it is possible for small QDs to increase the percentage of polarized light by increasing the external magnetic field.

For small QDs, $B = 0T$ and $e = 1.4$ the light is $\sim 18\%$ polarized and for $B = 16T$, $e = 1.4$ the light is $\sim 32\%$ polarized. The dependence of the light polarization on the separation distance in the presence of an external field is under investigation by our group.

Lastly, we mention that for other polarization directions other than [110], the light polarization has different behavior, as in the case of the absence of the the magnetic field. The circular polarization along the directions [100], [010] is zero and 1 along the direction [001] and appears to be independent on magnetic field.

5. Conclusions

Summing overall, we have researched a possible qubit made with SAQDs. Our investigation shows that for the case of absence of an external magnetic field, although the carriers are 100% polarized the emitted light is less than 100% polarized. The largest polarization ($P_{110} = 0.38$) was generated for small D, b=3 and e=1.4. Thus, the observed 1% polarization was generated by carriers that were $1/0.38 \sim 2.6\%$ polarized. The circular light polarization depends strongly on the interdot distance (Fig.5), the size of the dots and the elongation. On the other hand, in the presence of an external magnetic field, it has been shown that both the magnetic field and the geometry of the QD are parameters of special importance in order to control the light polarization along the direction of spin polarization in QD qubits (Fig.9). For small dots, circular light polarization along the direction [110] increases by increasing the magnetic field and in the case of large dots it decreases as the field increases. Lastly, other polarization directions like [001] and [100] lead to circular light polarization value 1 and 0 respectively.

6. Acknowledgment

The authors would like to thank Prof. C. Pryor, for useful discussions and Commander of the Hellenic Navy D. Filinis for his support. The author V.N.S. would like also to acknowledge the financial support given by the Training Mobility of Researchers (TMR), NSA and ARDA under ARO Contract No. DAAD19-03-1-0128. and the University of Iowa under the grand No. 52570034.

7. References

Artús, L., Cuscó, R., Hernández, S., Patanè, A., Polimeni, A., Henini, M. & Eaves, L. (2000). Quantum-dot phonons in self-assembled InAs/GaAs quantum dots: Dependence on the coverage thickness, *Appl. Phys. Lett.*, 77, (3556-3558).

Bahder, T.,B. (1990). Eight-band k.p model of strained zinc-blende crystals, *Phys. Rev.* B 41, (11992-12001).

Bahder, T.,B. (1992). Analytic dispersion relations near the Γ point in strained zinc-blende crystals, *Phys. Rev.* B 45, (1629-1637).

Bányai, L. & Koch, S.W. (1993). *Semiconductor Quantum Dots* (World Scientific).

Barabási, A.,L & Stanley, H.,E. (1995). *Fractal Concepts in Surface Growth*, (Cambridge University Press, Cambridge).

Barabási, A.,L, (1997). Self-assembled island formation in heteroepitaxial growth, *Appl. Phys. Lett.*, 70, (2565-2567).

Baruffa, F., Stano, P. & Fabian, J. (2010). Theory of Anisotropic Exchange in Laterally Coupled Quantum Dots, *Phys. Rev. Lett*, 104, (126401-126401).

Bertoni, A., Rontani, M., Goldoni, G. & Molinari, E. (2005). Reduced Electron Relaxation Rate in Multielectron Quantum Dots, *Phys. Rev. Lett.*, 95, (066806-066809)

Biehl, M. & Much, F. (2003). Simulation of wetting-layer and island formation in heteroepitaxial growth, *Europhys. Lett.*, 63,(14-20).

Biehl, M. & Much, F. (2005). Off-lattice Kinetic Monte Carlo simulations of Stranski-Krastanov-like growth, *B. A. Joyce et al. (eds.), Quantum Dots: Fundamentals, Applications, and Frontiers*, (89-102) (Springer).

Blackburn, J.,L., Ellingson, R.J., Micic, O.,I. & Nozik, A.,J. (2003) Electron Relaxation in Colloidal InP Quantum Dots with Photogenerated Excitons or Chemically Injected Electrons, *J. Phys. Chem. B* 107, (102-109).

Bimberg, D., Grundmann, M. & Ledentsov, N.,N. (1999). *Quantum Dot Heterostructures* (Wiley-VCH Verlag, Germany).

Burkard, G., Loss, D. & DiVincenzo, D.P. (1999). Coupled quantum dots as quantum gates, *Phys. Rev. B*, 59, (2070-2078).

Cantele, G., Ninno, D. & Iadonisi, G. (2001). Calculation of the Infrared Optical Transitions in Semiconductor Ellipsoidal Quantum Dots, *Nano Lett.*, 1, (121-124).

Cantele, G., Piacente, G., Ninno, D. & Iadonisi, G. (2002). Optical anisotropy of ellipsoidal quantum dots *Phys. Rev.*, B 66, (113308-1 - 113308-4).

Chestnoy, N., Hull, R. & Brus, L.,E. (1986). Higher excited electronic states in clusters of ZnSe, CdSe, and ZnS: Spin-orbit, vibronic, and relaxation phenomena, *J. Chem. Phys.* B 85, (2237-2242).

Chye, Y., White, M.E., Johnston-Halperin, E., Gerardot, B.D., Awschalom D.D. & Petroff, P.M. (2002). Spin injection from (Ga,Mn)As into InAs quantum dots, *Phys. Rev. B*, 66, (201301(R)-201304(R)).

Darula, I. & Barabási, A.,L, (1997). Island Formation and Critical Thickness in Heteroepitaxy *Phys. Rev. Lett.*, 78, (3027-3027).

Daudin, B., Widmann, F., Feuillet, G., Samson, Y., Arlery, M. & Rouvière, J.L. (1997). Stranski-Krastanov growth mode during the molecular beam epitaxy of highly strained GaN, *Phys. Rev. B*, 56, (R7069-R7072).

Dawson, P., Göbel, E. O., Pierz, K., Rubel, O., Baranovskii, S. D. & Thomas, P. (2007). Relaxation and recombination in InAs quantum dots, Phys. Stat. Sol. (b) 244, (2803-2815).

Eaglesham, D.J. & Cerullo, M. (1990). Dislocation-free Stranski-Krastanow growth of Ge on Si(100) *Phys. Rev. Lett.*, 64, (1943-1946).

Fessatidis, V., Horing, N. J. M. & Kashif S. (1999) Retarded Green's function for an electron in a parabolic quantum dot subject to a constant uniform magnetic field and an electric field of arbitrary time dependence and orientation *Philosophical Magazine Part B*, 79, (77-89).

Foulkes, W.M.C., Mitas, L., Needs, R.J. & Rajagopal G. (2001). Quantum Monte Carlo simulations of solids. *Rev. Mod. Phys.*, 73, (33-83).

Fujisawa, T., Oosterkamp, T. H., Kouwenhoven, P. & Wilfred G.(1998). Spontaneous Emission Spectrum in Double Quantum Dot Devices, *Science* , 282 (932-935).

Fujisawa T., Hayashi T. & Sasaki S. (2006) Time-dependent single-electron transport through quantum dots, *Rep. Prog. Phys.*, 69 (759-796).

Gershoni, D., Henry, C.,H. & Baraff, G.,A. (1993) Calculating the optical properties of multidimensional heterostructures: Application to the modeling of quaternary quantum well lasers, *IEEE J. Quant. Elec.*, 29 (2433-2450).

Ghaisas, S.,V. & Madhukar, A. (1986). Role of surface molecular reactions in influencing the growth mechanism and the nature of nonequilibrium surfaces: a Monte Carlo study of molecular-beam epitaxy, *Phys. Rev. Lett.*, 56, (1066-2078).

Gross, E., K., U., Runge, E. & Heinonen, O. (1991). *Many-Particle Theory* (Taylor & Francis).

Hayashi, T., Fujisawa, T., Cheong, H.D., Jeong, Y.H. & Y. Hirayama (2004). Coherent Charge Oscillation in a Semiconductor Double Quantum Dot, *IEEE Transactions on Nanotechnology*, 3 (300-303).

Hohenberg, P. & Kohn, W. (1964). Inhomogeneous Electron Gas, *Phys. Rev.*, 136, (864-871).

Hu, X. & Sarma,S., Das (2000). Hilbert-space structure of a solid-state quantum computer: Two-electron states of a double-quantum-dot artificial molecule, *Phys. Rev.* A 61, (062301, 062319).

Hu, X. & Sarma,S., Das (2001). Spin-based quantum computation in multielectron quantum dots, *Phys. Rev.* A 64, (042312, 042316).

Imamoglu, A., Awschalom, D.D., Burkard, G., DiVincenzo, D.P. Loss, D., Sherwin, M. & Small, A. (1999). Quantum Information Processing Using Quantum Dot Spins and Cavity QED, *Phys. Rev. Lett.*, 83, (4204-4207).

Jackson, J.D. (1998). *Classical Electrodynamics*, (Wiley, 3 edition).

Kew, J., Wilby, M.,R. & Vvedensky, D.,D. (1993). Continuous-space Monte Carlo simulations of epitaxial growth., *J. Cryst. Growth*, 127, 508.

Khaetskii, A.V. & Nazarov, Y.V. (2001). Spin-flip transitions between Zeeman sublevels in semiconductor quantum dots, *Phys. Rev. B*, 64, (125316-125321).

Khor, K.,E. & Sarma, S.,D. (2000). Quantum dot self-assembly in growth of strainedlayer thin films: a kinetic Monte Carlo study. *Phys. Rev. B.*, 62, (16657-16664).

Kohn, W. & Sham, L.J. (1965). Self-Consistent Equations Including Exchange and Correlation Effects, *Phys. Rev.*, 140, (A1133-A1138).

Lam, C.,H., Lee, C.,K. & Sander, L.,M. (2002). Competing roughening mechanisms in strained heteroepitaxy: A fast kinetic Monte Carlo study, *Phys. Rev. Lett.* 89, (216102-216105).

Lebens, J.,A., Tsai, C.,S., Vahala, K.,J. & Kuech, T.,F. (1990). Application of selective epitaxy to fabrication of nanometer scale wire and dot structures, *Appl. Phys. Lett.* 56, (2642, 2644).

Lee, S., Daruka, I., Kim, C.,S., Barabási, A.,L,, Merz, J.,L. & Furdyna, J.,K. (1998). Dynamics of Ripening of Self-Assembled II-VI Semiconductor Quantum Dots *Phys. Rev. Lett*, 81, (3479-3482).

Lee, S., Lazarenkova, O.L., Allmen, P, Oyafuso, F. & Klimeck, G. (2004). Effect of wetting layers on the strain and electronic structure of InAs self-assembled quantum dots, *Phys. Rev. B*, 70, (125307-125313).

Madhukar, M. (1983) Far from equilibrium vapor phase growth of lattice matched IIIV compound semiconductor interfaces: some basic concepts and Monte-Carlo computer simulations, *Surf. Sci.*, 132, 344-374.

Mohanty, P. (2000). Notes on decoherence at absolute zero, *Physica* B. 280, (446-452) (and references therein).

Much, F., Ahr, M., Biehl, M. & Kinzel, W. (2001). Kinetic Monte Carlo simulations of dislocations in heteroepitaxial growth, *Europhys. Lett.*, 56,(791-796).

Nielsen, M. and Chuang, I.L. (2000). *Quantum Computation and Quantum Information* (Cambridge University Press, Cambridge,).

Orr, B.,G., Kessler, D., Snyder, C.,W. & Sander, L. M. (1992). A model for strain-induced roughening and coherent island growth, *Europhys. Lett.*, 19,(33).

Owen, G. (1985). Electron lithography for the fabrication of microelectronic devices , *Rep. Prog. Phys.* 48, 795.

Pryor, C., Pistol, M.E. & Samuelson, L. (1997). Electronic structure of strained $InP/Ga_{0.51}In_{0.49}P$ quantum dots, *Phys. Rev.* B 56, (10404-10411).

Pryor, C. (1998). Eight-band calculations of strained InAs/GaAs quantum dots compared with one-, four-, and six-band approximations, *Phys. Rev.* B 57, (7190-7195).

Pryor, C.E., & Flatté, M.E. (2003). Accuracy of Circular Polarization as a Measure of Spin Polarization in Quantum Dot Qubits, *Phys. Rev. Lett.* 91, (257901-257904).

Räsänen, E., Saarikoski, H., Stavrou, V. N., Harju, A., Puska, M. J. & Nieminen, R. M. (2003). Electronic structure of rectangular quantum dots, *Phys. Rev.* B 67, (235307-1, 045320-8).

Rastelli, A., Stoffel, M., Tersoff, J., Kar, G.,S. & Schmidt, O.,G. (2005) Kinetic Evolution and Equilibrium Morphology of Strained Islands, *Phys. Rev. Lett.* 95, (026103-026106).

Saada, A.,S. (1989). *Elasticity:Theory and applications* (R. E. Krieger Publishing Co.).

Sadiku, M.,N.,O. (2001). *Numerical Techniques in Electromagnetics* (CRC Press).

Sham, L.J. & Kohn, W. (1966). One-Particle Properties of an Inhomogeneous Interacting Electron Gas, *Phys. Rev.*, 145, (561-567).

Simserides, C. D., Hohenester, U., Goldoni, G. & Molinari, E. (2000). Local absorption spectra of artificial atoms and molecules. *Phys. Rev.* B 62, (13657-13666).

Simserides, C., Zora, A. & Triberis, G. (2006). Near-field magnetoabsorption of quantum dots. *Phys. Rev.* B 73, (155313-1, 155313-13).

Simserides, C., Zora, A. & Triberis, G. (2006). Magneto-Optics of Quantum Dots in the Near Field. *International Journal of Modern Physics B* 21, (1649-1653).

Sitter, H. (1995). MBE growth mechanisms; studies by Monte-Carlo simulation, *Thin Solid Films* 267, 37.

Stano, P. & Fabian, J. (2005). Spin-orbit effects in single-electron states in coupled quantum dots. *Phys. Rev.* B 72, (155410-1, 155410-14).

Stano, P. & Fabian, J. (2006). Orbital and spin relaxation in single and coupled quantum dots. *Phys. Rev.* B 74, (045320-1, 045320-12).

Stavrou, V.N. & Hu, X. (2005). Charge decoherence in laterally coupled quantum dots due to electron-phonon interactions. *Phys. Rev.* B 72, (075362-1, 075362-8).

Stavrou, V.N. & Hu, X. (2006). Electron relaxation in a double quantum dot through two-phonon processes. *Phys. Rev.* B 73, (205313-1, 205313-5).

Stavrou, V.N. (2007) Suppression of electron relaxation and dephasing rates in quantum dots caused by external magnetic fields. *J. Phys.: Condens. Matter* 19, (186224-1, 186224-9).

Stavrou, V.N. (2008). Circularly polarized light in coupled quantum dots. *J. Phys.: Condens. Matter* 20, (395222-1, 395222-4).

Stavrou, V.N. (2009). Polarized light in quantum dot qubit under an applied external magnetic field. *Phys. Rev.* B 80, (153308-1, 153308-4).

Temkin, H., Dolan, G.,J., Panish, M.,B. & Chu, S.,N.,G. (1987). Low-temperature photoluminescence from InGaAs/InP quantum wires and boxes, *Appl. Phys. Lett.* 50, (413-415).

Tersoff, J. (1995). Enhanced Solubility of Impurities and Enhanced Diffusion near Crystal Surfaces. *Phys. Rev. Lett.* 74, (5080-5083).

Tersoff, J. (1998). Enhanced Nucleation and Enrichment of Strained-Alloy Quantum Dots. *Phys. Rev. Lett.* 81, (3183-3186).

Thijssen, J.,M. (1998). *Computational Physics* (Cambridge University Press).

Thorwart, M., Eckel, J. & Mucciolo, E.R. (2005). Non-Markovian dynamics of double quantum dot charge qubits due to acoustic phonons. *Phys. Rev.* B 72, (235320-1, 235320-6).

Tu, Y., & Tersoff, J. (2007). Coarsening, Mixing, and Motion: The Complex Evolution of Epitaxial Islands *Phys. Rev. Lett.* 98, (096103-096106).

Varga, Kálmán & Driscoll, Joseph A. (2011). *Computational Nanoscience* (Cambridge University Press, Cambridge).

Vurgaftman,I, Meyer, J.R. & Ram-Mohan, L.R. (2001). Band parameters for III-V compound semiconductors and their alloys. *J. A. Appl. Phys.* 89, (5815-5875).

Wang,Y. & Herron, N. (1991) Nanometer-Sized Semiconductor Clusters: Materials Synthesis, Quantum Size Effects, and Photophysical Properties *J. Phys. Chem.* 95, (525-532).

Witzel, W. M., Hu, X. & Sarma,S. Das (2007). Decoherence induced by anisotropic hyperfine interaction in Si spin qubits, *Phys. Rev.* B 76, (035212-1, 035212-8).

Woods, L.M., Reinecke, T.L. & Lyanda-Geller, Y. (2002). Spin relaxation in quantum dots, *Phys. Rev.* B 66, (161318-1, 161318-4).

Part 4

Quantum Dots in Biology

11

Quantum Dots-Based Biological Fluorescent Probes for *In Vitro* and *In Vivo* Imaging

Yao He*

*Institute of Functional Nano & Soft Materials (FUNSOM) and
Jiangsu Key Laboratory of Carbon-based Functional Materials & Devices
Soochow University, Suzhou, Jiangsu
China*

1. Introduction

The 2008 Nobel Prize in chemistry was awarded to three distinguished scientists who discovered and developed the green fluorescent protein (GFP), which has proved to be a powerful and versatile fluorescent probe for biological and biomedical researches [1]. High-performance fluorescent biological probes are expected to possess several important properties including excellent water-dispersibility, high photoluminescent quantum yield (PLQY), robust photostability, and favorable biocompatibility. A consensus has been reached that the GFP and organic dyes, recognized as the well-established fluorescent probes, are to some extent inappropriate for long-time in bioimaging because of their photobleaching property (Figure 1). In this respect, II-VI QDs have attracted intensive attentions due to their unique optical and electronic properties (e.g. size-tunable emission, broad photoexcitation, narrow and symmetrical emission spectra, strong fluorescence, and robust photostability). Indeed, the QDs are currently considered as most promising fluorescent probes, which open new opportunities for real-time and long-term monitoring and imaging both *in vitro* and *in vivo* [2].

1.2 Synthesis of II-VI QDs

It is well-known that there are two rudimentary approaches to prepare II-VI QDs: one is the organometallic route, the other is the aqueous method.

The organic synthesized QDs (orQDs) with excellent spectral properties have been achieved through the well-established organic strategy in elegant work [3]. As far as 1993, Bawendi's group developed a simple strategy, which was based on the pyrolysis of organometallic reagents by injection into a hot coordinating solvent, to production of serial highly monodispersed II-VI QDs (e.g., CdS, CdSe, and CdTe QDs). Such resultant QDs possessed uniform size, shape, and sharp absorption and emission features [3a]. Alivisatos et al. studied the growth kinetics of the CdSe QDs in a detailed way, providing a useful guidance for testing theories of quantum confinement, and also for designing high-quality QDs of controllable shapes [3b]. Peng and coworkers, by using CdO as precursor, developed more

---

* Corresponding Author

green-chemistry methods to prepare II-VI QDs [3c]. On the basis of which, they further optimized experiment conditions to synthesiz CdSe QDs with excellent optical properties, whose PLQY value reached 85% [3d]. Thereafter, different kinds of core-shell structured QDs (e.g., CdTe-CdS and CdTe-ZnS QDs) were achieved via organic synthesis [3e,f]. It is worth noting that, these orQDs cannot be directly used in bioapplications due to their hydrophobic character. A general strategy is to transfer the hydrophobic nanocrystals from the organic phase to aqueous solution by wrapping an amphiphilic polymer around the particles [4]. Although this method is efficient, it is relatively complicated and requires addition steps. Another effective strategy is to substitute the hydrophilic molecules, which have strong polar groups such as carboxylic acid or reactive groups such as Si-O-R, for surface-binding TOPO [5]. Nevertheless, in addition to relatively complicated manipulations, the treatment may depress optical properties and stability of the QDs (e.g, the PLQY may decrease when the orQDs are transferred into water because the polarity of water is too strong to break various equilibriums related to the nanocrystals) [6].

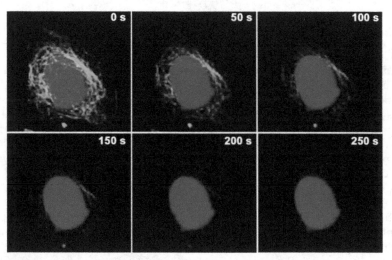

Fig. 1. Photobleaching of organic dyes. Microtubules were labeled with fluoresceinisothiocyanate (FITC, one typical kind of commercial organic dyes). Green fluorescence signals of FITC were rapidly disappeared in 250-s irradiation owing to severe photobleaching.

Aqueous synthesis is an alternative strategy to directly prepare water-dispersed QDs, which is relatively simpler, cheaper, less toxic, and more environmentally friendly. More importantly, the aqueous synthesized QDs (aqQDs) are naturally water-dispersed without any posttreatment due to a large amount of hydrophilic ligand molecules (e.g., 3-Mercaptopropionic acid, thioglycolic acid, et al.) covered on their surface. Rogach and coworkers reported aqueous synthesis of CdTe aqQDs those are directly prepared in water by using thiol molecules as ligands. However, the PLQY of aqQDs was generally lower than 30% [7]. Systematic investigations reveal that, large amount of surface defects, which are often generated due to the long reaction time in aqueous phase, results in low PLQY [8]. Tremendous effort has been devoted to improve the spectral properties of QDs directly prepared in the aqueous phase. Zhang et al developed a hydrothermal method for synthesis of highly luminescent aqQDs (PLQY: ~50%). Notably, the growth rate of QDs was greatly accelerated under high reaction temperature (e.g, 200 °C), leading to effective reduce of surface defects [9].

More recently, microwave-assisted method has been developed as novel promising strategies for synthesizing high-quality QDs. Compared with conventional heating method, there are four main advantages about microwave irradiation as heating method: (a) temperature can be rapidly raised due to the high utilization factor of microwave energy, and the kinetics of the reaction are increased by 1-2 orders of magnitude, (b) stirring effect on molecular level can be realized through microwave irradiation, which are favorable for uniform heating, (c) stagnant phenomenon can be effectively avoided compared with traditional heating method because microwave energy vanishes once closing microwave device, and (d) the initial heating is rapid which can lead to energy savings [10]. Ren et al reported CdTe aqQDs with PLQY of 40-60%, which were rapidly prepared via microwave synthesis [11]. He and coworkers developed a facile microwave-assisted method for synthesizing a variety of highly luminescent aqQDs (e.g., CdTe/CdS core-shell QDs, CdTe/CdS/ZnS core-shell-shell QDs). Notably, such resultant aqQDs feature excellent optical properties (PLQY: ~50-80%) due to the above-mentioned unique merits of microwave irradiation (Figure 2) [12]. Very recently, they further synthesized the near-infrared (NIR)-emitting QDs via microwave synthesis. Significantly, the prepared NIR-emitting QDs possessed excellent aqueous dispersibility, ultrasmall size (~4 nm), robust storage-, chemical-, and photo-stability, and finely-tunable emission in the NIR range (700-800 nm) [13].

1.3 II-VI QDs for biological fluorescence imaging

Biological fluorescence imaging is one of most important and interdisciplinary research involving chemistry, biology, life science, and biomedicine, which has been widely used for various *in vivo* and *in vitro* studies [1]. Particularly, fluorescent biological probes are essential tools for bioimaging. For optimum imaging and tracking of biological cells, these probes should be water-dispersible, anti-bleaching, luminescent, and biocompatible. In the last century, organic dyes and fluorescent proteins were mostly used as fluorescent probes in biological research; however they suffer from severe photobleaching that restricts their applications for long-term *in vitro* or *in vivo* cell imaging. This shortcoming has led to the intense interest in II-VI QDs due to their many unique merits. Particularly, compared to fluorescent dyes, QDs possess the unique advantages such as size-tunable emission color, broad photoexcitation, narrow emission spectra, strong fluorescence, and high resistance to photobleaching. Consequently, QDs have been widely used as a new class of fluorescent probes in biological research, particularly for *in vitro* and *in vivo* imaging [1,2].

In 1998, the groups of Alivisatos and Nie independently reported the first examples of QDs-based bioprobes, suggesting great promise of II-VI QDs for cell imaging [14]. Thereafter, the II-VI QDs have been extensively developed and optimized to be a kind of well-established fluorescent biological probes for a variety of bioimaging studies. Wu et al used CdSe/ZnS QDs liked to immunoglobulin G (IgG) and streptavidin to label the breast cancer marker Her2 on the surface of fixed and live cancer cells, to stain actin and microtubule fibers in the cytoplasm, and to detect nuclear antigens inside the nucleus. Notably, taking advantage of the emission flexibility of QDs, one single cell was dual-color labeled the QDs of different

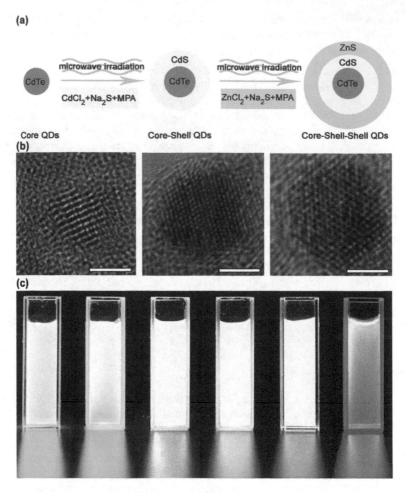

Fig. 2. (a) Schematic illustration of microwave-assisted synthesis of aqQDs. (b) High-resolution transmission electron microscopy (HRTEM) images of CdTe (left), CdTe-CdS core-shell (middle), CdTe-CdS-ZnS core-shell-shell (right) aqQDs. (c) Photograph of the wide spectral range of bright luminescence from aqQDs aqueous solution under irradiation with 365-nm ultraviolet light from a UV lamp. (Reprinted with permission from [12d], 2008 Wiley).

luminescence. They further compared the photostability of the QDs and Alexa 488 (one kind of commercial organic dyes). Significantly, labeling signals of Alexa 488 faded quickly and became undetectable within 2 min; in striking contrast, the QDs preserved stable and bright fluorescence for 3 min illumination period, which provides powerful demonstration of superior photostability of QDs compared to organic dyes [15]. Meanwhile, Simon and coworkers developed procedures for using QDs to label live cells and demonstrated their use for long-term multicolor imaging of live cells. The live cells were labeled via two approaches, i.e., (1) endocytic uptake of QDs and (2) selective labeling of cell surface proteins with antibodies-conjugated QDs. The QDs-labeled cells were readily long-time monitored via tracking QDs fluorescence due to robust photostability of QDs [16]. Wang et al employed aqQDs-based nanospheres for *in vitro* imaging, demonstrating that the aqQDs were superbly efficacious for long-term and high-specificity immunofluorescent cellular labeling, and multi-color cell imaging [12e]. Based on progresses of QDs-based cellular imaging, the QDs bioprobes were further utilized for *in vivo* imaging. Typically, Nie et al designed a triblock polymer-encapsulated and bioconjugated QDs with excellent aqueous dispersibility and strong luminescence. The prepared QDs were effectively targeted tumors both by the enhanced permeability and retention (EPR) of tumor sites and by antibody binding to cancer-specific cell surface biomarkers. Moreover, multicolor fluorescence imaging of cancer cells under *in vivo* conditions was achieved by using both subcutaneous injection of QD-tagged cancer cells and systemic injection of QD probes. Their studies offer high promise for ultrasensitive and multiplexed imaging of tumor targets *in vivo* [17]. It is worthwhile to point out that, near-infrared (NIR) fluorescence imaging is widely recognized as an effective method for high-resolution and high-sensitivity bioimaging due to minimized biological autofluorescence background and increased penetration of excitation and emission light through tissues in the NIR wavelength window (700-900 nm) [18]. More recently, He et al developed a kind of NIR-emitting, ultrasmall-sized aqQDs, and further employed the prepared QDs for highly spectrally and spatially resolved imaging in cells and animals. The NIR QDs were specifically highly accumulated in the tumor region through a passive targeting process caused by an enhanced permeability and retention (EPR) effect. Significantly, the fluorescent signals of QDs were distinctively bright, clearly spectrally and spatially resolved, despite the strong autofluorescence background in mouse. This study clearly demonstrates the advantages of NIR QDs for in-vivo imaging for which the QDs fluorescence and biological autofluorescence of the mouse are spectrally separated, and that the NIR-emission is less absorbed by tissues as compared to visible luminescence [13].

1.4 Biosafety assessment of II-VI QDs

While fluorescent II-VI QDs are recognized as novel high-performance biological probes and at the forefront of nano-biotechnology research, sufficient and objective assessment of QDs-relative biosafety is necessary for their wide ranging bioapplications. To meet requirement of practical biological and biomedical applications, a large amount of studies on biosafety assessment of the QDs have been carried out.

Bhatia et al shown that surface oxidation of QDs led to the formation of reduced Cd on the QD surface and release of free cadmium ions, and correlated with cell death [19]. Yamamoto et al found that the cytotoxicity of QDs was not only caused by the nanocrystalline particle itself, but also by the surface-covering molecules of QDs, i.e., surface-covered functional groups (e.g., -NH2 and -COOH) covering on the surface of QDs [20]. Parak et al demonstrated that, in addition to the release of Cd ions from the surface of QDs, QDs precipitation on the cell surface could also impaire cells. They further suggested that cytotoxic effects were different in the case that QDs are ingested by the cells compared to the case that QDs were just present in the medium surrounding the cells [21]. Fan et al presented systematic cytotoxicity assessment of a series of aqQDs, i.e., thiols-stabilized CdTe, CdTe/CdS core-shell structured and CdTe/CdS/ZnS core-shell-shell structured QDs. They demonstrated that the CdTe aqQDs were highly toxic for cells; epitaxial growth of a CdS layer reduced the cytotxicity of QDs; and the presence of a ZnS outlayer greatly improved the biocompatibility of aqQDs (Figure 4) [22]. In their following investigation, they further revealed relationship between the cytotoxicity of aqQDs and free cadmium ions. Significantly, they found that the CdTe aqQDs were more cytotoxic than $CdCl_2$ solutions even when the intracellular Cd^{2+} concentrations were identical in the treated cells, implying the cytotoxicity of aqQDs cannot be attributed solely to the toxic effect of free Cd2+, but also dependent on the concentration of total aqQDs ingested by cells [23]. These studies are useful for understanding the *in vitro* toxicity of QDs, and for systematical assassment of cytotxicity of QDs.

(a)

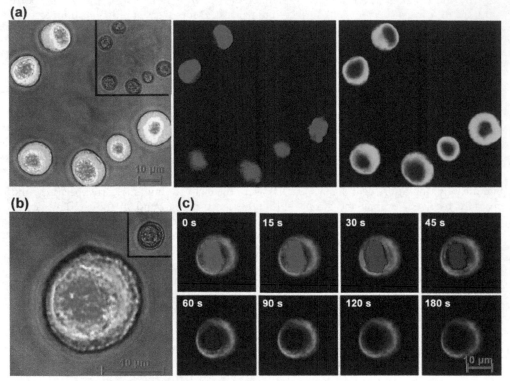

Fig. 3. Two-color staining of fixed cells. a, Fluorescent microscopy images of leukaemia K562 cells stained with the aqQDs-based nanospheres and PI dye. b, Fluorescent microscopy image of a single K562 cell labeled with the aqQDs-based nanospheres and PI dye. Inset displays the corresponding bright-field image. c, Temporal evolution of fluorescent signals of the K562 cells labeled with the aqQDs-based nanospheres (green fluorescent signal) and PI (red fluorescent signal). Images are captured with a cooled CCD camera at 15 s intervals automatically. Images at 0, 15, 30, 45, 60, 90, 120, and 180 s are shown. (Reprinted with permission from [12e], 2011 Biomaterials).

Such *in vitro* achievements are useful for biosafety assessment of the QDs; notwithstanding, comprehensive studies concerning *in vivo* toxicity are superior, since the results will assist in pinpointing the potential target organ and cells involved [24]. As a result, in the past several years, *in vivo* toxicity of QDs has been extensively studied. Particularly, Chan's group presented the first quantitative report on the *in vivo* biodistribution of QDs in 2006 [25]. In their latest study, they further systematically studied short- and long-term toxicity, animal survival, hematology, biochemistry, and organ histology of the QDs. Significantly, they demonstrated that the QDs did not cause appreciable toxicity *in vivo* even over long-term periods (e.g., 80 days), which differs from *in vitro* results (e.g, obvious QDs-induced cytotoxicity) [26]. Besides, Choi et al recently revealed that *in vivo* behavior of QDs are greatly dependent on their hydrodynamic diameters. Of particularle note, they found that, compared

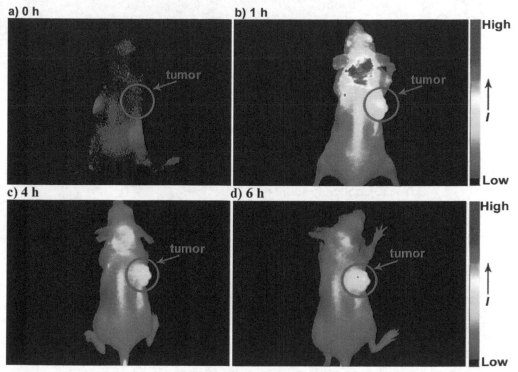

Fig. 4. *In vivo* tumor targeting of NIR QDs. Spectrally unmixed *in vivo* fluorescence images of KB tumor bearing nude mice at 0, 1 h, 4 h, and 6 h post injection of the prepared QDs. Mouse autofluorescence was removed by spectral unmixing in the above images. High tumor uptake of QDs was observed for the tumor models. (Reprinted with permission from [13], 2011 Wiley).

to those of large hydrodynamic diameter (> 15 nm), the QDs with smaller hydrodynamic diameter were more rapidly and efficiently eliminated from the mice through renal clearance. On the basis of which, they suggested that QDs with final hydrodynamic diameter < 5.5 nm produce feeble *in vivo* toxicity, and thus are more favorable for further bioapplications [27]. It is worthwhile to point out that, the QDs studied in all these mentioned publications are prepared via organometallic routes (organic synthesized QDs, orQDs). By using CdTe aqQDs as models, He et al carried out a comprehensive investigation on *in vivo* behaviors of the aqQDs, including their short- and long-term *in vivo* biodistribution, pharmacokinetics, and toxicity. They found that the biodistribution was largely dependent on hydrodynamic diameters of the QDs, blood circulation time, and types of organs. Based on histological and biochemical quantitative analysis, and real-time body weight measurement, we further revealed that, the aqQDs produced no overt adverse effect to mice [28]. These studies are important for understanding the *in vivo* toxicity of QDs, and for designing QDs with acceptable biocompability for biomedical applications.

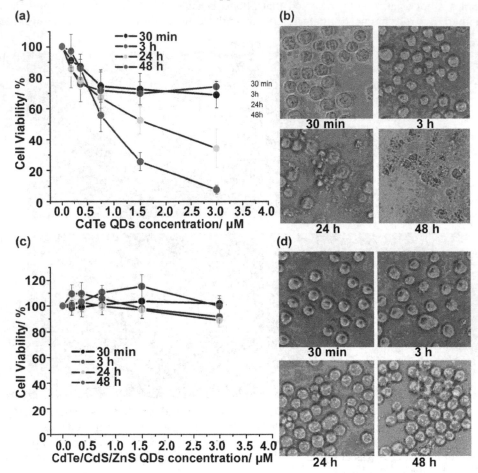

Fig. 5. Cytotoxicity of CdTe QDs (a,b) and CdTe-CdS-ZnS (c,d) QDs with different concentrations and incubation time with K562 cells. (a,c) Viability of K562 cells after treat with CdTe (a) and CdTe-CdS-ZnS (c) QDs. (b,d) Morphology of K562 cells after incubated with 3 μM CdTe (b) and CdTe-CdS-ZnS (d) QDs for 0.5, 3, 24, 48 h, respectively. (Reprinted with permission from [22], 2009 Biomaterials).

2. Carbon/silicon quantum dots-based biological fluorescent probes

2.1 Introduction

Two kinds of fluorescent QDs, i.e., carbon QDs (CQDs) and silicon QDs (SiQDs), have been recently developed as promising fluorescent biological probes for *in vivo* and *in vitro* imaging. Carbon/silicon-based nanostructures, such as nanoribbons, nanowires, nanotubes, and nanodots, are being intensely investigated and utilized for various applications ranging from electronics to biology [29]. Particularly, the quantum confinement phenomenon in CQDs and SiQDs is the focus of research since it would increase the probability of irradiative recombination via indirect-to-direct band gap transition, leading to enhanced fluorescent intensity and the prospect of long-awaited optical applications. In the past several years, exciting progress on development of CQDs/SiQDs-based fluorescent biological probes have been achieved.

2.2 Carbon quantum dots-based biological fluorescent probes

Carbon nanostructures, such as fullerenes, carbon nanotubes (CNTs), graphene, and carbon quantum dots (CQDs) have been utilized in a broad range of technological applications due to their many attractive advantages, including high natural abundance of carbon, low specific weight, favorable biocompatibility, as well as the chemical and thermal robustness, et al [30]. Of particular note, recent report revealed that CQDs could produce fluorescence in some specific conditions, which is attributed to passivated defects on the surface of carbon nanostructures acting as excitation energy traps [31]. As a result, CQDs have been extensively explored as novel fluorescent bioprobes owing to their strong fluorescence and low toxicity.

Sun's group reported a kind of carbon nanodots that were produced via laser ablation of carbon target in the presence of water vapor with argon as carried gas. The prepared nanodots yield bright luminescence (PQLY: 4-10%) upon the surface passivation by attaching organic species to their surface. such resultant carbon nanodots preserved stable fluorescence with respect to photoirradiation, exhitiging no meaningful reduction in the observed intensities under continuously long-time excitations [32]. On the basis of which, they further developed the carbon nanodots exhibiting strong luminescence with two-photon excitation. The prepared nanodots were thus utilized for cell imaging with two-photo luminescence microscopy. The labeled cell membrane and the cytoplasm of MCF-7 cells showed distinct green signals of the nanodots [33]. Liu et al presented multicolor fluorescent, ultrasmall (sizes: < 2 nm) and water-dispersible CQDs obtained from the combustion soot of candles. The CQDs fluoresce with different color (the emission-peak wavelengths ranging from 415 (violet) to 615 nm (orange-red)) under a single-wavelength UV excitation. Moreover, the CQDs contain carboxylic acid groups on their surface, allowing functionalization with biomolecules through N-hydroxysuccinimide (NHS) chemistry to construct CQDs-based bioprobes [34]. Thereafter, Sun et al demonstrated the first study of carbon nanodots for optical imaging *in vivo*. Particularly, the nanodots injected in various ways (e.g., subcutaneous injection, interdermal injection, and intravenous injection) into mice remain strongly fluorescent *in vivo* (Figure 7). They further revealed that no animal exhibited any sign of acute toxicological responses due to favorable biocompatibility of CQDs [35]. More recently, Lee's group developed a facile one-step alkali-assisted electrochemical fabrication of highly luminescence (PLQY: ~12%) CQDs. The CQDs with sizes of 1.2-3.8 nm displayed controllable fluorescence via adjustment of current density. In addition to the strong and size-controllable luminescence, the prepared CQDs featured excellent aqueous dispersibility and upconversion luminescence properties, offering great potential for optical bioimaging and related biomedical applications [36].

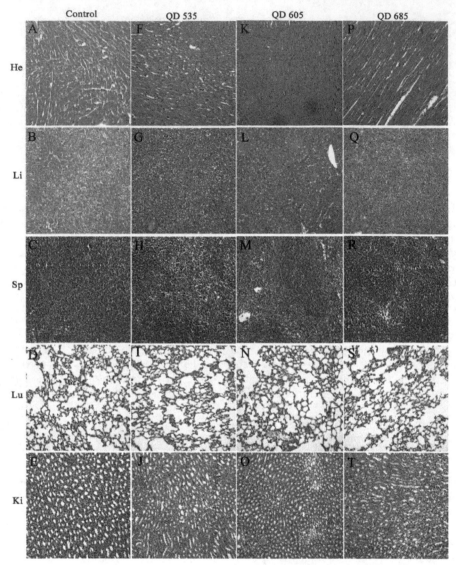

Fig. 6. Representative organ histology for control and treated animals. For control (A-E) and injected with aqQD535 (F-J), aqQD605 (K-O) and aqQD685 (P-T) animals, heart (He), liver (Li), spleen (Sp), lung (Lu) and kidney (Ki) are shown. Our analysis shows that organs did not exhibit signs of toxicity. (Reprinted with permission from [28], 2011 Biomaterials).

On the other hand, fluorescent nanodiamons, as another typical kind of carbon nanostructures, have also been explored as promising fluorescent biomarkers for *in vitro* and *in vivo* studies. Due to page limitation, this chapter will not discuss their bioimaging applications in a detailed way. The corresponding progresses could be referred in previous reports [37-39].

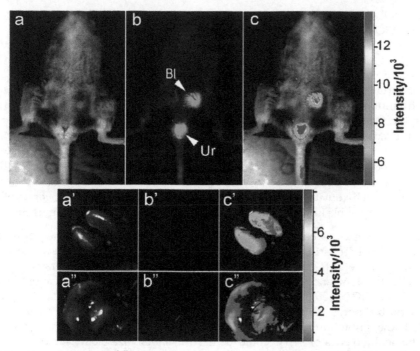

Fig. 7. Intravenous injection of fluorescent carbon nanodots: (a) bright field, (b) as-detected fluorescence (BI, bladder; Ur, urine), and (c) color-coded images. The same order is used for the images of the dissected kidneys (a' – c') and liver (a'' – c''). (Reprinted with permission from [35], 2009 American Chemical Society).

2.3 Silicon quantum dots-based biological fluorescent probes

Silicon nanomaterials are a type of important nanomaterials with attractive properties including excellent electronic/mechanical properties, favorable biocompatibility, huge surface-to-volume ratios, surface tailorability, improved multifunctionality, as well as their compatibility with conventional silicon technology [29,40]. Consequently, there has been great interest in developing functional silicon nanomaterials for various applications ranging from electronics to biology. To meet increasing demands of silicon-based applications, silicon materials of various nanostructures (e.g., nanodot [41], nanowire [29a,42], nanorod [43], and nanoribbon [44]) have been developed. Particularly, previous research reveal that quantum confinement phenomenon in SiQDs can increase the probability of irradiative recombination via direct band gap transition [45], leading to improvement of fluorescent intensity and the prospect of optical applications. Due to their excellent biocompatibility and noncytotoxic property, SiQDs are considered as promising fluorescent biological probes for *in vivo* and *in vitro* imaging.

Since the first reports of room-temperature light emission from porous silicon in the early 1990s [46], a variety of fabrication techniques have been developed for preparation of silicon quantum dots, including solution-phase reductive [47], plasma-assisted aerosol precipitation [48], microemulsion [49], mechanochemical [50], laser ablation [51], and sonochemical synthesis [52]. While luminescent intensity of silicon is enhanced at the nanoscale, SiQDs usually possess inferior optical properties (quantum yield < 10%) to semiconductor II-VI QDs (e.g. CdSe, CdSe/ZnS QDs) with direct band gaps (quantum yield 80-90%) [3,12]. Recent theoretical studies reveal that optical properties of SiQDs are significantly influenced by surface oxidation. Particularly, oxygen bonded to the surface of silicon quantum dots often reduces the band gap and limits luminescent emission, resulting in low quantum yield [53]. Consequently, it is essential to avoid surface oxidation to improve the quantum yields of SiQDs [53,54]. Indeed, Kortshagen et al. recently reported successful preparation of SiQDs with remarkably high ensemble quantum yields exceeding 60% by using plasma-assisted synthesis with strict removal of oxygen and elaborate surface passivation [48b], which provides an excellent example. It demonstrates that SiQDs could possess quantum yields as high as II-VI QDs under optimum conditions. Lee et al. recently developed a polyoxometalate-assisted electrochemical etching method for synthesizing SiQDs with controllable luminescent colors (e.g., blue, orange, and red) (Figure 8) [55].

Despite these progresses on SiQDs synthesis, most as-prepared SiQDs are not well water-dispersible since their surfaces are covered by hydrophobic moieties (e.g., styrene, alkyl, and octene). Extensive efforts have been undertaken to realize aqueous dispersibility of SiQDs. In 2004, Ruckenstein and Li developed a UV-induced graft polymerization for surface modification of SiQDs [56]. These SiQDs became well water-dispersible with the grafting of a water-soluble poly (acrylic acid) (PAAc) layer. In addition, the grafted PAAc also improved the photoluminescence stability of the SiQDs. Moreover, high density of carboxylic acid moieties of PAAc could be used to immobilize biomolecules (e.g., protein). Photostability comparison of SiQDs and four types of organic dyes (e.g., Alexa 488, Cy5, fluorescein isothiocyanate (FITC), and laser dye styryl (LDS751)) demonstrated superior resistance to photobleaching than the conventional organic dyes. Such modified SiQDs with quantum yield of 24% were employed as biological probes for cell imaging, suggesting potential bioimaging applications of modified SiQDs. Tilley and co-workers later reported a room-temperature synthesis for preparing water-dispersed SiQDs that exhibited strong blue photoluminescence [57]. In their method, a platinum chemical was utilized as catalyst for initiating reaction between Si-H surface bonds of the SiQDs and C=C bonds of allylamine. The resultant blue-emitting SiQDs became hydrophilic because their surfaces were modified with allylamine. In addition to the good aqueous dispersibility as well as relatively high quantum yield (~10%), the allylamine-capped SiQDs possessed robust storage- and photo-stability. They kept stable optical properties for several-month and long-time (more than 1 h) UV irradiation. As a comparison, the photoluminescence from rhodamine 6G dropped by 60% under the same illumination conditions. Sato and Swihart utilized photoinitiated hydrosilylation to successfully attach propionic acid (PA) to the surface of SiQDs, thereby producing water-dispersible, PA-terminated SiQDs with average diameter of less than 2.4 nm [58]. Compared to the former two reports showing water-dispersed SiQDs of a single emission color (red or blue), this work is significant because the size and corresponding PL emission color of SiQDs could be controlled by varying conditions. PA-terminated SiQDs with continuous luminescent color from yellow to green were readily synthesized in their work. Recently, Erogbogbo, Swihart et al. revealed that most of the modified SiQDs often showed obvious PL degradation especially in biological media with different pH, despite their high storage- and photo-stability in water [59]. Low pH stability would severely hinder their broad applications in biology. For example, conjugation of SiQDs with antibodies would be technically difficult if they were instable at neutral and alkaline pH environment. They further developed a new kind of SiQDs encapsulated by phospholipid micelles. Significantly, such micelle-encapsulated SiQDs kept stable optical properties under various biologically relevant conditions of pH values (4-10) and temperatures (20-70 ℃) [59]. The micelle-encapsulated SiQDs were further used in multiple cancer-related *in vivo* applications, including tumor vasculature targeting, sentinel lymph node mapping, and multicolor imaging in live mice [60]. Tilley and coworkers systematically investigated the chemical reactions on molecules attached to the surface of SiQDs, and further developed a multi-stepped chemical method for surface modification of SiQDs [61]. This stepwise approach offers new opportunities to prepare the SiQDs with diverse and desirable functionalities. This study sheds new insight into biological applications of silicon quantum dots.

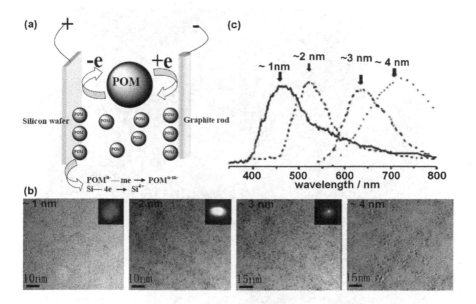

Fig. 8. (a) Schematics for the POMs-assisted electrochemical etching process. (b) TEM picture of serial sizes of SiQDs and their corresponding luminescence colors under UV irradiation (inset). (c) Typical PL spectra of SiQDs with sizes from ~1 to ~4 nm. (Reprinted with permission from [55], 2007 American Chemical Society.)

Lee and coworkers recently presented an EtOH/H2O2-assisted oxidation method to synthesize water-dispersed Si/SiOxHy core/ shell quantum dots with a Si core of different controlled diameters [62]. Significantly, this method allows for fine tuning emission wavelengths of QDs, producing seven luminescent colors from blue to red. On the basis of such studies [55,62] and theoretical prediction [63], they developed a new class of fluorescent silicon nanospheres (SiNSs) that each containing several hundreds of SiQDs (Figure 5). The as-prepared nanospheres, featuring excellent aqueous dispersibility, strong fluorescence, robust photo stability, and favorable biocompatibility, were further utilized for long-term cellular imaging [64]. Very recently, they developed a new microwave-assisted method for one-pot synthesis of high-quality SiQDs, which were facilely and rapidly prepared in short reaction times (e.g. 15 min). Remarkably, the ~4 nm SiQDs featured excellent aqueous dispersibility, robust photo- and pH-stability, strong fluorescence (PLQY: ~15%), and favorable biocompatibility (Figure 9). They further demonstrated that the prepared SiQDs were suitable for long-term immunofluorescent cellular imaging as biological probes. Particularly, the SiQDs-labeled microtubules yielded very stable fluorescent signals during continuous 240-min observation. In sharp contrast, the signals of the control groups using CdTe QDs (recognized as photostable fluorescent labels) or FITC (one typical kind of conventional fluorescent dyes) as fluorescent labels almost completely disappeared in 25-min irradiation under the same conditions (Figure 10) [65]. These studies well demonstrated the great promise for real-time and long-term bioimaging with the SiQDs-based fluorescent probes.

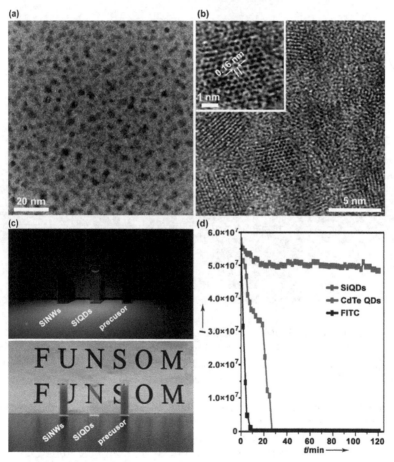

Fig. 9. TEM and HRTEM images (a,b), Inset in (b) presents the HRTEM image of a single SiQD. (c) Temporal evolution of fluorescence of the SiQDs under various pH values. (d) Photostability comparison of FITC, CdTe QDs, and as-prepared SiQDs. All samples are continuously irradiated by a 450 W xenon lamp. (Reprinted with permission from [65], 2011 American Chemical Society.)

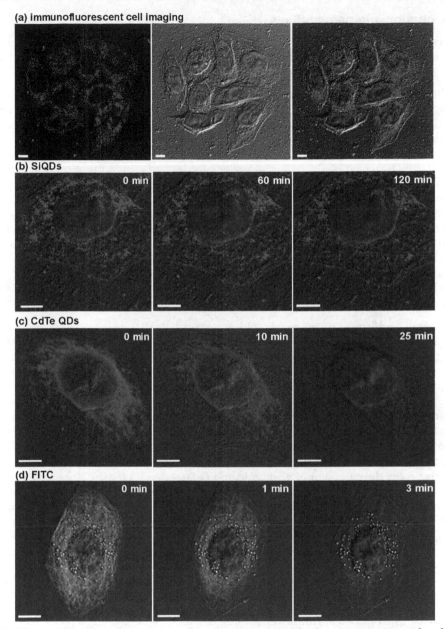

Fig. 10. Photos of immunofluorescent cell imaging captured by laser-scanning confocal microscopy. (a) Left: microtubules of Hela cells are distinctively labeled by the SiQDs/protein bioconjugates. Middle: bright field image. Right: superposition of fluorescence and transillumination images. (b)-(d) Stability comparison of fluorescence signals of Hela cells labeled by SiQDs (b), CdTe QDs (c), and FITC (d). Scale bar = 5 μm. (Reprinted with permission from [65], 2011 American Chemical Society.)

3. Conclusion and perspectives

In the past two decades, there have been considerable advances in the development of QDs-based fluorescent biological probes. Various kinds of high-quality QDs, i.e., II-VI QDs, CQDs, and SiQDs, have been developed for biological imaging.

II-VI QDs, serving as high-performance bioprobes, have been widely used for a variety of bioimaging applications, such as cell labeling, and tracking cell migration, tumor targeting, ect. Along with wide ranging bioapplications, concerns about their biosafety have attracted increasingly intensive attentions. *In vitro* studies have suggested that cytotoxicity of QDs is ascribed to release of toxic metals, production of reactive oxygen species, and hydrodynamic size of the QDs, which could be largely alleviated by surface modification (e.g., epitaxial growth of ZnS shell). On the other hand, *in vivo* experiments indicate that QDs of proper concentrations were not toxic to the animals (e.g., No apparent histopathological abnormalities or lesions are observed in QDs-treated mice). Therefore, while II-VI QDs are not completely innocuous, a safe range likely exists in which they can accomplish their task without major interference with the process under study. Notwithstanding, to meet requirement of practical biological and biomedical applications, further efforts are still urgently required to fully address the potential toxicity problem of the II-VI QDs, including the evaluation of QDs composition, surface chemistry, diameter, as well as the effect of their byproducts on biodistribution, toxicity, and pharmacokinetics.

The inherent problems, i.e. severe photobleaching or potential toxicity, associated with the traditional dyes or the fluorescent II/VI QDs remain completely unsolved, and have fueled a continual and urgent pursuit for new fluorescent bioprobes that are more photostable and biocompatible. Despite exciting progress on this area, extensive efforts (e.g., systematical assessment of QDs-relative biosafety, optimizing optical properties of CQDs/SiQDs, enhancing photo/chemical stability of QDs, investigating QDs behaviors in complicated biological environment, etc) are necessary to fit the demands of biological and biomedical applications. While there have been several exciting reports on the utility of CQDs/SiQDs for biological imaging, further efforts are necessary to modify and tailor CQDs/SiQDs architectures to fit the demands of biological and biomedical applications. One big challenge remaining is the development of economic and facile strategies for the large-scale synthesis of highly luminescent CQDs/SiQDs with controllable colors, which is the fundamental basis for their widespread bioapplications. Furthermore, effective methods of surface modification are required to further improve aqueous dispersibility and optical properties of CQDs/SiQDs. Moreover, a number of biological parameters have to be satisfactorily addressed before eventual *in vivo* and *in vitro* applications. While carbon and silicon are expected to be biocompatible, it is critically important to carry out systematic studies to assess their biosafety, including biodistribution and interactions between CQDs/SiQDs and biomolecules, cells and animals for *in vivo* bioimaging [1,2,66].

Key Words: Quantum dots; Fluorescent biological probes; Bioimaging; Silicon; Carbon

4. References

[1]   G. U. Nienhaus, *Angew. Chem. Int. Ed*. 2008, *47*, 8992-8994.

[2]   a) M. P. Bruchez, M. Moronne, P. Gin, S.Weiss, A. P. Alivisatos, *Science* 1998, *281*, 2013-2016. b) W. C. W. Chan, S. Nie, *Science* 1998, *281*, 2016-2018. c) X. Y. Wu, H. J. Liu, J.

Q. Liu, K. N. Haley, J. A. Treadway, J. P. Larson, N. F. Ge, F. Peale, M. P. Bruchez, *Nature Biotech.* 2003, *21*, 41-46. d) X. Michalet, F. F. Pinaud, L. A. Bentolila, J. M. Tsay, S. Doose, J. J. Li, G. Sundaresan, A. M. Wu, S. S. Gambhir, S. Weiss, *Science* 2005, *307*, 538-544.

[3] a) C. B. Murray, D. J. Noms, M. G. Bawendi, *J. Am. Chem. Soc.* 1993, *115*, 8706-8715. b) X. G. Peng, L. Manna, W. D. Yang, J. Wickham, E. Scher, A. Kadavanich, A. P. Alivisatos, *Nature* 2000, *404*, 59-61. c) Z. A. Peng, X. G. Peng, *J. Am. Chem. Soc.* 2001, *123*, 183-184. d) L. H. Qu, X. G. Peng, *J. Am. Chem. Soc.* 2002, *124*, 2049-2055. e) J. J. Li, Y. A. Wang, W. Z. Guo, J. C. Keay, T. D. Mishima, M. B. Johnson, X. G. Peng, *J. Am. Chem. Soc.* 2003, *125*, 12567-12575. f) J. M. Tsay, M. Pflughoefft, L. A. Bentolil, S. Weiss, *J. Am. Chem. Soc.* 2004, *126*, 1926-1927.

[4] T. Pellegrino, L. Manna, S. Kudera, T. Liedl, D. Koktysh, A. L. Rogach, S. Keller, J. Radler, G. Natile, W. J. Parak, *Nano Lett.* 2004, *4*, 703-707.

[5] a) D. Gerion, F. Pinau, S. C. Williams, W. J. Parak, D. Zanchet, S. Weiss, A. P. Alivisatos, *J. Phys. Chem. B* 2001, *105*, 8861-8871. b) H. Borchert, D. V. Talapin, N. Gaponik, C. McGinley, S. Adam, A. Lobo, T. Mo1ller, H. Weller, *J. Phys. Chem. B* 2003, *107*, 9662-9668.

[6] a) S. F. Wuister, I. Swart, F. V. Driel, S. G. Hickey, C. D. M. Donega, *Nano Lett.* 2003, *3*, 503-507. b) H. B. Bao, Y. J. Gong, Z. Li, M. Y. Gao, *Chem. Mater.* 2004, *16*, 3853-3859. c) Y. Wang, Z. Y. Tang, M. A, C. Duarte, I. P. Santos, M. Giersig, N. A. Kotov, L. M. L. Marza, *J. Phys. Chem. B* 2004, *108*, 15461-15469.

[7] N. Gaponik, D. V. Talapin, A. L. Rogach, K. Hoppe, E. V. Shevchenko, A. Kornowski, A. Eychmu1ller, H. Weller, *J. Phys. Chem. B* 2002, *106*, 7177-7185.

[8] a) M. T. Crisp, N. A. Kotov, *Nano Lett.* 2003, *3*, 174-177. b) J. Guo, W. L. Yang, C. C. Wang, *J. Phys. Chem. B* 2005, *109*, 17467-17473.

[9] H. Zhang, L. P. Wang, H. M. Xiong, L. H. Hu, B. Yang, W. Li, *Adv. Mater.* 2003, *15*, 1712-1715.

[10] a) D. M. P, Mingos, D. R. Baghurst, *Chem. SOC. Rev.* 1991, *20*, 1-47. b) S. A. Galema, *Chemical Society Reviews* 1997, *26*, 233-238. c) H. Grisaru, O. Palchik, A. Gedanken, V. Palchik, M. A. Slifkin, A. M. Weiss, *Inorganic Chemistry* 2003, *42*, 7148-7155. d) D. J. Brooks, R. Brydson, R. E. Douthwaite, *Adv. Mater.* 2005, *17*, 2474-2477. e) V. Swayambunathan, D. Hayes, K. H. Schmidt, Y. X. Liao, D. Meisel, *J. Am. Chem. Soc.* 1990, *112*, 3831-3837. f) J. A. Gerbec, D. Magana, A. Washington, G. F. Strouse, *J. Am. Chem. Soc.* 2005, *127*, 15791-15800.

[11] L. Li, H. F. Qian, J. Ren, *Chem. Commun.* 2005, 528-530.

[12] a) Y. He, H. T. Lu, L. M. Sai, W. Y. Lai, Q. L. Fan, L. H. Wang, W. Huang, *J. Phys. Chem. B* 2006, *110*, 13352-13356. b) Y. He, H. T. Lu, L. M. Sai, W. Y. Lai, Q. L. Fan, L. H. Wang, W. Huang, *J. Phys. Chem. B* 2006, *110*, 13370-13374. c) Y. He, L. M. Sai, H. T. Lu, M. Hu, W. Y. Lai, Q. L. Fan, L. H. Wang, W. Huang, *Chem. Mater.* 2006, *19*, 359-365. d) Y. He, H. T. Lu, L. M. Sai, Y. Y. Su, M. Hu, C. H. Fan, W. Huang, L. H. Wang, *Adv. Mater.* 2008, *20*, 3416-3421. e) Y. He, H. T. Lu, Y. Y. Su, L. M. Sai, M. Hu, C. H. Fan, L. H. Wang, *Biomaterials* 2011, *32*, 2133-2140.

[13] Y. He, Z. H. Kang, Q. S. Li, Chi Him A. Tsang, C. H. Fan, S. T. Lee, *Angew. Chem. Int. Ed.* 2009, *48*, 128-132.

[14] a) M. B. Jr, M. Moronne, P. Gin, S. Weiss, A. P. Alivisatos, *Science* 1998, *281*, 2013-2016. b) W. C. W. Chan, S. Nie, *Science* 1998, *281*, 2016-2018.

[15] X. Y. Wu, H. J. Liu, J. Q. Liu, K. N. Haley, J. A. Treadway, J. P. Larson, N. F. Ge, F. Peale, M. P. Bruchez, *Nat. Biotechnol.* 2003, *21*, 41-46.

[16] J. K. Jaiswal, H. Mattoussi, J. Mattoussi, J. M. Mauro, S. M. Simon, *Nat. Biotechnol.* 2003, *21*, 47-51.

[17] X. H. Gao, Y. Y. Cui, Richard M Levenson, L. W. K. Chung, S. Nie, *Nat. Biotechnol.* 2004, *22*, 969-976.

[18] a) R. Weissleder, Nat. Biotechnol. 2001, 19, 316-317. b) R. Weissleder, V. Ntziachristos, *Nat. Med.* 2003, *9*, 123-128. c) S. Lee, E. J. Cha, K. Park, S. Y. Lee, J. K. Hong, I. C. Sun, S. Y. Kim, K. Choi, I. C. Kwon, K. Kim, C. H. Ahn, *Angew. Chem. Int. Ed.* 2008, *47*, 2804-2807.

[19] A. M. Derfus, W. C. W. Chan, S. N. Bhatia, *Nano Lett.* 2004, *4*, 11-18.

[20] A. Hoshino, K. Fujioka, T. Oku, M. Suga, Y. F. Sasaki, T. Ohta, M. Yasuhara, K. Suzuki, K. Yamamoto, *Nano Lett.* 2004, *4*, 2163-2169.

[21] C. Kirchner, T. Liedl, S. Kudera, T. Pellegrino, A. M. J. H. E. Gaub, S. Stolzle, N. Fertig, W. J. Parak, *Nano Lett.* 2005, *5*, 331-338.

[22] Y. Y. Su, Y. He, H. T. Lu, L. Sai, Q. N. Li, W. X. Li, L. H. Wang, P. P. Shen, Q. Huang, C. H. Fan, *Biomaterials* 2009, *30*, 19-25.

[23] Y. Y. Su, M. Hub, C. H. Fan, Y. He, Q. N. Li, W. X. Li, L. H. Wang, P. P. Shen, Q. Huang, *Biomaterials* 2010, *31*, 4829-4834.

[24] J. Edward, B. S. J, C. Meghan, D. A. M, V. Yuri, G. k. Y. K, K. N. A, *ACS Nano* 2008, *2*, 928-932.

[25] H. C. Fischer, L. C. Liu, K. S. Pang, W. C. W. Chan, *Adv. Funct. Mater.* 2006, *16*, 1299-1305.

[26] T. S. Hauck, R. E. Anderson, H. C. Fischer, S. Newbigging, W. C. W. Chan, *Small* 2010, *1*, 138-144.

[27] H. S. Choi, W. H. Liu, P. Misra, E. Tanaka, J. P. Zimmer, B. I. Ipe, M. G. Bawendi, J. V. Frangioni, *Nat. Biotechnol.* 2007, *25*, 1165-1170.

[28] Y. Y. Su, F. Peng, Z. Y. Jiang, Y. L. Zhong, Y. M. Lu, X. X. Jiang, Q. Huang, C. H. Fan, S. T. Lee, Y. He, *Biomaterials* 2011, *32*, 5855-5862.

[29] a) D. D. D. Ma, C. S. Lee, F. C. K. Au, S. Y. Tong, S. T. Lee, *Science* 2003, *299*, 1874-1877. b) F. Lu, L. Gu, M. J. Meziani, X. Wang, P. G. Luo, L. M. Veca, L. Cao, Y. P. Sun, *Adv. Mater.* 2009, *21*, 139-152. c) Y. He, C. H. Fan, S. T. Lee, *Nano Today* 2010, *5*, 282-295. d) Y. He, S. Su, T. T. Xu, Y. L. Zhong, J. A. Zapien, J. Li, C. Fan, S. T. Lee, *Nano Today* 2011, *6*, 122-130. e) Y. He, Y. L. Zhong, F. Peng, X. P. Wei, Y. Y. Su, S. Su, W. Gu, L. S. Liao, S. T. Lee, *Angew. Chem. Int. Ed.* 2011, *50*, 3080-3083.

[30] T. N. Hoheisel, S. Schrettl, R. Szilluweit, H. Frauenrath, *Angew. Chem. Int. Ed.* 2010, *49*, 6496-6515.

[31] X. Y. Xu, R. Ray, Y. L. Gu, H. J. Ploehn, L. Gearheart, K. Raker, W. A. Scrivens, *J. Am. Chem. Soc.* 2004, *126*, 12736-12737.

[32] Y. P. Sun, B. Zhou, Y. Lin, W. Wang, K. A. S. Fernando, P. Pathak, M. J. Meziani, B. A. Harruff, X. Wang, H. F. Wang, P. G. Luo, H. Yang, M. E. Kose, B. Chen, L. M. Veca, S. Y. Xie, *J. Am. Chem. Soc.* 2006, *128*, 7756-7757.

[33] L. Cao, X. Wang, M. J. Meziani, F. Lu, H. F. Wang, P. G. Luo, Y. Lin, B. A. Harruff, L. M. Veca, D. Murray, S. Y. Xie, Y. P. Sun, *J. Am. Chem. Soc.* 2007, *129*, 11318-11319.

[34] H. Liu, T. Ye, C. Mao, *Angew. Chem. Int. Ed.* 2007, *46*, 6473-6475.

[35] S. T. Yang, L. Cao, P. G. Luo, F. Lu, Xin Wang, H. F. Wang, M. J. Meziani, Y. F. Liu, G. Qi, Y. P. Sun, *J. Am. Chem. Soc.* 2009, *131*, 11308-11309.

[36] H. T. Li, X. D. He, Z. H. Kang, H. Huang, Y. Liu, J. L. Liu, S. Y. Lian, C. H. A. Tsang, X. B. Yang, S. T. Lee, *Angew. Chem. Int. Ed.* 2010, *49*, 4430-4434.

[37] S. J. Yu, M. W. Kang, H. C. Chang, K. M. Chen, Y. C. Yu, *J. Am. Chem. Soc.* 2005, *127*, 17604-17605.

[38] C. C. Fu, H. Y. Lee, K. Chen, T. S. Lim, H. Y. Wu, P. K. Lin, P. K. Wei, P. H. Tsao, H. C. Chang, W. Fann, *PNAS* 2007, *104*, 727-732.

[39] N. Mohan, C. S. Chen, H. H. Hsieh, Y. C. Wu, H. C. Chang, *Nano Lett.* 2010, *10*, 3692-3699.

[40] a) L. Pavesi, L. D. Negro, C. Mazzoleni, G. Franzo, F. Priolo, *Nature* 2000, *408*, 440-444. b) Z. F. Ding, B. M. Quinn, Haram, K. P. Santosh, E. Lindsay, Korgel, A. Brian, A. JBard, *Science* 2002, *296*, 1293-1298. c) J. E, Allen, E. R. Hemesath, D. E. Perea, J. L. Lenschfalk, Z. Y. Li, F. Yin, M. H. Gass, P. Wang, A. L. Bleloch, R. E. Palmer, L. J. Lauhon, *Nature nanotechnology* 2008, *3*, 168-173. d) G. F. Grom, D. J. Lockwood, J. P. McCaffrey, H. J. Labbé, P. M. Fauchet, J. B. White, J. Diener, D. Kovalev, F. Koch, L. Tsybeskov, *Nature* 2000, *407*, 358-361.

[41] a) J. D. Holmes, K. J. Ziegler, R. C. Doty, L. E. Pell, K. P. Johnston, B. A. Korge, *J. Am. Chem. Soc.* 2001, *123*, 3743-3748. b) M. Cavarroc, M. Mikikian, G. Perrier, L. Boufendi, *Appl. Phys. Lett.* 2006, *89*, 013107−013110.

[42] V. Schmidt, J. V. Wittemann, S. Senz, U. Gosele, *Adv. Mater.* 2009, *21*, 2681-2702.

[43] a) J. G. Fan, X. J. Tang, Y. P. Zhao, *Nanotechnology* 2004, *15*, 501-504. b) Hawker, T. P. Russell, M. Steinhart, U. Gosele, *Nano Lett.* 2007, *7*, 1516-1520.

[44] a) A. J. Baca, M. A. Meitl, H. C. Ko, S. Mack, H. S. Kim, J. Y. Dong, P. M. Ferreira, J. A. Rogers, *Adv. Funct. Mater.* 2007, *17*, 3051-3062. b) H. C. Ko, A. J. Baca, J. A. Rogers, *Nano Lett.* 2006, *6*, 2318−2324.

[45] a) W. L. Wilson, P. F. Szajowski, L. E. Brus, *Science* 1993, *262*, 1242-1244. b) N. M. Park, C. J. Choi, T. Y. Seong, S. J. Park, *Phys. Rev. Lett.* 2001, *86*, 1355-1357.

[46] a) Canham, L. T. *Appl. Phys. Lett.* 1990, *57*, 1046-1048. b) Cullis, A. G.; Canham, L. T. *Nature* 1991, *335*, 335-338.

[47] a) C. S. Yang, R. A. Bley, S. M. Kauzlarich, H. W. H. Lee, G. R. Delgado, *J. Am. Chem. Soc.* 1999, *121*, 5191-5195. b) R. K. Baldwin, K. A. Pettigrew, E. Ratai, M. P. Augustine, S. M, Kauzlarich, *Chem. Commun.* 2002, 1822-1823.

[48] a) L. Mangolini, E. Thimsen, U. Kortshagen, *Nano Lett.* 2005, *5*, 655-659. b) D. Jurbergs, E. Rogojina, *Appl. Phys. Lett.* 2006, *88*, 233116-1-233116-2.

[49] R. D. Tilley, K. Yamamoto, *Adv. Mater.* 2006, *18*, 2053-2056.

[50] A. S. Heintz, M. J. Fink, B. S. Mitchell, *Adv. Mater.* 2007, *19*, 3984-3988.

[51] D. Riabinina, C. Durand, M. Chaker, F. Rosei, *Appl. Phys. Lett.* 2006, *88*, 073105-1-073105-3.

[52] N. A. Dhas, C. P. Raj, A. Gedanken, *Chem. Mater.* 1998, *10*, 3278-3281.

[53] Z. Zhou, L. Brus, R. Friesner, *Nano Lett.* 2003, *3*, 163-167.

[54] A. Puzder, A. J. Williamson, J. C. Grossman, G. Galli, *J. Am. Chem.Soc.* 2003, *125*, 2786-2791.

[55] Z. H. Kang, C. H. A. Tsang, Z. D. Zhang, M. L. Zhang, N. B. Wong, J. A. Z. Shan, S. T. Lee, *J. Am. Chem. Soc.* 2007, *129*, 5326-5327.

[56] Z. F. Li, E. Ruckenstein, *Nano Lett.* 2004, *4*, 1463-1467.

[57] J. H. Warner, A. Hoshino, K. Yamamoto, R. D. Tilley, *Angew. Chem. Int. Ed.* 2005, *44*, 4550-4554.

[58] S. Sato, M. T. Swihart, *Chem. Mater.* 2006, *18*, 4083-4088.

[59] F. Erogbogbo, K. T. Yong, I. Roy, G. X. Xu, P. N. Prasad, M. T. Swihart, *ACS Nano* 2008, 2, 873-878.

[60] F. Erogbogbo, K. T. Yong, I. Roy, R. Hu, W. C. Law, W. W. Zhao, H. Ding, F. Wu, R. Kumar, M. T. Swihart, P. N. Prasad, *ACS Nano* 2011, 5, 413-423.

[61] A. Shiohara, S. Hanada, S. Prabakar, K. Fujioka, T. H. Lim, K. Yamamoto, P. T. Northcote, R. D. Tilley, *J. Am. Chem. Soc.* 2010. *132*, 248-253.

[62] Z. H. Kang, Y. Liu, C. H. A. Tsang, D. D. D. Ma, X. Fan, N. B. Wong, S. T. Lee, *Adv. Mater.* 2009, *21*, 661-664.

[63] a) Q. S. Li, R. Q. Zhang, S. T. Lee, T. A. Niehaus, T. Frauenheim, *Appl. Phys. Lett.* 2008, *92*, 053107-1-053107-3. b) Q. S. Li, R. Q. Zhang, T. A. Niehaus, T. Frauenheim, S. T. Lee, *J. Chem. Theory Comput.* 2007, *3*, 1518-1526. c) X. Wang, R. Q. Zhang, T. A. Niehaus, T. Frauenheim, *J. Phys. Chem. C* 2007, *111*, 2394-2400.

[64] a) Y. He, Z . H. Kang, Q. S. Li, C. H. A. Tsang, C. H. Fan, S. T. Lee, *Angew. Chem. Int. Ed.* 2009, *48*, 128-132. b) He, Y. Y. Su, X. B. Yang, Z. H. Kang, T. T. Xu, R. Q. Zhang, C. H. Fan, S. T. Lee, *J. Am. Chem. Soc.* 2009, *131*, 4434-4438.

[65] Y. He, Y. L. Zhong, F. Peng, X. P. Wei, Y. Y. Su, Y. M. Lu, S. Su, W. Gu, L. S. Liao, S. T. Lee, *J. Am. Chem. Soc.* 2011, *133*,14192-14195.

[66] S. P. Song, Y. Qin, Y. He, Q. Huang, C. H. Fan, H. Y. Chen, *Chem. Soc. Rev.* 2010, *39*, 4234-4243.

Energy Transfer-Based Multiplex Analysis Using Quantum Dots

Young-Pil Kim

Dept. of Life Science, Hanyang University,
Republic of Korea

1. Introduction

Semiconductor nanocrystals, also known as quantum dots (QDs), have emerged as a significant new class of materials over the past decade. The capabilities of QDs – high quantum yield, improved sensitivity, high photostability, and size-tunable colors have paved the way for numerous studies including imaging, sensing and targeting biomolecules. Thus QDs are now rapidly replacing traditional fluorophores in almost all fluorescence-based applications. Unlike organic dyes and fluorescent proteins, QDs are size-tunable with non-overlapping emission band profiles due to their narrow and symmetric emission bands (full width at half maximum of 25–40 nm) that can span the light spectrum from the ultraviolet even to the infrared. As illustrated in **Figure 1**, this property enables the QDs to be useful for multiplexing assay in a single run (Chan et al., 2002; Jaiswal et al., 2003; Wu et al., 2003). Moreover, QDs typically have very broad absorption spectra with very large molar extinction coefficients (0.5–5×10^6 M^{-1} cm^{-1}) (Hawrylak et al., 2000; Moreels and Hens, 2008). This makes QDs absorb 10-50 times more photons than organic dyes at the same excitation photon flux, providing a sufficient brightness for the sensing system (Gao et al., 2004). Owing to high photostability, QD-based sensing and imaging are favorable for continuous tracking studies over a long period of time. Most importantly, when the QD is harnessed in fluorescence resonance energy transfer (FRET), several advantages over organic dye-based probes have been acquired. Multiple binding of an energy acceptor per a single QD is expected to increase the overall energy transfer efficiency (Clapp et al., 2004; Zhang et al., 2005). Additionally, large Stokes shift of the QD can avoid the crosstalk between the donor QD and the acceptor counterpart because its broadband absorption allows excitation at a short wavelength that does not directly excite the acceptor. The continuously tunable emissions that can be matched to any desired acceptor, makes it possible to use many fluorophores for multiplexed assay (Medintz et al., 2003). The quencher (organic (Mauro et al., 2003) or metal (Kim et al., 2008a; Oh et al., 2005; Oh et al., 2006; Wargnier et al., 2004) substances) or emissive fluorescent molecules (fluorophores, proteins, or other QDs) (Wang et al., 2002) can be promising acceptors. While QDs are frequently used as donors in FRET, they may also play a critical role as energy acceptors either in bioluminescence resonance energy transfer (BRET)(Rao et al., 2006a; Rao et al., 2006b) or chemiluminescene resonance energy transfer (CRET)(Huang et al., 2006) as the energy donor (**Figure 2**). To this end, energy transfer system allows QDs to be suitable for many biological applications, such as the analyses of enzyme activity, protein-protein interactions, and other environmental conditions (pH, ion concentration and so forth). To avoid redundancy in a myriad of applications of QDs, the energy transfer-based detection will be focused here between QDs and other binding partners. A short overview regarding QD-FRET, QD-BRET, and QD-CRET will be demonstrated.

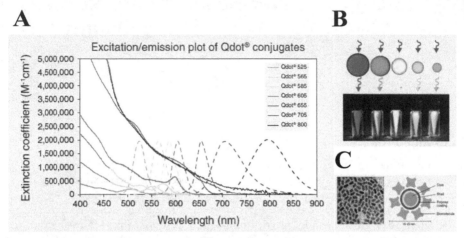

Fig. 1. Characters of Quantum Dot (QD). (A) Excitation (solid line) and emission spectra (dotted line) of CdSe quantum dots. (B) Size tuneability and emission color of five different QDs with the same long-wavelength UV lamp. (C) Structure of a QD nanocrystal: transmission electron microscope image of core shell QD at 200,000× magnification (left) and schematic of the overall structure (right). Figures are obtained from Invitrogen website (www.invitrogen.com).

2. Fluorescence resonance energy transfer (FRET) system using QDs

FRET is the most commonly utilized technique in these applications because of the high sensitivity, good reproducibility, and real-time monitoring capabilities. General configuration of FRET consists of chromophores with different combinations, such as auto-quenched probes (Kim et al., 2009; Weissleder et al., 1999), dual chromophore probes (Kircher et al., 2004), or multiphoton FRET-based probes (Stockholm et al., 2005). However, such organic fluorophores often have problems such as photobleaching, susceptibility to environment, difficulty in multiplexed analysis by specific paring between donor and acceptor. As aforementioned, these problems in FRET assays can be overcome when appropriate fluorophore or quencher is used in conjunction with quantum dots (QDs) (Medintz et al., 2006; Shi et al., 2006). In FRET, QDs are typically used as fluorescence donors while the fluorescent (or quenchable) acceptors are used as acceptors by appropriated labeling with biomolecules (DNA, aptamer, peptide and protein).

FRET-based QD-DNA nanosensor allows for detecting low concentrations of DNA, where the target strand binds to a dye-labelled reporter strand thus forming a FRET between QD and dye (Zhang et al., 2005). The QD also functions as a concentrator that amplifies the target signal by confining several targets in a nanoscale domain. They applied the nanosensors in combination with the oligonucleotide ligation assay to the detections of Kras point mutations (codon 12 GGT to GTT mutation) in clinical samples from patients with ovarian serous borderline tumours. Unbound nanosensors produce near-zero background fluorescence, but on binding to even a small amount of target DNA (~50 copies or less) they generate a very distinct FRET signal.

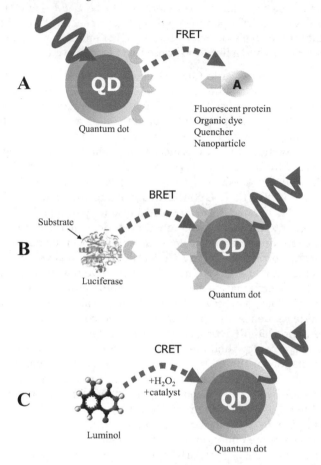

Fig. 2. Representative schematic of resonance energy transfer-based QD probes. QD-FRET (A), QD-BRET (B), and QD-CRET (C). A indicates energy acceptor.

QD-based FRET sensor has been increasingly combined with aptamers which take a great deal of advantages including high affinity, easy development through systematic evolution of ligands by exponential enrichment (SELEX), great stability and immuno-resistance. It has been reported that QD-aptamer (QD-Apt) conjugate can image and deliver anticancer drugs to prostate cancer (PCa) cells to the targeted tumor cells based on the mechanism of FRET(Bagalkot et al., 2007), in which RNA aptamers covalently attached to the surface of QD serves a dual function as targeting molecules and as drug carrying vehicles. When doxorubicin (Dox) intercalated within the A10 aptamer, a donor-acceptor quenching between QD and Dox was induced by the energy transfer mechanism. Specific uptake of QD-Apt(Dox) conjugates into target cancer cell through PSMA mediate endocytosis evokes the release of Dox from the QD-Apt(Dox) conjugates, resulting in the recovery of fluorescence from both QD and Dox and the intracellular delivery of Dox inside cancer cells. Levy et al. demonstrated an aptamer beacon-based approach for detecting the protein thrombin with QD-based fluorescence readout, in which the detection limit was determined to be 1 µM (Levy et al., 2005). Most recently, Cheng et al. Once MUC1 peptide binds to the aptamer strand on the surface of QD, fluorescence intensity successively decreases through FRET effect, showing low detection limit for MUC1 (nanomolar level) and a linear response found in blood serum.

Much attention has also been paid to QD-FRET for *in vitro* and *in vivo* assay of proteases. The ability of enzymes to target the specific substrate in complex milieu is essential for understanding the fidelity of most biological functions. Among enzymes, proteases are particularly significant because proteolytic processing is a final step to establish the functional roles of many expressed proteins and most of proteases are involved in major human cancers. Thus, methods to assay proteases and their inhibitors using QD-FRET have been of great interest in diagnosis of protease-relevant diseases and development of potential drugs. Assaying proteolytic activity can be easily designed by combining such peptide substrates with QD-FRET. Upon cleavage of the substrates by the protease, the resulting signals are regenerated from quenched or energy-trapped QDs because of the distruption of the energy transfer. Although conventional methods to adapt this FRET-based principle for monitoring *in vivo* protease activity has been made by using genetically engineered fluorescence protein (typically CFP-YFP), the real analysis of energy transfer is often hindered by cross-talk and bleed-through between two fluorescent proteins due to the narrow Strokes shift. By conjugating the energy acceptor groups (organic fluorophore or quencher (Chang et al., 2005; Kim et al., 2008b; Medintz et al., 2006; Shi et al., 2006)) with a QD through a peptide sequence, this problem can be overcome. A recent study by Shi *et al.* has shown a ratiometric QD-FRET probe to measure protease activity *in vivo*, consisting of the donor QD and the acceptor rhodamine labeled peptide (Rosenzweig et al., 2007; Shi et al., 2006). Local excitation of the QDs is able to induce efficient energy transfer into the adjacent rhodamine dye. Approximately 48 numbers of a rhodamine dye labeled substrate (RGDC) for collagenase were conjugated on a single QD through a sulfhydryl group of cysteine residue. When the probes were first tested for trypsin (500 µg/mL for 15 min) in solution, they gave rise to 60% increase in the photoluminescence of the QDs and a corresponding decrease in the emission of the rhodamine molecules, based on FRET signal changes (Rosenzweig et al., 2007). Similarly, upon the cleavage by collagenase, fluorescence from QD at 545 nm was recovered up to 60% while that of rhodamine at 560 nm were diminished. Thus, ratiometric change in fluorescence emissions of QD and rhodamine allowed for the real-time detection and quantification of collagenase activity. The activity of collagenase, matrix metalloproteinases (MMPs) was monitored in normal (HTB 125) and cancerous breast cells (HTB 126) in which collagenase is negligible. This result clearly showed that QD-dye FRET sensors could be very useful to detect protease activity by measuring the ratiometric change. Mattoussi group has reported the similar approach to measure the activities of different proteases by tuning the appropriate peptide substrate (Medintz et al., 2006). Modular peptides were rationally designed to have four different parts; i) an N-terminal hexahistidine(His$_6$) domain for self-assembly with QD, (ii) a helix-linker spacer region, (iii) an exposed protease recognition/cleavage sequence, and (iv) a C-terminal site-specific location (cysteine thiol) for dye attachment. In particular, the artificial residue alpha-amino isobutyric acid (Aib) and alanine were incorporated into the helix linker to provide rigidity, and an organic fluorophore or a quencher was attached to the C-term via thiol chemistry. For the detection of caspase-1 activity, dihydrolipoic acid (DHLA)-capped QD538 and the Cy3 was employed as the FRET pair, and QD-FRET probes for three other proteases (thrombin, collagenase and chymotrypsin) were also designed to have the quencher (QXL-520) instead of Cy3. They described that the relative FRET efficiency within each QD nano-assembly can be controlled through the number of peptides assembled per QD. Therefore, these assays provide quantitative data for protolytic activity by measuring enzymatic velocity, Michaelis–Menten kinetic parameters, and mechanisms of enzymatic inhibition. Unlike the conventional protease assay format in which substrate concentration has to be used in the micromolar range, they demonstrated that the developed format can be consistently used with lower substrate concentrations (200nM QD and peptide concentration of 0.2–1.0 μM) over a diverse selection of enzymes that manifest different specific activities. Moreover, it is likely that the substrates incorporated on the QD complex allow easy access to the desired protease and provide the high sensitivity and low background by being easily tuned. More recently, they utilized this system to detect caspase-3 activity, a key downstream effector of apoptosis. Along with that, the presence of calcium ions with the acceptor CaRbCl can increase the FRET efficienty, enabling calcium-sensitizing sensors (Medintz et al., 2010). Also they have reported this QD-dye FRET system to serve as a papain or proteinase K sensor, based on tunable coupler, 520QD and Cy3 (Clapp et al., 2008). In a similar way, Biswas *et al.* has showed that a genetically programmable protein module was designed to have a His$_6$, a cleavage site labeled with a Alexa dye via cysteine residue, an elastin-like peptide (ELP) domain for purification, and a flanking TAT peptide (Chen et al., 2011). This QD-dye FRET module was used for the detection of HIV-1 Pr activity *in vitro* and in cancer cells, which particularly takes responsibility for drug-resistance against rapidly mutating viruses such as HIV-1. Analysis of enzymatic inhibition was also performed in the presence of specific inhibitors. QD-FRET assay system to measure protease activity has been applied to chip-based format by Kim *et al* (Kim et al., 2007). While the photoluminescence (PL) of donor streptavidin-QD525 immobilized on a surface was quenched due to the presence of an energy acceptor (peptide substrates modified with TAMRA and biotin at N- and C-terminus, respectively) in close proximity, the protease activity caused modulation in the efficiency of the energy transfer between the acceptor and donor, thus enabling the highly sensitive detection of MMP-7 activity. In contrast to a solution-based analysis, the chip-based format allowed more reliable analysis, with no aggregation of QDs. Plus, this format required a much smaller reaction volume. This method is likely to have a potential to screen the activity of disease-associated proteases for the development of therapeutics and diagnostics in a high-throughput manner. In addition to organic fluorophores, a fluorescent protein was easily designed as an energy acceptor against QD donor. Boeneman *et al* (Medintz et al., 2009) has demonstrated that a red fluorescent protein (mCherry) expressing the caspase 3 cleavage site and a His$_6$ sequence were self-assembled to the surface of CdSe–ZnS DHLA QDs via metal affinity coordination, leading to FRET quenching of the QD and sensitized emission from the mCherry acceptor. Caspase-3 activity caused the FRET efficiency to be reduced. Owing to the favorable spectral overlap (Förster distance R_0=4.9 nm) between QD550 and mCherry, considerable loss in QD PL was observed along with an increase in sensitized mCherry emission. A FRET efficiency of approximately 50% was measured when the number of mCherry per a single QD was six. Caspase 3-induced changes in FRET efficiency were comparable to those observed in fluorescent protein sensors. However, compared to two fluorescent proteins, some advantages of QD-fluoresecent protein encompass 5–10 times less substrate and ~3 orders of magnitude less enzyme in terms of quantity to be used. As a result, they were able to detect enzymatic activity for caspase 3 concentrations as low as 20 pM. This capability seems to be mainly due to multivalent effect of QDs.

In order to construct QD-FRET probe, quenching groups (organic quenchers or metal nanoparticles) (Chang et al., 2005; Kim et al., 2008b; Medintz et al., 2006; Shi et al., 2006) can be bound to the surface of a QD through a peptide sequence. Unlike QD-dye FRET system based on the ratiometry of dual emission, this close proximity causes only quenching of the QD emission via the resonance energy transfer, while subsequent cleavage of the peptide sequence by the corresponding protease led to a recovery of the QD fluorescence. The quenching ability of the gold nanoparticle (AuNP) has been known to be much higher than that of organic quenchers as described elsewhere (Oh et al., 2006). As such, the use of AuNPs as energy acceptors enables the energy transfer to be valid even in the excess distance of the traditional FRET. One of the feasible mechanisms might be associated with the property of metal surface; it was reported that the metal surface extended the effective energy transfer distance up to 22 nm, resulting in a high energy transfer efficiency (Dulkeith et al., 2002; Jennings et al., 2006a; Jennings et al., 2006b; Pons et al., 2007; Yun et al., 2005). The initial report of protease detection with a QD-AuNP system was made by Chang *et al* (Chang et al., 2005). A peptide substrate, GGLGPAGGCG, was employed to measure the activity of collagenase. The N-terminal amines of the peptides were coupled to the carboxylic acids on the QDs by EDC, and the cysteine was conjugated to the maleimide functionalized gold nanoparticles. When the AuNP level was six per QD, a quenching efficiency of 71% was observed. To be expected, the excess peptides on the QD possibly decrease the probe sensitivity to protease hydrolysis even if they cause the quenching efficiency to increase. By the release of the gold quencher in the presence of collagenase (0.2 mg mL^{-1}), QD fluorescence was recovered up to 51%, meaning that a number of enzyme molecules are not enough to be fully accessible between two large nanoparticles.

It is demonstrated by Suzuki et al (Suzuki et al., 2008a) that QD-based nanoprobes were used to detect multiple cellular signaling events including the activities of protease (trypsin), deoxyribonuclease, DNA polymerase, as well as the change in pH. This system was designed based on the FRET between the QD as donor and an appended fluorophore as acceptor; protease and deoxyribonuclease (DNase) induces the change in FRET efficiency between donor (QD) and acceptor (GFP or fluorophore-modified double-stranded DNA), whereas DNA polymerase action leads to the close proximity of fluorescently labeled nucleotides to the surface of the QD, and pH-sensitive fluorophore conjugated on the surface of QD produces changes in FRET efficiency by pH change (**Figure 3**). This mixture of modified QDs showed distinct changes in emission peaks before and after enzyme treatment by simultaneous-wavelength excitation in the same tube.

A multiplexed system to detect the activity and inhibitory effects of several proteases (MMPs, thrombin, and caspase-3) has been proposed by Kim *et al*, which is utilized by the principle of energy transfer between the AuNP and respective QDs on a glass slide (**Figure 4**) (Kim et al., 2008a). For construction of nanoprobes, the AuNP acceptors conjugated with a peptide substrate including cysteine and biotin were associated with streptavidin (SA)-

conjugated QDs (SA-QDs, energy donor) deposited on a glass slide, thus quenching the PL of the QD by the energy transfer. Upon addition of a protease to cleave the peptide substrate

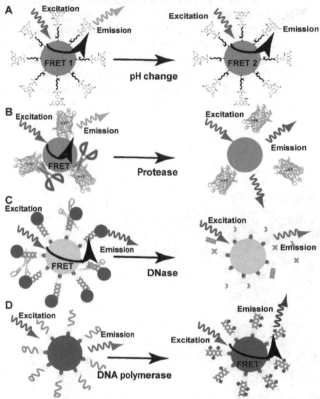

Fig. 3. FRET-based QD bioprobes designed to give FRET changes on (from top to bottom): (A) pH change via pH-sensing dyes attached to a QD; (B) cleavage of a GFP variant with an inserted sequence recognized by a protease (e.g., trypsin) to release GFP from the QD surface; (C) digestion by DNase of dsDNA (labeled with fluorescent dUTP) bound to a QD; (D) incorporation of fluorescently labeled dUTPs into ssDNA on a QD by extension with DNA polymerase.

on the AuNP–QD conjugates, there was a significant regeneration of the photoluminescence emission of the QDs. Protease inhibitors also prevented any recovery of the photoluminescence of QDs by inhibiting the protease activity. When three types of SA-QD (SA-QD525, SAQD605, and SA-QD655) were independently complexed with the AuNPs with different peptide substrates, a specific reaction of the protease induced strong photoluminescence intensity from each spot, at a specified wavelength. Marginal cross-reactional images of the protease against other peptide substrates were observed, thus confirming the multiplexed capability of this assay system (**Figure 4A**). Since the AuNPs can be employed as a common energy acceptor, a variety of QDs with different colors could be used as the energy donor (**Figure 4B and 4C**), thus enabling a multiplexed assay. Moreover, high quenching efficiency of AuNPs allows application of the assay system to an extended separation distance between a donor and an acceptor. This developed system also overcome some drawbacks resulting from a solution-based format, including the aggregation of nanoparticles and the fluctuation in photoluminescence, and the consumption of large amounts of reagents. This multiplexed assay using a QD-FRET nanosensor has also been applied to different types of enzymes other than proteases. Suzuki *et al* (Suzuki et al., 2008b). designed QD-based nanoprobes on a FRET with QD as donor and an appended fluorophore as acceptor in order to detect multiple cellular signaling events, including the activities of protease (trypsin), deoxyribonuclease, and DNA polymerase, as well as changes in the pH. Notably, this mixture of modified QDs showed distinct changes in emission peaks before and after enzyme treatment by simultaneous-wavelength excitation in the same tube.

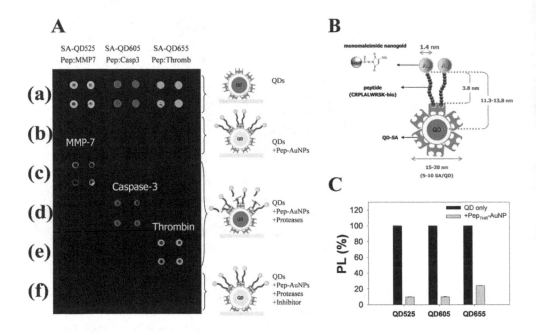

Fig. 4. Development of multiplexed system to detect protease activity with QDs. (A) Multiplexed assay of proteases by using QDs with different colors. SA-QD525, SA-QD605, and SA-QD655 were used (from left to right). Biotinylated peptide substrates for MMP-7, caspase-3, and thrombin were conjugated to the AuNPs, and then the resulting Pep-AuNPs were associated with SA-QD525, SA-QD605, and SA-QD655, respectively. (a) SA-QDs only. (b) SA-QDs + respective Pep-AuNPs. (c) SA-QDs + Pep-AuNPs + MMP-7. (d) SA-QDs + Pep-AuNPs + caspase-3. (e) SA-QDs + Pep-AuNPs + thrombin. (f) QDs + Pep-AuNPs + mixture of the respective protease and its inhibitor. (B) Configuration of QD-AuNP nanosensor to detect the activity of MMP-7 (SA: streptavidin). (C) Changes in the : photoluminescence (PL) intensities of QDs having different colors (SA-QD525, SA-QD605, and SA-QD655) in the presences of biotinylated Pep_{THR}-AuNPs in solution. The molar ratios of Pep-AuNPs to respective QDs were equally maintained at 50.

3. Bioluminescence resonance energy transfer (BRET) system using QDs

In FRET, QDs are not able to function as effective acceptors (Nabiev et al., 2004) because a direct excitation of donor fluorophores has to accompany by unavoidable excitation of the QD with a broad absorption. However, when QDs can serve as an energy acceptor with a light-emitting protein, luciferase, this problem can be overcome; since the bioluminescence energy of a luciferase-catalyzed reaction occurs only in its active site and cannot excite the acceptor QD, it can be successfully transferred to the QDs to produce quantum dot light emission. Upon addition of the luciferase substrate (coelenterazine) to QD-luciferase conjugate, a short blue light emission at 475 nm is transferred the QDs. Main advantage of QD-luciferase system is to eliminate the need for excitation light which causes inevitable background autofluorescence. In the case of QD-FRET that has been widely employed as activity-based probes, the excited illumination can partly increase the background noise level especially in serum sample, thus diminishing the value of acquired information as a result of the false-positive signal. Moreover, a common luciferase protein serves as the BRET donor for several QDs with different colors because QDs have similar absorption spectra and absorb blue light efficiently. Especially in contrast to QD-FRET system where the FRET efficiency improves as the number of FRET acceptors per QD increases, the BRET ratio with varying numbers of luciferase on the surface of single QD were quite similar although the intensity of both luciferase and QD emissions varied by approximately100-fold. This means that the QD-BRET system is more dependent on the donor-to-acceptor distance rather than varying donor number, thus being applied to detect the distance-dependent assay with high fidelity. Rao group has initially demonstrated the feasibility of using QDs as the acceptor in a bioluminescence resonance energy transfer (BRET) system (So et al., 2006). This conjugate was known as "self-illuminating quantum dots." in the sense that no external illumination light source is needed for the QDs to fluoresce (**Figure 5**). The self-illuminating QDs are advantageous over conventional QDs for the *in vivo* animal imaging purposes, due to extremely high sensitivity (an *in vivo* signal-to-background ratio of >1000 for 5 pmole of conjugates subcutaneously injected) and capability to look into deep tissue without considering external light delivery.

This QD-BRET based probe can be applied to the detection of protease activity. Yao *et al.* has focused on the detection of the activity of MMPs that is a promising cancer biomarker enzyme. Since the secretion level of MMPs in human serum is of very interest as a promising prognostic marker (Lein et al., 1997; Nikkola et al., 2005) due to the up-regulation of MMPs in almost all human cancerous cells or in malignant tissues, the designed probe to measure MMP activity and related inhibitory effect is valuable for discovering drug candidates for anticancer therapeutics (Coussens et al., 2002; Denis and Verweij, 1997; Hidalgo and Eckhardt, 2001). For the construction of QD-BRET energy donor, a bioluminescent protein fused to the MMP-2 substrate (GGPLGVR) and a hexahistidine tag at its C terminus was genetically expressed in *E. coli*. The bioluminescent protein is a *Renilla reniformis* luciferase mutant dubbed Luc8, which has eight mutations and shows higher stability and improved catalytic efficiency than the wild-type luciferase (Loening et al., 2006). Since the simultaneous coordination is generated between carboxyl groups on the QD surface and the Luc8 His tag in the presence of Ni^{2+}, a strong BRET signal was observed just immediately after luciferase-GGPLGVRGGH$_6$ is mixed with carboxyl QD655 and Ni^{2+}. Upon cleavage of the flanking peptide region by MMP-2, the His tag was released from the fusion Luc8 and the BRET signal decreased. In comparison to FRET-based QD sensors,

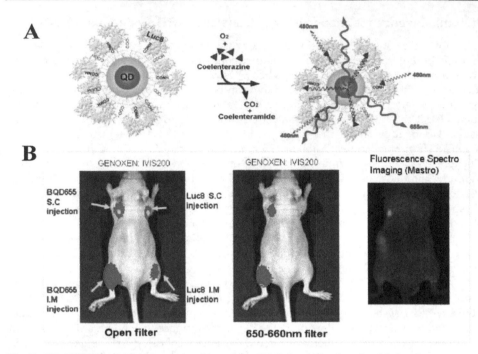

Fig. 5. QD-BRET based in vivo animal imaging. (A) Schematic showing bioluminescence resonance energy transfer between *Renilla* luciferase mutant, Luc8 protein (donor) and QD (acceptor) The bioluminescence energy of Luc8-catalyzed oxidation of coelenterazine is transferred to QDs in close proximity, resulting in the QD emission (655 nm); (B) In vivo imaging using QD-BRET conjugates. Luc8-QD bioconjugates were injected into nude mice either intramuscularly or subcutaneously and imaged bioluminescently (Xenogen, open filter (left image) and QD filter (middle image)) or fluorescently (Mastro, right image, QD signals shown in bright white). Luc8 alone were also injected into the same animal for comparison.

BRET based QD biosensors have several attractive features. Most significantly, the large spectral separation between the BRET donor and acceptor emissions makes it easy to detect both emissions. In this case, a ratiometric measurement is far more accurate and reliable than that of QD-FRET because the integrated intensities from two separable spectra was favorable to reflect the energy-transferred efficiency (Rao et al., 2006a). In addition, considering that the BRET ratio for a concentration of MMP-2 of 2 ng mL^{-1} (~30 pM) decreased by about 10%, and by 15% for a concentration of MMP-2 of 5 ng mL^{-1} (~75 pM), the sensitivity is high due to a low background emission. By taking advantage of effective ratiometry where the spectral separation between the BRET donor and acceptor emissions is large, a QD-BRET nanosensor with a stable covalent linkage has been developed that allows the detection of protease activity in mouse sera and tumor lysates(Xia et al., 2008). MMP-2, MMP-7, and urokinase-type plasminogen activator (uPA). Most importantly, the nanosensors were not only capable of detecting these proteases in complex biological media (e.g., mouse serum and tumor lysates) with sensitivity down to 1 ng mL^{-1}, but could also detect multiple proteases present in one sample. To accomplish the site-specific conjugation, the carboxylated QDs had been functionalized with adipic dihydrazide and subsequently treated with the luciferase–protease substrate recombinant protein with an additional intein segment. Since hydrazides are excellent nucleophiles to attack the thioester intermediate of inteins that are natural protein ligation mediators. The reaction proceeded rapidly when the two components were mixed together, and resulted in cleavage of the intein and ligation of the C terminus of the recombinant protein to the QDs.

It has been demonstrated that luciferase enzyme can be directly used as the template to generate new near-infrared (NIR) QD-BRET nanosensor via a biominealization process (Ma et al., 2010). The synthesis was accomplished by incubating luciferase with lead acetate $(Pb(Ac)_2)$ at ambient conditions to allow the binding of Pb^{2+} to Luc8 and facilitate the heterogeneous nucleation. Then a sodium sulfide (Na_2S) was added into the $Luc8-Pb^{2+}$ to produce a Luc-PbS hybrid nanostructure with a mean hydrodynamic diameter of 19.9 nm. Despite the very low quantum yield (~3.6%) of the Luc-PbS complex, the BRET signal was comparable to other NIR QDs. The small size of the complex is expected to be more advantageous for *in vivo* imaging and other sensing applications than that (30–40 nm in diameter) of the conjugates between luciferase and as-prepared QDs. However, the long-term stability in luminescence intensity, which might be due to the QD surface oxidation, should be overcome for more practical applications of this protein-nanomateral hybrid.

4. Chemiluminescence resonance energy transfer (CRET) system using QDs

Similar to bioluminescence, chemiluminescence offers many advantages such as high detection sensitivity, a wide linear range for quantification, and no requirement for a light source. Typically, chemiluminescence can be produced when luminol is mixed with oxidizing agents (e.g. hydrogen peroxide) in an alkaline solution in the presence of metal catalysts (iron or copper), leading to a strong blue emission at around 425 nm. A horseradish peroxidase enzyme (HRP) has been also used for catalyzing the oxidation of luminol in the presence of hydrogen peroxide via several intermediates, and the light emitted can be enhanced up to 1000-fold with enhancers such as *p*-iodophenol. This process is known to be the enhanced chemiluminescence. Based on this principle, nonradiative energy transfer can be accomplished between a chemiluminescent (CL) donor and a fluorophore acceptor (or QD), which is called chemiluminescence RET (CRET).

Ren and co-workers investigated the chemiluminescence resonance energy transfer (CRET) by using luminol–H_2O_2 system as energy donor and HRP-conjugated CdTe QD as acceptor (Huang et al., 2006). By the CL generation among the luminol, H_2O_2, and HRP, QD could absorb part of the excited-state luminol energy and re-emit it at longer wavelengths without external light source (**Figure 6**). A comparison of CL spectra between HRP–QDs conjugates and a mixture of HRP and QDs showed that the CL intensity ratio of QDs:luminol from the mixture of QDs and HRP was very low, which was attributed to the low adsorption of QDs onto HRP and to the resultant longer distance. The multiplexed CRET was able to be realized similar to the multiplexed BRET and a proof-of-concept experiment for multiplexed immunoassay was also explored using different QD-antibody-HRR complex. Although the QD-CRET efficiency is dependent on several factors, it is worth noting that the quantum yield of the acceptor QD in different status is the crucial factor to the CRET efficiency; that is, high quantum yield of QDs brings in a high CRET efficiency (Wang et al., 2008).

To avoid the requirement of labeling the energy acceptor with HRP and H_2O_2, which may limit the application of this CRET system, enzyme-free QD-CRET system was developed using a new oxidizing reagent, NaBrO (Zhao et al., 2010). The CRET ratio calculated by dividing the acceptor emission by the donor emission was comparable to that observed from the luminol–H_2O_2–HRP–conjugated QD system (~30%). A feasible explanation for this mechanism remains unclear, but an intermediate complex might be formed between BrO- anions and CdTe QDs, which bring the energy donor (luminol molecules) and the acceptor (QDs) very close to each other at the time of oxidation to produce chemiluminescence emission. Main advantage of this system is that highly stable QD-CRET sensor can be achieved because QD emission in HRP-based QD-CRET system is critically affected by free H_2O_2, leading to the instability of QD sensor. Therefore, this new luminol–NaBrO–QD CRET system is expected to have a great potential for simultaneous prognosis and diagnosis.

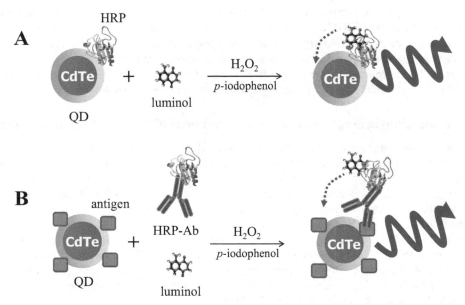

Fig. 6. (A) Schematic of QD-CRET based on luminol donors and HRP-labeled CdTe QD accepters, (B) Schematic of QD-CRET for luminol donors and QD accepters based on the immuno-reaction of QD–antigen and antibody–HRP.

5. Conclusion

The intrinsic properties of QDs have encouraged scientists to further develop this material for many biological applications. One of the most emerging uses of QDs lies in QD-based energy transfer systems, primarily consisting of QD-FRET, QD-BRET and QD-CRET. Many imaging and biosensing applications of QD-RET are now becoming very popular as a powerful tool for dissecting complex biomolecular detection and trafficking. Moreover, the ability to create different colors with QDs will certainly allow for multiplexed analysis both *in vitro* and *in vivo* applications. Given the rapid growth in new QD based materials, the improved interdisciplinary achievement will certainly be bright, but a lot of techniques about QDs are still required to be improved with the development of new QDs in the near future as follows; (i) introduction of multiple functional groups to a single QDs for multimodal monitoring, (ii) improved properties of QD: reduced size and removing blinking phenomenon, (iii) need of toxicity and kinetics study in human before extensive application for clinical diagnosis and therapy, and (iv) development of near infrared QDs for in *vivo* imaging.

6. Acknowledgement

This research was supported by Basic Science Research Program through the National Research Foundation of Korea(NRF) funded by the Ministry of Education, Science and Technology (2011-0022757)

7. References

Bagalkot, V., Zhang, L., Levy-Nissenbaum, E., Jon, S., Kantoff, P.W., Langer, R., and Farokhzad, O.C. (2007). Quantum dot-aptamer conjugates for synchronous cancer imaging, therapy, and sensing of drug delivery based on bi-fluorescence resonance energy transfer. *Nano Lett.* 7, 3065-3070.

Chan, W.C.W., Maxwell, D.J., Gao, X.H., Bailey, R.E., Han, M.Y., and Nie, S.M. (2002). Luminescent quantum dots for multiplexed biological detection and imaging. *Curr. Opin. Biotech.* 13, 40-46.

Chang, E., Miller, J.S., Sun, J., Yu, W.W., Colvin, V.L., Drezek, R., and West, J.L. (2005). Protease-activated quantum dot probes. *Biochem. Biophys. Res. Commun.* 334, 1317-1321.

Chen, W., Biswas, P., Cella, L.N., Kang, S.H., Mulchandani, A., and Yates, M.V. (2011). A quantum-dot based protein module for in vivo monitoring of protease activity through fluorescence resonance energy transfer. *Chem. Commun.* 47, 5259-5261.

Clapp, A.R., Goldman, E.R., Uyeda, H.T., Chang, E.L., Whitley, J.L., and Medintz, I.L. (2008). Monitoring of Enzymatic Proteolysis Using Self-Assembled Quantum Dot-Protein Substrate Sensors. *J. Sensors* 2008, 1-10.

Clapp, A.R., Medintz, I.L., Mauro, J.M., Fisher, B.R., Bawendi, M.G., and Mattoussi, H. (2004). Fluorescence resonance energy transfer between quantum dot donors and dye-labeled protein acceptors. *J. Am. Chem. Soc.* 126, 301-310.

Coussens, L.M., Fingleton, B., and Matrisian, L.M. (2002). Matrix metalloproteinase inhibitors and cancer: trials and tribulations. *Science* 295, 2387-2392.

Denis, L.J., and Verweij, J. (1997). Matrix metalloproteinase inhibitors: present achievements and future prospects. *Invest. New Drugs* 15, 175-185.

Dulkeith, E., Morteani, A.C., Niedereichholz, T., Klar, T.A., Feldmann, J., Levi, S.A., van Veggel, F.C., Reinhoudt, D.N., Moller, M., and Gittins, D.I. (2002). Fluorescence quenching of dye molecules near gold nanoparticles: radiative and nonradiative effects. *Phys. Rev. Lett.* 89, 203002.

Gao, X., Cui, Y., Levenson, R.M., Chung, L.W., and Nie, S. (2004). In vivo cancer targeting and imaging with semiconductor quantum dots. *Nat Biotechnol* 22, 969-976.

Hawrylak, P., Narvaez, G.A., Bayer, M., and Forchel, A. (2000). Excitonic absorption in a quantum Dot. *Phys. Rev. Lett.* 85, 389-392.

Hidalgo, M., and Eckhardt, S.G. (2001). Development of matrix metalloproteinase inhibitors in cancer therapy. *J. Natl. Cancer Inst.* 93, 178-193.

Huang, X.Y., Li, L., Qian, H.F., Dong, C.Q., and Ren, J.C. (2006). A resonance energy transfer between chemiluminescent donors and luminescent quantum-dots as acceptors (CRET). *Angew. Chem. Int. Ed.* 45, 5140-5143.

Jaiswal, J.K., Mattoussi, H., Mauro, J.M., and Simon, S.M. (2003). Long-term multiple color imaging of live cells using quantum dot bioconjugates. *Nat. Biotechnol.* 21, 47-51.

Jennings, T.L., Schlatterer, J.C., Singh, M.P., Greenbaum, N.L., and Strouse, G.F. (2006a). NSET molecular beacon analysis of hammerhead RNA substrate binding and catalysis. *Nano Lett.* 6, 1318-1324.

Jennings, T.L., Singh, M.P., and Strouse, G.F. (2006b). Fluorescent lifetime quenching near d = 1.5 nm gold nanoparticles: probing NSET validity. *J Am Chem Soc* 128, 5462-5467.

Kim, H.S., Kim, Y.P., Oh, Y.H., and Oh, E. (2007). Chip-based protease assay using fluorescence resonance energy transfer between quantum dots and fluorophores. *Biochip J.* 1, 228-233.

Kim, H.S., Kim, Y.P., Oh, Y.H., Oh, E., Ko, S., and Han, M.K. (2008a). Energy transfer-based multiplexed assay of proteases by using gold nanoparticle and quantum dot conjugates on a surface. *Anal. Chem.* 80, 4634-4641.

Kim, K., Lee, S., Ryu, J.H., Park, K., Lee, A., Lee, S.Y., Youn, I.C., Ahn, C.H., Yoon, S.M., Myung, S.J., et al. (2009). Polymeric Nanoparticle-Based Activatable Near-Infrared Nanosensor for Protease Determination In Vivo. *Nano Lett.* 9, 4412-4416.

Kim, Y.P., Oh, Y.H., Oh, E., Ko, S., Han, M.K., and Kim, H.S. (2008b). Energy transfer-based multiplexed assay of proteases by using gold nanoparticle and quantum dot conjugates on a surface. *Anal. Chem.* 80, 4634-4641.

Kircher, M.F., Weissleder, R., and Josephson, L. (2004). A dual fluorochrome probe for imaging proteases. *Bioconju. Chem.* 15, 242-248.

Lein, M., Nowak, L., Jung, K., Koenig, F., Lichtinghagen, R., Schnorr, D., and Loening, S.A. (1997). Analytical aspects regarding the measurement of metalloproteinases and their inhibitors in blood. *Clin. Biochem.* 30, 491-496.

Levy, M., Cater, S.F., and Ellington, A.D. (2005). Quantum-dot aptamer beacons for the detection of proteins. *Chembiochem* 6, 2163-2166.

Loening, A.M., Fenn, T.D., Wu, A.M., and Gambhir, S.S. (2006). Consensus guided mutagenesis of Renilla luciferase yields enhanced stability and light output. *Protein. Eng. Des. Sel.* 19, 391-400.

Ma, N., Marshall, A.F., and Rao, J.H. (2010). Near-Infrared Light Emitting Luciferase via Biomineralization. *J. Am. Chem. Soc.* 132, 6884-6885.

Mauro, J.M., Medintz, I.L., Clapp, A.R., Mattoussi, H., Goldman, E.R., and Fisher, B. (2003). Self-assembled nanoscale biosensors based on quantum dot FRET donors. *Nat. Mater.* 2, 630-638.

Medintz, I.L., Boeneman, K., Mei, B.C., Dennis, A.M., Bao, G., Deschamps, J.R., and Mattoussi, H. (2009). Sensing Caspase 3 Activity with Quantum Dot-Fluorescent Protein Assemblies. Journal of the American Chemical Society 131, 3828-+.

Medintz, I.L., Clapp, A.R., Brunel, F.M., Tiefenbrunn, T., Uyeda, H.T., Chang, E.L., Deschamps, J.R., Dawson, P.E., and Mattoussi, H. (2006). Proteolytic activity monitored by fluorescence resonance energy transfer through quantum-dot-peptide conjugates. *Nat. Mater.* 5, 581-589.

Medintz, I.L., Clapp, A.R., Mattoussi, H., Goldman, E.R., Fisher, B., and Mauro, J.M. (2003). Self-assembled nanoscale biosensors based on quantum dot FRET donors. *Nat. Mater.* 2, 630-638.

Medintz, I.L., Prasuhn, D.E., Feltz, A., Blanco-Canosa, J.B., Susumu, K., Stewart, M.H., Mei, B.C., Yakovlev, A.V., Loukov, C., Mallet, J.M., et al. (2010). Quantum Dot Peptide Biosensors for Monitoring Caspase 3 Proteolysis and Calcium Ions. *ACS Nano* 4, 5487-5497.

Moreels, I., and Hens, Z. (2008). On the interpretation of colloidal quantum-dot absorption spectra. *Small* 4, 1866-1868; author reply 1869-1870.

Nabiev, I., Sukhanova, A., Devy, M., Venteo, L., Kaplan, H., Artemyev, M., Oleinikov, V., Klinov, D., Pluot, M., and Cohen, J.H.M. (2004). Biocompatible fluorescent nanocrystals for immunolabeling of membrane proteins and cells. *Anal. Biochem.* 324, 60-67.

Nikkola, J., Vihinen, P., Vuoristo, M.S., Kellokumpu-Lehtinen, P., Kahari, V.M., and Pyrhonen, S. (2005). High serum levels of matrix metalloproteinase-9 and matrix metalloproteinase-1 are associated with rapid progression in patients with metastatic melanoma. *Clin. Cancer. Res.* 11, 5158-5166.

Oh, E., Hong, M.Y., Lee, D., Nam, S.H., Yoon, H.C., and Kim, H.S. (2005). Inhibition assay of biomolecules based on fluorescence resonance energy transfer (FRET) between quantum dots and gold nanoparticles. *J. Am. Chem. Soc.* 127, 3270-3271.

Oh, E., Lee, D., Kim, Y.P., Cha, S.Y., Oh, D.B., Kang, H.A., Kim, J., and Kim, H.S. (2006). Nanoparticle-based energy transfer for rapid and simple detection of protein glycosylation. *Angew. Chem. Int. Ed.* 45, 7959-7963.

Pons, T., Medintz, I.L., Sapsford, K.E., Higashiya, S., Grimes, A.F., English, D.S., and Mattoussi, H. (2007). On the quenching of semiconductor quantum dot photoluminescence by proximal gold nanoparticles. *Nano Lett.* 7, 3157-3164.

Rao, J.H., So, M.K., Loening, A.M., and Gambhir, S.S. (2006a). Creating self-illuminating quantum dot conjugates. *Nat. Protoc.* 1, 1160-1164.

Rao, J.H., So, M.K., Xu, C.J., Loening, A.M., and Gambhir, S.S. (2006b). Self-illuminating quantum dot conjugates for in vivo imaging. *Nat. Biotechnol.* 24, 339-343.

Rosenzweig, Z., Shi, L.F., and Rosenzweig, N. (2007). Luminescent quantum dots fluorescence resonance energy transfer-based probes for enzymatic activity and enzyme inhibitors. *Anal. Chem.* 79, 208-214.

Shi, L., De Paoli, V., Rosenzweig, N., and Rosenzweig, Z. (2006). Synthesis and application of quantum dots FRET-based protease sensors. *J. Am. Chem. Soc.* 128, 10378-10379.

So, M.K., Xu, C., Loening, A.M., Gambhir, S.S., and Rao, J. (2006). Self-illuminating quantum dot conjugates for in vivo imaging. *Nat. Biotechnol.* 24, 339-343.

Stockholm, D., Bartoli, M., Sillon, G., Bourg, N., Davoust, J., and Richard, I. (2005). Imaging calpain protease activity by multiphoton FRET in living mice. *J. Mol. Biol.* 346, 215-222.

Suzuki, M., Husimi, Y., Komatsu, H., Suzuki, K., and Douglas, K.T. (2008a). Quantum dot FRET biosensors that respond to pH, to proteolytic or nucleolytic cleavage, to DNA synthesis, or to a multiplexing combination. *J. Am. Chem. Soc.* 130, 5720-5725.

Suzuki, M., Husimi, Y., Komatsu, H., Suzuki, K., and Douglas, K.T. (2008b). Quantum dot FRET Biosensors that respond to pH, to proteolytic or nucleolytic cleavage, to DNA synthesis, or to a multiplexing combination. *J. Am. Chem. Soc.* 130, 5720-5725.

Wang, H.Q., Li, Y.Q., Wang, J.H., Xu, Q., Li, X.Q., and Zhao, Y.D. (2008). Influence of quantum dot's quantum yield to chemiluminescent resonance energy transfer. *Anal. Chim. Acta* 610, 68-73.

Wang, S.P., Mamedova, N., Kotov, N.A., Chen, W., and Studer, J. (2002). Antigen/antibody immunocomplex from CdTe nanoparticle bioconjugates. *Nano Lett.* 2, 817-822.

Wargnier, R., Baranov, A.V., Maslov, V.G., Stsiapura, V., Artemyev, M., Pluot, M., Sukhanova, A., and Nabiev, I. (2004). Energy transfer in aqueous solutions of oppositely charged CdSe/ZnS core/shell quantum dots and in quantum dot-nanogold assemblies. *Nano Lett.* 4, 451-457.

Weissleder, R., Tung, C.H., Mahmood, U., and Bogdanov, A. (1999). In vivo imaging of tumors with protease-activated near-infrared fluorescent probes. *Nat. Biotechnol.* 17, 375-378.

Wu, X.Y., Liu, H.J., Liu, J.Q., Haley, K.N., Treadway, J.A., Larson, J.P., Ge, N.F., Peale, F., and Bruchez, M.P. (2003). Immunofluorescent labeling of cancer marker Her2 and other cellular targets with semiconductor quantum dots (vol 21, pg 41, 2003). *Nat. Biotechnol.* 21, 452-452.

Xia, Z., Xing, Y., So, M.K., Koh, A.L., Sinclair, R., and Rao, J. (2008). Multiplex detection of protease activity with quantum dot nanosensors prepared by intein-mediated specific bioconjugation. *Anal. Chem.* 80, 8649-8655.

Yun, C.S., Javier, A., Jennings, T., Fisher, M., Hira, S., Peterson, S., Hopkins, B., Reich, N.O., and Strouse, G.F. (2005). Nanometal surface energy transfer in optical rulers, breaking the FRET barrier. *J. Am. Chem. Soc.* 127, 3115-3119.

Zhang, C.Y., Yeh, H.C., Kuroki, M.T., and Wang, T.H. (2005). Single-quantum-dot-based DNA nanosensor. *Nat. Mater.* 4, 826-831.

Zhao, S.L., Huang, Y., Liu, R.J., Shi, M., and Liu, Y.M. (2010). A Nonenzymatic Chemiluminescent Reaction Enabling Chemiluminescence Resonance Energy Transfer to Quantum Dots. *Chemistry-a European Journal* 16, 6142-6145.

13

II-VI Quantum Dots as Fluorescent Probes for Studying Trypanosomatides

Adriana Fontes[1], Beate. S. Santos[2],
Claudilene R. Chaves[2] and Regina C. B. Q. Figueiredo[3]
[1]Departamento de Biofísica e Radiobiologia,
Universidade Federal de Pernambuco, Recife
[2]Departamento de Ciências Farmacêuticas,
Universidade Federal de Pernambuco, Recife
[3]Centro de Pesquisa Aggeu Magalhães,
Fundação Oswaldo Cruz, Recife
Brazil

1. Introduction

Fluorescence provides a unique method for understanding how biomolecules interact with each other in many levels, from single cell to whole organisms. Researchers commonly use fluorescence-based techniques mainly due to its high specificity and sensitivity (capable of detecting even a single molecule) (Michalet *et al.*, 2006; Giepmans *et al.*, 2006). The evolution of tools based on fluorescence including new techniques (like multiphoton microscopy), new lasers and new fluorescent probes (presenting new features when compared to the traditional organic fluorophores), allowing us to take advantages of the full potential of fluorescence. This is the case of Quantum dots (QDs) that since the end the 90´s have been increasingly used in biological and biomedical field.

Quantum dots are semiconductor nanoparticles, with typical dimensions ranging from 2 to 10 nm. QDs are often referred as artificial atoms not only because electrons are dimensionally confined in these nanoparticles (just like in a real atom), but also because the QDs also present discrete energy levels and can give rise to fluorescence when excited by light. In general, QDs have been successfully used as fluorescent labels for imaging fixed and live cells and also small animals, immunoassays (Tian *et al.*, 2010), diagnostic methodologies (Yezhelyev *et al.*, 2006) and recently also for photodynamic therapy (Samia *et al.*, 2006).

For an optimal performance in biological application, QDs have been synthesized in order to optimize their fluorescence, active surface and chemical stability. These necessary characteristics result in a nanostructured complex multilayered chemical assembly where: 1. The chemical nature of the core is the main responsible for the fluorescence emission color. 2. The passivation shell determines its brightness and photostability and 3. The organic capping determines its stability and biological functionality, as shown in Figure 1. The passivation shell is defined as a chemical coating at the surface of the nanoparticle. This coating, composed of a few monolayers of another semiconductor material, helps to decrease the surface defects characteristic of a high surface/volume ratio of these nanometric systems. When this passivation shell is applied nanoparticles are called "core-shell" systems.

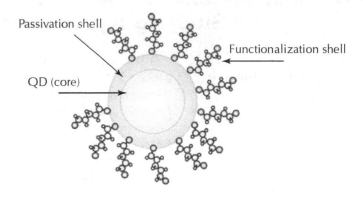

Fig. 1. Schematic representation of a semiconductor QD for biomedical application purposes (Santos *et al.*, 2008).

Both core and shell usually are made of elements from II B and IV A group giving rise to core/shell nanoparticles such as CdTe/CdS, CdSe/ZnSe and ZnSe/ZnS. For active optical properties, the band gap of the semiconductor present in the shell is usually higher than the band gap of the core. QDs can be synthesized in organic or aqueous medium. When compared to organometallic routes, water-based QDs synthesis is cheaper, less toxic and intrinsically biocompatible for applications in biological fields (Santos *et al.*, 2008). The stability of water dispersed QDs is generally accomplished by using charged organic molecules or polymers. A commonly used class of organic molecules is alkyl-thiol molecules (such as mercaptopropionic acid - MPA, mercaptoacetic acid - MAA and mercaptosuccinic acid - MSA) and more recently cysteine. These stabilizing agents have carboxyl or amine terminals that can be conjugated to biomolecules (such as proteins and antibodies) rendering more specificity for biological applications.

Despite the success of QDs as fluorescent probes, some drawbacks still exist (mainly when the application is focused in live cells and small animals). Synthesis of high quality water dispersed QDs, high stability in aqueous media at physiological pH, full biocompatibility; narrow size distribution and (only) specific interactions with target molecules. Taken together these features are still hard tasks to be achieved. Moreover, little is still known about the mechanisms of QDs interactions with live biological systems and how these mechanisms can affect cellular functions. As in other classes of nanoparticles, the cellular uptake of QDs, for example, depends on the surface coating, the cell type and QDs size (Kelf *et al.*, 2010). The comprehension of the interconnections existing between the cellular dynamics and the QDs kinetics (including the mechanisms of uptake and intracellular delivery), is an important step to engineering more functional QDs for biomedical research and also to take advantage of the QDs – cells interactions to understand basic cellular biology.

In this context, the purpose of this Chapter is to discuss important characteristics that have to be considered when using QDs for live cells applications and to highlight the use of water dispersed CdS/Cd(OH)$_2$ and CdTe/CdS QDs as a bioimaging tool to improve the comprehension of cell biology aspects of protozoa. Here we will focus on the *Leishmania sp.* and *Trypanosoma cruzi* protozoa which are the etiological agents of Leishmaniasis and Chagas disease respectively.

2. QDs for life sciences applications

QDs are colloidal semiconductor nanocrystals which have unique optical properties due to their 3D dimensional quantum confinement regime. These properties give to QDs some advantages over the conventional fluorescent dyes. Compared to organic fluorophores, the major advantages of interest to biologists offered by QDs are:

1. A broad absorption band, allowing a flexible cross section for multiphoton microscopy – as shown in Figure 2;
2. A size tunable emission wavelength which means that QDs of the same material, but with different sizes, can emit light of different colors (quantitatively speaking, the energy gap determines the wavelength of the fluorescence). Larger QDs emit fluorescence towards the red region as shown in Figure 3;
3. An active surface for chemical conjugation. QDs conjugated to proteins or antibodies play important roles, because they are inorganic-biological hybrids nanoparticles that combine characteristics of both materials, that is, the fluorescence properties of QDs with the biochemical functions of the proteins and antibodies.
4. High resistance to photobleaching: the most important advantage of QDs over organic dyes.

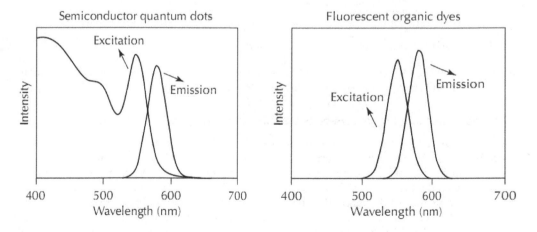

Fig. 2. A general comparative excitation and emission spectrum profile of QDs versus organic dyes (Santos *et al.*, 2008).

The feasibility of using QDs for antigen detection in fixed cells and tissues has been demonstrated in several studies. Specific genomic sequences and antigens could be labeled in fixed samples with great success by using QDs (Say *et al.*, 2011; Barroso, 2011). On the other hand, the use of QDs for live cells imposes some challenges when compared to fixed samples and special care must be taken to guarantee that cell remain alive and healthy along all the experiment. Water dispersed QDs usually are less fluorescent than organic synthesized QDs. Conversely, due to their hydrophilic characteristic they are intrinsically more biocompatible than organic QDs. Before applying QDs as fluorescent probes for live cells it is necessary to analyze:

- QDs physico-chemical aspects (such as original emission/absorption and photophysical/chemical stability in biological media);
- QDs-cells interactions (for example: Do these cells have an active uptake? The size, surface electrical charges and the coating of QDs are important to QDs-cell interactions?);
- Intracellular labeling and non-specific interactions (which and where is the target biomolecule, inside the cell or on the cell membrane?);
- QDs toxicity (how and which QDs properties can affect cells? The influence of pH, osmolarity and composition).

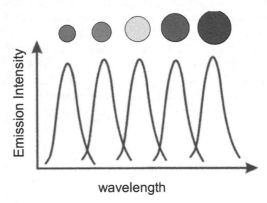

Fig. 3. Fluorescence maximum as a function of QD size (Santos *et al.*, 2008).

In this Section we will discuss some of these important characteristics in order to promote an adequate and efficient biological application.

2.1 General aspects about QDs-cells specific and non-specific interactions

The cell membrane functions as a semi-permeable barrier, allowing a very few molecules to cross it while the majority of large and particulate materials are impaired. Uptake of nutrients and all communication among or between the cells and their microenvironment occur through the plasma membrane. So, understanding QD-cell membrane interactions is a very important step to develop efficient QDs for many biological applications as well as to predict and overcome any potential toxicities. These interactions depend on both the composition and environmental conditions of membranes and the chemical and physical properties of QDs themselves. QDs can interact with cells in a non-specific or specific manner. Nonspecific adsorption of QDs to the cell membranes occurs mainly by electrostatic interactions. QDs can be also delivered into the cytosol or nucleus actively by transiently destabilizing the cell/nuclear membrane. These active interventions include micro/nanoinjection, electroporation and osmotic lyses. Although nonspecific binding of QDs to cell membrane has been exploited for extracellular labeling and intracellular delivery of QDs, it has become an undesirable property of QDs which impair QDs to be addressed to specific extracellular and intracellular sites.

For specific interactions a functionalizing agent intermediates the QDs and biomolecules as antibodies, proteins, nucleic acids, antigens, etc, directing them to specific targets in the cells. The process of binding the functionalizing agent to the biomolecule is called *Bioconjugation*. It can be accomplished by simple adsorption (of stabilizing agent such as: MPA, MSA or cysteine) to the biomolecule or by covalent attachment. The most common covalent bioconjugations occur between carboxyl-amine (trough EDC and Sulfo-NHS) groups, amine-amine (by glutaraldehyde), disulfide bonds and streptavidin-biotin interactions.

The bioconjugation process is not 100% effective, so it is important to remove non-bioconjugated QDs, QDs can non-specifically bind to the cell leading to misinterpretation of the biological labeling results. There are several purification procedures to remove non-bioconjugated QDs (ex. by exclusion chromatography). Some authors also state that the coating of QDs with polyethylene glycol decrease non-specific interactions (Bentzen *et al.*, 2005).

2.2 Dynamic of QDs-cell interactions

Researches have been assuming that after binding to the cell surface by specific or nonspecific interactions particles are taken up by cells through endocytosis, an energy-dependent process of internalization of macromolecules (Conner & Schmid, 2003). Based on the size and shape of the formed vesicle, the nature of internalized material and the molecules involved, endocytosis can be divided in three endocytic pathways: phagocytosis, macropinocytosis and pinocytosis (Fig. 4). Phagocytosis is mainly conducted by specialized mammalian cells (like monocytes, macrophages and neutrophils) to engulf solid particles with diameters > 750 nm by the cell membranes which form internal phagosomes. Smaller particles ranging from a few to several hundred nanometers are internalized by pinocytosis or macropinocytosis, which occur in almost all cell types. In macropinocytosis and phagocytosis, large vacuoles are formed by membrane protrusions towards the material to be endocytosed. These processes are highly dependent on the actin filament polymerization. In pinocytosis, small vesicles are formed by plasma membrane invagination. Pinocytic process can be further subdivided accordingly to the participation of cytoplasmic proteins clathrin, dynamin and caveolin. Clathrin-dependent endocytosis depends on the assembly of clathrin coat from cytoplasm to the cytoplasmatic side of the forming vesicle. This process involves the participation of GTPase, dynamin and can be mediated by binding of extracellular ligands to specific receptors on the cell surface. Another kind of pinocytosis depends on the participation of both caveolin and dynamin, with formation of caveosomes. Some cargoes can be endocytosed by mechanisms that are independent of the coat protein clathrin/caveolin and the fission GTPase, dynamin. Once inside the cell, most internalized cargoes are delivered to early endosomes via vesicular (clathrin- or caveolin-coated vesicles) or tubular intermediates that are derived from the plasma membrane. Some pathways may first traffic to intermediate compartments, such glycosyl phosphatidylinositol-anchored protein enriched early endosomal compartments (GEEC), before reaching the early endosome. Fusion with early endosomes delivers molecules to acidic environment that allows recycling of some molecules to plasma membrane. The remaining molecules are addressed to degradation and transported to late endosome or multivesicular body (MVB) where the process of degradation performed by hydrolytic enzymes come from Golgi complex takes place. The complete degradation is finished in lysosomes (Mayor & Pagano, 2007; De Souza *et al.*, 2009).

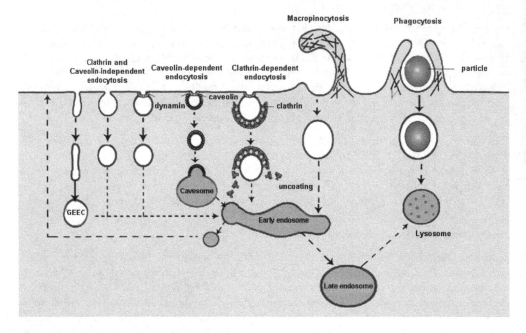

Fig. 4. Simplified scheme of endocytic pathway in mammal cells - adapted from (Mayor & Pagano, 2007) and (De Souza *et al.*, 2009).

It is well known that the source of QD materials and the nature of the QD surface ligand chemistry can be determinant for the intracellular fate of the nanoparticles. Typically size and shape are not the main factors that dominate the important biological process of cellular uptake of traditional matter or molecules. For most applications of nanoparticles, surface chemistry manipulations is critical for purposes like functionalization, reducing surface reactivity, reducing toxicity or enhancing chemical and optical stability.

2.3 Intracellular delivery of free QDs

The study of intracellular molecules and structures is still a challenge because in many cases QDs cannot reach freely the cell cytosol, limiting their application for experiments of intracellular labeling. Therefore, development of methods to cross the plasma membrane and/or escape from endo/lysosomal trapping after endocytosis is required for these purposes.

The mechanisms proposed for intracellular delivery of free QDs can generally be grouped into two categories: active and facilitated. Active delivery of QDs can be performed by microinjection and electroporation. However these methods are very invasive and often compromise the integrity of cellular structure and reduce the cellular viability (Biju *et al.*, 2010).

The facilitated delivery of QDs can be achieved by using peptides (such as cell-penetrating peptides, CPPs), viral vectors, some chemicals and more recently also by using liposomes (Jesorka & Orwar, 2008). Facilitated delivery can use agents such as cationic peptide complexed to QDs that can specifically induce their internalization. However, this procedure can also result in the trapping of nanoparticles within intracellular endosomal-lysosomal systems requiring further strategies to release the sequestered material to the cytosol when other intracellular targets are desired. The treatment of cells with chemicals, such as sucrose, chloroquine or polymers, can be applied to disrupt the endosomes and liberate QDs into cytosol (Biju et al., 2010). CPPs also known as protein transduction domains (PTDs) are short peptides containing significant amounts of basic amino acids that can permeate cell membrane. CPPs, originally derived from the Tat protein of the human immunodeficiency virus type 1.1, do not require receptors or energy-dependent pathway to penetrate the membrane being an attractive tool for delivering molecules (including QDs) into cells. Delehanty et al. (2006) demonstrated that efficient cellular uptake of CdSe/ZnS by HEK 293T/17 and COS cells were achieved when polyarginine-bearing cell-penetrating peptide was self-assembled onto the QD surface via non-covalent metal-affinity interactions. These CPP-QDs complexes were trapped into endosomal/lysosomal compartments for up three days in culture. The same group has performed an extensive investigation on active delivery methods based on CPPS and concluded that these peptides are able to bypass the endocytic system. They have found that Palm1-peptide was able to mediate both rapid endocytosis and a subsequent efficient cytosolic delivery of QDs conjugates 48 hours after initial exposure. The Palm-1 peptide appeared to be well tolerated with minimum toxicity (Delehanty et al., 2006, 2010).

2.4 QDs physico-chemical aspects

Although QDs present a broad absorption band, their application can be improved by knowing in which wavelength they can be excited to promote the highest emission intensity. A narrow first maximum band in the absorption spectrum is usually linked to a good size dispersion of the nanoparticles and a narrow (FWHM < 50 nm) and intense emission spectrum is usually linked to well passivated QDs. Carboxyl coated QDs are generally very stable in basic pH (higher than 8.5) and in their original medium that they have been synthesized. When QDs are in contact with the biological system the pH is near 7.5 and the stabilizing molecules may not be all deprotonated enabling aggregation of the QDs, influencing their intracellular fate. This can lead to an unsuccessfully biological application of QDs as fluorescent probes. Moreover, some biological media can not only promote aggregation but also the quenching of QDs fluorescence. For this reason it is important to study the spectroscopic behavior of QDs in the fluids where they will be dispersed. Chemical stability in biological media can be improved, for example, by coating QDs with silica.

2.5 General aspects of QDs toxicity

Nanoparticle toxicity is determined by its physicochemical characteristics. Size, charge, concentration, external coating bioactivity (capping material and functional groups) and oxidative, photolytic and mechanical stability can be determinant for QDs toxicity. Several studies have suggested that the cytotoxic mechanism of QDs involve photolysis and oxidation. Under oxidative and photolytic conditions, the coating of QDs is labile and may expose potentially toxic capping materials or intact core metalloid complexes or may result in the dissolution of core metal components (e.g. Cd, Se, etc) (Li & Chen, 2011).

Cadmium is probably carcinogen, having a biological half-life of 15-20 years in humans. Cd^{2+} bioaccumulates and crosses the blood-brain barrier and placenta, and is systematically distributed to all body tissues including liver and kidney. So, If Cd^{2+} ions dissociate from QDs, they can be eventually toxic to the organism. QDs could be coated with inert materials, such as silica or polymers, to prevent dissociation and render Cadmium bearing QDs are less toxic compared to conventional organic dyes. Despite their potential cytotoxic effects the CdSe and CdTe QDs are the most fluorescent and for this reason are the most used QDs in life sciences. Researchers are improving the synthesis of Mn^{2+} doped Zinc based QDs to overcome the toxicity and to produce red emitting nanoparticles. Excitation by using UV light can also underline side effects to the cell.

Lastly it is also important to pay attention in the stabilizing agent and in osmolarity; some stabilizing agents and ionic counter ions result in unbalanced media that can lyse labile cells such as erythrocytes. Cells and biomolecules integrity also depend on the QDs pH. When synthesis produces QDs with pH near 10 – 11, it is necessary to reduce the pH to near physiological values to use QDs in live cells for keeping metabolic cell and the antibodies/antigens functions.

3. Water dispersed QDs

In this Section we will discuss aspects concerning to the synthesis, characterization and functionalization of QDs applied by us in the staining of parasites.

Fluorescent QDs used for biological imaging are usually obtained by colloidal synthesis involving the chemical reaction of inorganic precursors. Water colloidal synthesis typically involves three main components: inorganic precursors, organic surfactants and water. The main parameters which need to be ajusted to optimize the synthesis are: initial pH, reaction temperature and the ratio among inorganic precursors and organic surfactants used. The surfactant helps to stabilize the nanoparticles (by preventing their aggregation) and to dissolute smaller nanocrystals in favor of larger ones. The key feature of this "bottom-up" approach is the ability to control precisely the final size of the QDs and consequently their optical properties.

3.1 QDs synthesis

$CdS/Cd(OH)_2$ nanoparticles were prepared in aqueous medium adapting the procedure described by Vossmeyer *et al.*, 1994. In a typical procedure $Cd(ClO)_4$ and H_2S were used as chemical precursors. Due to the intrinsic thermodynamically instability of colloidal suspensions, a stabilizing agent (sodium polyphosphate, $Na(PO_3)_9$) was added to the reacting system. Subsequent surface passivation with $Cd(OH)_2$ was carried out to improve luminescence (Figure 5).

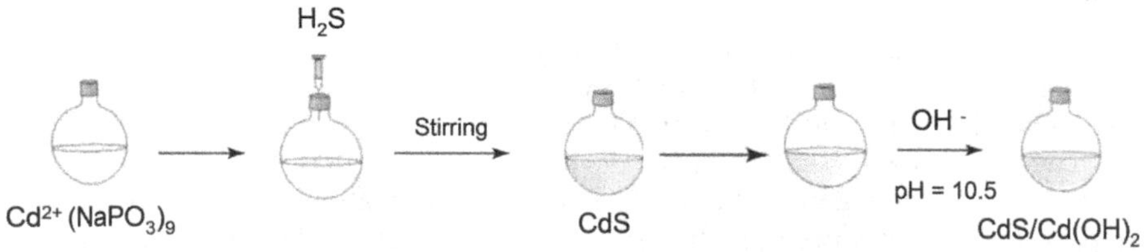

Fig. 5. Schematic illustration of $CdS/Cd(OH)_2$ nanoparticles synthesis.

CdTe/CdS nanoparticles were synthesized according to the previously reported methodology (Santos, 2008). Briefly, QDs were prepared by the addition of Te^{2-} solution in a Cd(ClO$_4$)$_2$ 0.01 M solution of pH above 10 in the presence of MPA (mercaptopropionic acid) or MSA (mercaptosuccinic acid) as stabilizing agents in a 2:1:6 ratio for Cd:Te:MPA and 2:1:2.4 of Cd:Te:MSA. The reaction proceeds under constant stirring and heating of 80 °C in argon or nitrogen (White Martins). The Te^{2-} solution was prepared using metallic tellurium and NaBH$_4$, under argon or nitrogen saturated inert atmosphere. At the end of the reaction, the colloidal dispersion showed a brownish color (Figure 6). The CdTe passivation shell is composed of a few monolayers of CdS grown during the post heating of the suspensions and monitored by the increase of the fluorescence intensity. The CdTe/CdS-MSA QDs were synthesized in PBS (Phosphate Buffered Saline) to improve biocompatibility.

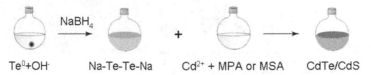

Te^0+OH$^-$ Na-Te-Te-Na   Cd^{2+} + MPA or MSA   CdTe/CdS

Fig. 6. Schematic illustration of the CdTe/CdS nanocrystals synthesis.

3.2 QDs structural characterizations

QDs structural properties were obtained by using Transmission Electron Microscopy (TEM) and X-ray Diffraction analysis (Figure 7). Samples for the microscopic analysis were prepared by drying a drop of the diluted colloidal suspension on a copper grid coated with carbon/formwar film. X-ray analysis of all QD employed show that they presented a cubic zinc blend phase structure.

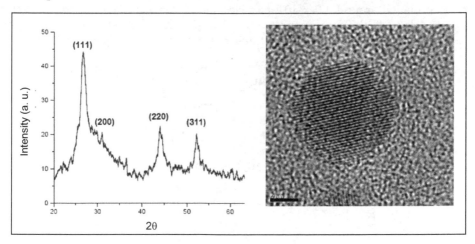

Fig. 7. A representative X-ray powder diffraction profile and TEM image of the CdS/Cd(OH)$_2$ QDs (bar = 2 nm).

By using the Scherrer formula, we can calculate the QDs mean diameters from the width of the most intense diffraction peak at half-maximum. The values obtained by X-ray diffraction were in accordance with TEM data for both QDs. According to these analysis, CdS/Cd(OH)$_2$ QDs have a mean diameter of 10 nm (Figure 7), CdTe/CdS-MPA QDs of 2.7 nm (Figure 8) and similarly CdTe/CdS-MSA QDs of 3.2 nm (data not shown).

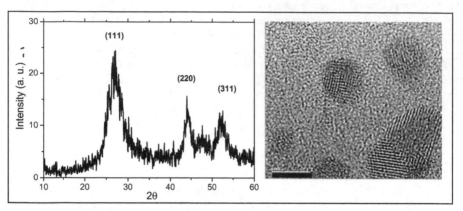

Fig. 8. A representative X-ray powder diffraction profile and TEM image of the CdTe/CdS QDs (bar = 3 nm).

3.3 QDs optical characterizations

Figures 9 and 10 present the absorption and emission spectra of CdS/Cd(OH)$_2$ and CdTe/CdS-MPA QDs. The CdS/Cd(OH)$_2$ shows maximum absorption at 475 nm. The mean diameter of CdS can be directly related to the value of the first maximum absorption wavelength. In 475 nm it was described a size of about 9.6 nm (Vossmeyer *et al.*, 1994). This synthesis provided around 1.0 x 10^{13} particles/mL. The mean diameters for CdTe/CdS-MPA and CdTe/CdS- MSA are respectively 2.7 nm (λ = 286 nm) and 3.5 nm (λ = 597 nm), also estimated using the spectrum first absorption maximum (Dagtepe *et al.*, 2007). The

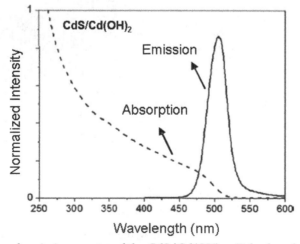

Fig. 9. Absorption and emission spectra of the CdS/Cd(OH)$_2$ – Polyphosphate QDs.

number of particles was 1.3 x 10^{15} particles/mL for CdTe/CdS-MPA and 5.6 x 10^{14} particles/mL for CdTe/CdS-MSA (Yu *et al.*, 2003). The mean diameters obtained by optical characterization are in accordance to the structural analysis.

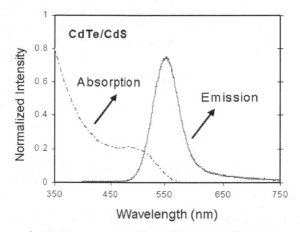

Fig. 10. Absorption and emission spectra of the CdTe/CdS-MPA QDs.

The emission band profile shows a maximum at 506 nm for CdS/Cd(OH)$_2$ QDs, 550 nm for CdTe/CdS-MPA QDs and 600 nm for CdTe/CdS-MSA QDs respectively. These CdTe/CdS-MPA showed intense luminescence after 3 hours of refluxing process at 80^{0}C and MSA after 7 hours. This heat treatment process, at high pH, promotes the hydrolysis of the MPA/MSA acid molecules releasing thiol group (SH-) which help to grow the passivation shell (Gaponik & Rogach, 2010). CdS/Cd(OH)$_2$ QDs presents final pH around 8.0 while CdTe/CdS QDs presents final pH equal to 10.5.

3.4 QDs preparation for applications

For application of QDs with biological systems, 1400 µL of the CdS/Cd(OH)$_2$ nanoparticles (1×10^{13} particles/ml) were functionalized with 100 µL PEG-1000 (0.0025 g/mL) in physiological medium. PEG is a water-soluble biocompatible polymer which possesses a low degradation rate and is commonly used to prevent the close contact of nanometric drug carrier systems with enzymes and other molecules present in the biological systems. The PEGylation of the particles is believed to decrease the cytotoxicity of the QDs allowing *in vitro* analysis of live cells for longer time periods (Chaves *et al.*, 2008).

The CdTe/CdS QDs were used without further bioconjugation. The MPA and MSA were used to promote the interactions between QDs and parasites.

4. Trypanosomatides

Protozoa of the Trypanosomatidae family are agents of parasitic disease that have a high incidence of occurrence with major negative economic impact on developing countries. *Trypanosoma cruzi*, the ethiologic agent of Chagas disease, is transmitted to human beings by either blood sucking triatomine insect vectors, blood transfusion, organ transplantation or congenital transmission. This disease is present in 18 tropical countries and has affected 17 million individuals. The life cycle of *T. cruzi* involves the obligatory passage through the vertebrate (mammals) and invertebrate (triatomine insects) hosts. Shortly after infection the individuals enter the acute phase of illness, which can extend for weeks or even months, being typically asymptomatic or associated with fever and other less specific clinical manifestations. In the absence of a specific treatment the rate of deaths associated to Chagas disease ranging from 10 to 15% and 10 to 50% of survivors evolve into a chronic phase, which is characterized by a potentially fatal cardiomyopathy, megaesophagus or megacolon[1].

In the case of Leishmaniasis, caused by different species of protozoa belonging to the *Leishmania* genus, about 16 million people are infected in Africa, Asia, parts of Europe and Latin America. Leishmaniasis is a poverty-related disease. It affects the poorest people and is associated with malnutrition, displacement, poor housing, illiteracy, gender discrimination, weakness of the immune system and lack of resources. Depending on the parasite species and immunologic response of host the disease can assume a wide range of clinical symptoms, which may be cutaneous, mucocutaneous or visceral. Cutaneous leishmaniasis is the most common form whereas visceral leishmaniasis is the most severe form, in which vital organs of the body are affected[2].

Trypanosomatides parasites are usually transmitted by insect vectors and invade a range or different tissues or cell types in mammalian These parasites undergo a number of distinct developmental and morphological changes during their complex life cycles in the insect vector and mammalian hosts (Figure 11).

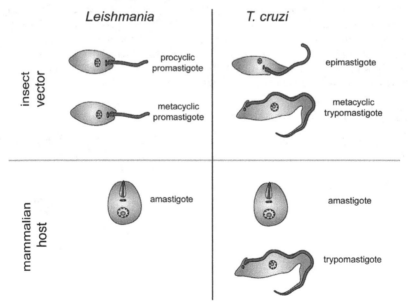

Fig. 11. Developmental stages of *T. cruzi* and *Leishmania spp.* during the life cycle in the vertebrate and invertebrate hosts.

---

[1] *http://www.fiocruz.br/chagas_eng/cgi/cgilua.exe/sys/start.htm?tpl=home*
[2] *http://www.who.int/leishmaniasis/en/*

In view of the importance of Trypanosomatides some efforts have been devoted to understand the interactions existing between the parasites and their hosts. In this way, several fluorescent probes, as acridine orange, Nile red, and GFP-transfected cells have been used for *T. cruzi* and *Leishmania* morphological and physiological characterization (Pereira *et al.*, 2011; Monte Neto *et al.*, 2010).

However, the organic fluorophores have limited use in live cells due to their instability, photobleaching effects and potential photocytotoxicity. In order to benefit from the advantageous optical properties of QDs as *in vivo* and *in vitro* probes for biological studies in these parasites, some issues must be addressed. Firstly: how do the parasites interact with different kind of QDs? How are they taken up by the parasites? Once inside the cell what is their intracellular fate? And finally, how are the QDs eliminated? Before discussing these questions we will briefly describe some particularities of the cell biology and the endocytic pathway of *T. cruzi* and *Leishmania*.

It is well established that to enter the cell and achieve their intracellular target QDs must first interact with the plasma membrane. The way by which QDs interact with the parasite surface is therefore crucial to determine their intracellular fate inside the cell.

Trypanosomatides are high polarized cells with organelles placed in specific regions of the cytoplasm. Although Trypanosomatides are similar in some aspects of endocytosis, they diverge as the structural organization of their endocytic pathways. The surface of all Trypanosomatides is composed by a functional assembly of two components: the plasma membrane and a stable corset of subpelicular microtubules connected with each other and with plasma membrane by short filaments. This arrangement forms a rigid structure that impairs endocytic vesicles to form along the whole parasite body. In *Leishmania* species the uptake of macromolecules is performed exclusively throughout the flagellar pocket, which is a plasma membrane invagination that is devoid of microtubules. Besides the flagellar pocket a highly specialized structure known as the cytostome is also involved in the ingestion of macromolecules in epimastigote forms of *T. cruzi*. The cytostome, present only in epimastigote and amastigote forms, is considered as the main site of endocytosis in epimastigotes but their functions in amastigotes is still unknown. In epimastigotes the cytostome is located near the flagellar pocket and extends deeply into the cytoplasm as a narrow tubule (the cytopharynx) that finishes in a dilatation from where endocytic vesicles pinch off.

The endocytic pathway of *Leishmania* promastigotes comprises at least three morphologically distinct compartments: a network of tubular endosomes that are localized near the flagellar pocket; a population of multivesicular bodies that are also located near the anterior end of promastigotes and an unusual multivesicular tubule (MVT)-lysosome that runs along the anterior-posterior axis of the parasite. It has been proposed that MVTs are the sites of accumulation of the protein and lipid that are ingested by the parasite and the final compartment of endocytic pathway in this evolutive form. The intracellular amastigote forms of *Leishmania* escape from host cell immune cell response by infecting macrophages. Intracellular amastigote are able to ingest macromolecules through the flagelar pocket which has access to the parasitophorous vacuole. This process seems to be mediated by clathrin molecules. The ingested material is degraded in the megasomes, large membrane-bounded structures with acidic environment. Megasomes are thought to be the lysosomal compartment of amastigotes (De Souza *et al.*, 2009).

As mentioned above *T. cruzi* endocytosis only occurs in the epimastigote forms. Low or lacking of endocytic activity can be observed in trypomastigote (the infective form of parasite) and amastigotes (the intracellular forms). Following binding to the cytostome and flagellar pocket, macromolecules are rapidly internalized in small uncoated vesicles which bud from these structures. The characterization of endocytic compartments in *T. cruzi* is still controversial (Porto-Carreiro *et al.*, 2000; Figueiredo & Soares, 2000). However, there is a consensus that the endocytic pathway seems to converge to large lysosome-related organelles called reservosomes. The reservosome has a protein electrodense matrix were small lipid inclusions are immersed. These organelles have been implicated mainly in macromolecule storage and degradation, although other important roles in the life cycle of parasite, as cell differentiation, autophagy and recycling processes have been also demonstrated (Figueiredo *et al.*, 2004; Soares *et al.*, 1992).

5. QDs applications for bioimaging of trypanosomatides

In this Section we will present some results obtained by our research group on the applications of CdS/Cd(OH)$_2$ and CdTe/CdS water dispersed QDs in bioimaging of *Trypanosoma cruzi* and *Leishmania amazonensis*. We have successfully demonstrated that different QDs can be efficiently used as fluorescent probes for bioimaging of live Trypanosomatides.

The amount of 250 µL of live epimastigote forms of *T. cruzi* (1.0 x 10^6) were incubated at room temperature with 350 µL of colloidal CdS/Cd(OH)$_2$–PEG QDs suspension (1 x 10^{13} particles/mL) for 180 minutes using Hank's solution (Invitrogen, USA). The choice of salt balanced Hank's solution as incubation medium was due to the fact that the medium commonly used for axenic cultivation of parasites suppresses the fluorescence of CdS/Cd(OH)$_2$ QDs. This probably happens because Cd(OH)$_2$ is a labile passivation shell. The Hank's solution has physiological pH (7.5) and keeps the parasites alive with minimum interference in the fluorescence signal.

Figure 12 shows fluorescence microscopy and transmission electron microscopy images of *T. cruzi* labeled with CdS/Cd(OH)$_2$. It is possible to observe that QDs entered the cytostome and could be further detected in small vesicles and finally in large compartments, which probably correspond to the reservosomes. As mentioned above the cytostome is the main site of receptor-mediated endocytosis in *T. Cruzi* (Figure 12 a). Previous studies demonstrated that proteins coupled to colloidal gold are efficiently endocytosed through the cytostome by a receptor-mediated mechanism. Vesicles pinch off from the bottom of these structures delivering their cargo directly into reservosomes (Figueiredo *et al.*, 2004; Figueiredo & Soares, 2000). It seems that the endocytosis of PEG-CdS/Cd(OH)$_2$ shares the same pathway.

An interesting advantage of QDs at biological point of view is that these particles are nanocrystals with high atomic number atoms (Cadmium, Tellurium, etc). It means that they are electron-dense and can be visualized by transmission electron microscopy (Figure 12 b). This feature is extremely useful considering that besides provide high quality fluorescence images QDs also allow the visualization and localization of particles at ultrastructural level. The transmission electron microscopy analysis of epimastigote incubated with CdS/Cd(OH)$_2$ corroborates the data obtained from fluorescence microscopy, showing a high concentration of QDs in endocytic vesicles and in tubules resembling cytosomes which are in close association with reservosomes (Figure 12 c). Additionally, the parasite had a well-preserved ultrastructure. All the data collected suggest that QDs are a useful tool for the bioimaging of *T.cruzi*.

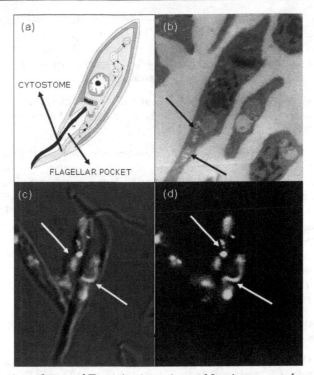

Fig. 12. (a) Endocytic pathway of *T. cruzi* epimastigote. Nutrients are taken up through the cytostome and the flagellar pocket and delivery into large membrane-bound structures, the reservosomes. (b) TEM of *T. cruzi* parasites showing the accumulation of CdS/Cd(OH)$_2$– PEG QDs (arrow) inside the tubular structure resembling the cytostome in close association with reservosomes. (c) Overlay of the image of fluorescence microscopy and differential interference contrast showing the strong QDs labeling in the cytostome. (d) The same image showing only fluorescence staining. Note the accumulation of tracer at the bottom of the cytostome (arrow). Excitation at 456 nm (Chaves *et al.*, 2008).

The analysis of *T. cruzi* by using fluorescent QDs can clarify the comprehension of the endocytic process as well as to clarify the cellular differentiation mechanism. Moreover, the images preserve the original fluorescence emission wavelength suggesting that the PEG layer protects the QDs passivation shell that can induce deactivation processes and loss of fluorescence. The chemical and optical stability of this inorganic–organic assembly inside the biological system show its great potential as functional fluorescent nanoprobes.

CdTe/CdS-MAA and CdSe/CdS-MAA nanocrystals were also used by other research groups to study *T. cruzi*–insect vector interactions (Feder *et al.* 2009). These QDs were synthesized by using mercaptoacetic acid (MAA). QDs were also conjugated with lectins to label specific carbohydrates involved in parasite-vector interaction. In this work parasites and intestinal cells were labeled with both green-emitting CdTe/CdS-MAA and yellow-emitting CdSe/CdS-MAA QDs. Parasites were maintained viable up to three day in culture medium containing of QDs. In experiments using QDs conjugated to *Sambucus nigra* lectin (SNA), which specifically bind to galactose and N-acetylgalactosamine carbohydrates residues, labeling was found in intracellular vesicles. So, it is likely that SNA-QDs conjugates are recognized by specific receptors at the parasite surface before to be interiorized in endocytic vesicles. No fluorescence could be detected in the cytostome suggesting that the chemical characteristic of QDs as well as their bioconjugation to specific ligands have influence in the uptake of these nanoparticles.

The quantity of 300 µL of CdS/Cd(OH)$_2$ QDs (at 1×10^{13} particles/mL) was also used by us to label *L. amazonensis* live parasites (100 µL – 2×10^5 cells/mL). Figure 13 shows the analysis of these parasites by fluorescence microscopy.

The living parasites showed no signal of damage after the incubation procedure and maintained their morphological integrity and viability even after 18 hours of incubation, demonstrating the isotonicity of the labeling procedure as well as the low toxicity of the QDs. The emission pattern (Figure 13 a and c), observed under 456 nm excitation, evidenciates the labeling of different internal structures present in the promastigote: nucleus, kinetoplast (specialized region of parasite mitochondria where the mitochondrial DNA, named k-DNA, is concentrated) and flagellar pocket (Figure 13 b). Contrary to *Trypanosoma cruzi*, no labeling in intracellular vesicles and in endocytic structure could be detected and the mechanism of uptake and delivery of QDs in nucleus and kinetoplast is still unknown. Further experiments will be conducted to better understand the routes of internalization and interaction of QDs-parasite.

We have also reported the labeling of *L. amazonensis* live parasites using CdTe/CdS with MPA and MSA agents that have carboxyl groups available.

The quantity of 350 µL of CdTe/CdS-MPA (at 1.3×10^{15} particles/mL) was incubated for 30 minutes with 250 µL of *L. amazonensis* live parasites (2.0×10^5). For the CdTe/CdS-MSA a similar procedure was applied by using 100 µL of cells and 300 µL of QDs (5.6×10^{14} particles/mL), in this experiment cells were incubated for 90 minutes.

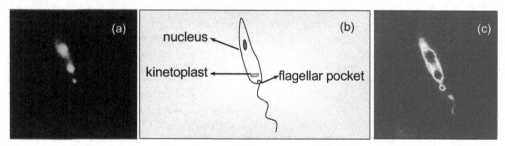

Fig. 13. (a) Confocal image of *L. amazonensis* promastigote labeled with CdS/Cd(OH)$_2$ QDs excited at 456 nm. (b) The Figure shows a schematic representation of a L. amazonensis promastigote parasite, showing three different internal structures, the nucleus, the kinetoplast and the flagellar pocket (c) Corresponding fluorescence intensity maps – blue represents the absence of fluorescence and red the highest intensity.

Incubation of parasites with CdTe/CdS-MPA QDs or CdTe/CdS-MSA showed a similar labeling pattern of CdS/Cd(OH)$_2$ QDs with strong labeling of flagellar pocket and nucleus (Figure 14 a). These results suggest that carboxylic groups of the stabilizing and functionalizing agents (present on the QDs surface) were responsible for non specific interaction with proteins in cells.

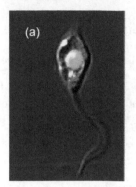 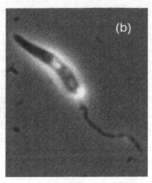

Fig. 14. (a) Confocal and (b) conventional fluorescence microscope images of the *L. amazonensis* incubated with CdTe/CdS-MPA (a) and CdTe/CdS-MSA (b).

When a study aims the labeling of live cells as *T. cruzi*, the potential cytotoxicity of QDs is an important issue to be addressed. We followed the growth and viability of epimastigote forms incubated with $CdS/Cd(OH)_2$ for 96 hours of cultivation, by light microscopy. *T. cruzi* epimastigotes were grown at 28°C in liver infusion tryptose medium supplemented with 10% bovine fetal serum (Chiari & Camargo, 1984; Jaffe *et al.*, 1984). In order to determine the effect of QDs on viability and growth, *T. cruzi* epimastigotes (1×10^6/mL) were incubated at different $CdS/Cd(OH)_2$ QDs concentrations and cell density was estimated by directly cell counting in a hemocytometric chamber. Our results showed that no significant difference in the parasite growth and viability could be detected at 24 hours of incubation with QDs at any concentration tested. However, after 48 hours of cultivation the number of parasites decreases in a time and concentration dependent manner (Figure 15).

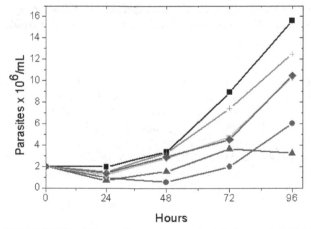

Hours

Fig. 15. Effects of $CdS/Cd(OH)_2$ QDs on *T. cruzi* growth. (■) Control; (▲) 0,5 x 10^{13}; (●) 2,5 x 10^{12}; (▼) 2,5 x 10^{11}; (♦) 2,5 x 10^9; (X) 2,5 x 10^7 particles/mL.

Similarly, Vieira *et al.*, 2011 analyzed the nanotoxic effects of CdTe/CdS-MAA QDs in *T. cruzi*. These authors found that high doses of QDs 20 µM and 200 µM/mL led to a decrease of *T. cruzi* growth patterns. Contrary, low concentrations of CdTe/CdS-MAA QDs did not affect parasite integrity, division and motility up to seven days of culture in presence of QDs. Transmission electron microscopy demonstrated that CdTe/CdS-MAA intracellular localization differs significantly from those found for CdS/Cd(OH)$_2$. CdTe/CdS-MAA labeling was present in cytoplasmic vesicles and near reservosomes, but not inside this latter organelle, suggesting that different mechanisms mediate the uptake and accumulation of CdS/Cd(OH)$_2$ and CdTe/CdS-MAA in this parasite. Since the QDs were also detected at the plasma membrane a mechanism for elimination of these nanoparticles could be through the fusion of intracellular vesicle containing CdTe to plasma membrane for shedding and elimination (Vieira et al., 2011).

6. General conclusions

In summary, our results have shown the possibility of applying fluorescent QDs as probes to label living T. cruzi and Leishmania and to study cellular processes in these parasites. Even though QDs have been presented as fluorescent probes in the late 90′s they are not yet conventional probes. There are still some important aspects to consider, specially those related to the maintenance of their optical properties and chemical stability in contact with the biological systems. Beyond the great optical properties of this inorganic-organic ensemble, QDs are electron-dense and can also be explored at ultrastructural level. This feature helps the monitoring of the cellular uptake of QDs and their intracellular fate. The investigation of QDs-trypanosomatides interactions can lead to a better comprehension of the endocytic routes providing new insights of the parasites cellular biology, which can help to understand the cellular mechanisms involved in the Chagas disease and Leishmaniasis.

7. Acknowledgement

The authors are grateful to CAPES, CNPq, FACEPE, HEMOPE, L′óreal, Brazilian Academy of Sciences and UNESCO. This work is also linked to the National Institute of Photonics.

8. References

Barroso, M. M. (2011) Quatum dots in cel biology. J Histochem Cytochem., Vol. 59, pp. 237-251.

Bentzen, E. L.; Tomlinson, I. D.; Mason, J.; Gresch, P.; Warnement, M. R.; Wright, D.; Sanders-Bush, E.; Blakely, R. & Rosenthal, S. J. (2005). Surface Modification To Reduce Nonspecific Binding of Quantum Dots in Live Cell Assays. Bioconjugate Chem., Vol. 16, pp. 1488-1494.

Biju, V.; Itoh, T. & Ishikawa, M. (2010). Delivering quantum dots to cell: bioconjugated quantum dots for target and nonspecific extracellular and intracellular imaging. Chem. Soc. Rev., Vol. 39, pp. 3031-3056.

Conner, S. D. & Schmid, S. L. (2003). Regulated portals of entry into the cell. Nature, Vol. 422, pp. 37 – 44.

Chaves, C. R.; Fontes, A.; Farias, P. M. A.; Santos, B.S.; Menezes, F.D.; Ferreira, R. C.; César, C. L. César, Galembeck, A. & R.C.B.Q. Figueiredo. (2008). Application of core–shell PEGylated CdS/Cd(OH)$_2$ quantum dots as biolabels of Trypanosoma cruzi parasites. Applied Surface Science, Vol. 225, pp. 728–730.

Chiari, E. & Camargo, E. P. (1984). Culturing and cloning of Trypanosoma cruzi. Genes and antigens of parasites, A laboratory manual, (2a ed), In Morel CM, Fundação Oswaldo Cruz, Rio de Janeiro, p. 23-26.

Dagtepe, P.; Chikan, V.; Jasinski, J. & Leppert, V. J. (2007). Quantized Growth of CdTe Quantum Dots, Observation of Magic-Sizes CdTe Quantum Dots. *Journal of Physical Chemistry*, Vol. 111, pp. 14977-14983.

De Souza, W.; SantAnna, C.; Cunha-e-Silva, N. L. & Narcisa, L. (2009). Electron microscopy and cytochemistry analysis of the endocytic pathway of pathogenic protozoa. *Progess in Histochemistry and Cytochemistry*. Vol. 44, pp. 67-124.

Delehanty, J. B.; Medintz, I. L.; Pons, T.; Dawson, P. E,; Brunel, F. M.; Mattoussi, H. (2006). Self-assembled quantum dot-peptide bioconjugates for selective intracellular delivery. *Bioconjugate Chem.*, Vol. 17, pp. 920-927.

Delehanty, J. B.; Bradburne, C. E.; Boeneman, K.; Susumu, K.; Farrell, D.; Mei, B. C.; Blanco-Canosa, J. B.; Dawson, G.; Dawson, P. E.; Mattoussi, H. & Medintz, I. L. (2010). Delivering quantum dot-peptide bioconjugates to the cellular cytosol: escaping from the endolysosomal system. *Integr Biol* (Camb), Vol. 2, pp. 265-277.

Feder, D.; Gomes, S. A.; de Thomaz, A. A.; Almeida, D. B.; Faustino, W. M.; Fontes, A.; Stahl, C. V.; Santos-Mallet, J. R. & César, C. L. (2009). In vitro and in vivo documentation of quantum dots labeled *Trypanosoma cruzi*--Rhodnius prolixus interaction using confocal microscopy. *Parasitol Res.*, Vol. 106, pp. 85-93.

Figueiredo, R. C. & Soares, M. J. (2000). Low temperature blocks fluid-phase pinocytosis and receptor-mediated endocytosis in *Trypanosoma cruzi* epimastigotes. *Parasitol Res.*, Vol. 86, pp. 413-418.

Figueiredo, R. C.; Rosa, D. S.; Gomes, Y. M.; Nakasawa, M.; Soares, M. J. (2004). Reservosome: an endocytic compartment in epimastigote forms of the protozoan *Trypanosoma cruzi* (Kinetoplastida: Trypanosomatidae). Correlation between endocytosis of nutrients and cell differentiation. *Parasitology*, Vol. 129, pp. 431-438.

Gaponik, N.; Rogach, A. (2010). L.Thiol-capped CdTe nanocrystals: progress and perspectives of the related research fields. *Phys. Chem. Chem. Phys.*, Vol. 12, pp. 8685-8693.

Giepmans, B. N. G.; Adams, S. R.; Ellisman, M. E. & Tsien, R. Y. (2006). The Fluorescent Toolbox for Assessing Protein Location and Function. *Science*, Vol. 312, pp. 217-224.

Jaffe, C. L.; Grimaldi, G.; McMahon-Pratt, D. (1984). The cultivation and cloning of Leishmania. Genes and antigens of parasites. A laboratory manual, 2nd ed., In Morel CM, Fundação Oswaldo Cruz, Rio de Janeiro, p. 48-91.

Jesorka, A.; Orwar, O. (2008). Liposomes: Technologies and Analytical Applications. *Annu. Rev. Anal. Chem.* Vol. 1, pp. 801-32.

Kelf, T. A.; Sreenivasan, V. K. A.; Sun, J.; Kim, E. J.; Goldys, E. M. & Zvyagin, A. V. (2010). Non-specific cellular uptake of surface-functionalized quantum dots. *Nanotechnology*, Vol. 21, pp. 1-8.

Li, Yu-Feng & Chen, C. (2011). Fate and Toxicity of Metallic and Metal-Containing Nanoparticles for Biomedical Applications. Small, pp. 1-16.

Mayor, S. & Pagano, R. E. (2007). Pathways of clatrin-independent endocytosis. Nature Reviews Molecular Cell Biology, Vol. 8, pp. 603-612.

Michalet, X.; Weiss, S. & Jäger, M. (2006). Single-Molecule Fluorescence Studies of Protein Folding and Conformational Dynamics. *Chemical Reviews*, Vol. 106, pp. 1785-1813.

Monte Neto, R. L.; Sousa, L. M.; Dias, C. S.; Barbosa Filho, J. M.; Oliveira, M. R. & Figueiredo, R. C. (2010). Morphological and physiological changes in *Leishmania* promastigotes induced by yangambin, a lignan obtained from Ocotea duckei. *Exp Parasitol.*, Vol. 127, pp. 215-221.

Pereira, M. G.; Nakayasu, E. S.; Sant'Anna, S.; De Cicco, N. N. T.; Atella, G. T.; De Souza, W. Almeida, I. C. & Cunha-Silva, N. (2011). Trypanosoma cruzi epimastigotes are able to store and mobilize high amounts of cholesteral in reservossome lipid inclusions. *PLOS One*, Vol. 6, pp. 222359.

Porto-Carreiro, I.; Attias, M.; Miranda K.; De Souza, W.; Cunha-e-Silva, N. (2000). Trypanosoma cruzi epimastigote endocytic pathway: cargo enters the cytostome and passes through an early endosomal network before storage in reservosomes. *European Journal of Cell Biology*. Vol. 79, pp. 858-869.

Samia, A. C. S., Dayal, S. & Burda, C., 2006. Quantum Dot-based Energy Transfer: Perspectives and Potential for Applications in Photodynamic Therapy. *Photochemistry and Photobiology*, Vol. 82, pp. 617-625.

Santos, B. S.; Farias, P.M.A. Farias & Fontes, A. (2008). Semiconductor Quantum Dots for Biological Applications. *Handbook of Self Assembled Semiconductor Nanostructures for Novel Devices in Photonics and Electronics*. pp. 773-799.

Say, R.; Kiliç, G. A.; Ozcan A. A,; Hür, D.; Yilmaz, F.; Kutlu, M.; Yazar, S.; Denizli, A.; Diltemiz, S. E. & Ersöz, A. (2011). Investigation of photosensitively bioconjugated target quantum dots for labeling of Cu/Zn superoxide dismutase in fixed cel and tissue sections. *Histochem Cell Biol.*, Vol. 135, pp. 523–530.

Soares, M. J.; Souto-Padrón, T.; De Souza, W. (1992). Identification of a large pre-lysosomal compartment in the pathogenic protozoon *Trypanosoma cruzi*. *J Cell Sci.*, Vol. 102, pp. 157-167.

Tian, J., Liu, R., Zhao, Y., Peng, Y., Hong, X., Xu, Q. & Zhao, S. (2010). Synthesis of CdTe/CdS/ZnS quantum dots and their application in imaging of hepatocellular carcinoma cells and immunoassay for alpha fetoprotein. *Nanotechnology*, Vol. 21, pp. 1-8.

Vieira, C. S.; Almeida, D. B.; Thomaz, A. A.; Menna-Barreto, R. F. S; Santos-Mallet, J. R; Cesar, C. L.; Gomes, S. A. O. G; Feder, D. (2011). Studying nanotoxic effects of CdTe quantum dots in *Trypanosoma cruzi*. *Mem Inst Oswaldo Cruz*, Rio de Janeiro, Vol. 106, pp. 158-165.

Vossmeyer, T.; Katsikas, L.; Giersig, M.; Popovic, IG, Diesner, K.; Chemseddine, A.; Eychmüller, A.; Weller, H. J. (1994). CdS Nanoclusters: Synthesis, Characterization, Size Dependent Oscillator Strength, Temperature Shift of the Excitonic Transition Energy, and Reversible Absorbance Shift. *The Journal of Physical Chemistry*, Vol. 98, pp. 7665.

Yezhelyev, M. V., Gao, X., Xing, Y., Al-Hajj, A., Nie, S. & O'Regan, R. M. (2006). Emerging use of nanoparticles in diagnosis and treatment of breast cancer. *The lancet oncology*, Vol. 7, pp. 657-667.

Yu, W. W.; Qu, L.; Guo, W. & Peng, X. (2003). Experimental Determination of the Extinction Coefficient of CdTe, CdSe, and CdS Nanocrystals. *Chem. Mater*, Vol. 15, pp. 2854-2860.r

Permissions

The contributors of this book come from diverse backgrounds, making this book a truly international effort. This book will bring forth new frontiers with its revolutionizing research information and detailed analysis of the nascent developments around the world.

We would like to thank Ameenah N. Al-Ahmadi, PhD, for lending her expertise to make the book truly unique. She has played a crucial role in the development of this book. Without her invaluable contribution this book wouldn't have been possible. She has made vital efforts to compile up to date information on the varied aspects of this subject to make this book a valuable addition to the collection of many professionals and students.

This book was conceptualized with the vision of imparting up-to-date information and advanced data in this field. To ensure the same, a matchless editorial board was set up. Every individual on the board went through rigorous rounds of assessment to prove their worth. After which they invested a large part of their time researching and compiling the most relevant data for our readers. Conferences and sessions were held from time to time between the editorial board and the contributing authors to present the data in the most comprehensible form. The editorial team has worked tirelessly to provide valuable and valid information to help people across the globe.

Every chapter published in this book has been scrutinized by our experts. Their significance has been extensively debated. The topics covered herein carry significant findings which will fuel the growth of the discipline. They may even be implemented as practical applications or may be referred to as a beginning point for another development. Chapters in this book were first published by InTech; hereby published with permission under the Creative Commons Attribution License or equivalent.

The editorial board has been involved in producing this book since its inception. They have spent rigorous hours researching and exploring the diverse topics which have resulted in the successful publishing of this book. They have passed on their knowledge of decades through this book. To expedite this challenging task, the publisher supported the team at every step. A small team of assistant editors was also appointed to further simplify the editing procedure and attain best results for the readers.

Our editorial team has been hand-picked from every corner of the world. Their multi-ethnicity adds dynamic inputs to the discussions which result in innovative outcomes. These outcomes are then further discussed with the researchers and contributors who give their valuable feedback and opinion regarding the same. The feedback is then collaborated with the researches and they are edited in a comprehensive manner to aid the understanding of the subject.

Apart from the editorial board, the designing team has also invested a significant amount of their time in understanding the subject and creating the most relevant covers. They scrutinized every image to scout for the most suitable representation of the subject and create an appropriate cover for the book.

The publishing team has been involved in this book since its early stages. They were actively engaged in every process, be it collecting the data, connecting with the contributors or procuring relevant information. The team has been an ardent support to the editorial, designing and production team. Their endless efforts to recruit the best for this project, has resulted in the accomplishment of this book. They are a veteran in the field of academics and their pool of knowledge is as vast as their experience in printing. Their expertise and guidance has proved useful at every step. Their uncompromising quality standards have made this book an exceptional effort. Their encouragement from time to time has been an inspiration for everyone.

The publisher and the editorial board hope that this book will prove to be a valuable piece of knowledge for researchers, students, practitioners and scholars across the globe.

List of Contributors

Viviane Pilla, Noelio O. Dantas, Anielle C. A. Silva and Acácio A. Andrade
Federal University of Uberlândia–UFU, Uberlândia, MG, Brazil

Egberto Munin
University Camilo Castelo Branco- UNICASTELO, São José dos Campos, SP, Brazil

Georg Pucker, Enrico Serra and Yoann Jestin
Bruno Kessler Foundation, Center for Materials and Microsystems, Italy

Alice Hospodková
Institute of Physics AS CR, v. v. i., Prague, Czech Republic

Hussein B. AL-Husseini
Nassiriya Nanotechnology Research Laboratory (NNRL), Physics Department, Science College, Thi-Qar University, Nassiriya, Iraq

Irati Ugarte and Roberto Pacios
Ikerlan S. Coop, Spain

Ivan Castelló and Emilio Palomares
ICIQ Institut Català d'Investigació Química, Spain

S. Valdueza-Felip, F. B. Naranjo and M. González-Herráez
Photonics Engineering Group (GRIFO), Electronics Dept., University of Alcalá, Spain

E. Monroy
CEA-Grenoble, INAC / SP2M / NPSC, France

J. Solís
Instituto de Óptica, C.S.I.C., Spain

Hirotaka Sakaue, Akihisa Aikawa and Yoshimi Iijima
Japan Aerospace Exploration Agency, Japan

Takuma Kuriki and Takeshi Miyazaki
The University of Electro-Communications, Japan

Diana Nesheva, Irina Bineva and Emil Manolov
Institute of Solid State Physics, Bulgarian Academy of Sciences, Sofia, Bulgaria

Nikola Nedev and Benjamin Valdez
Institute of Engineering, Autonomous University of Baja California, Mexicali, B. C., Mexico

Mario Curiel
Centro de Nanociencias y Nanotecnología, Universidad Nacional Autónoma de México
Ensenada, B. C., Mexico

Faxian Xiu
Electrical and Computer Engineering, Iowa State University, Ames, IA, USA

Yong Wang
Division of Materials, The University of Queensland, Brisbane, Australia
Materials Science and Engineering, Zhejiang University, Hangzhou, China

Kang L. Wang
Device Research Laboratory, Department of Electrical Engineering, University of California,
Los Angeles, California, USA

Jin Zou
Division of Materials, The University of Queensland, Brisbane, Australia

V. N. Stavrou
Department of Physics and Astronomy, University of Iowa, Iowa City, USA

G. P. Veropoulos
Division of Physics, Hellenic Naval Academy, Hadjikyriakou and Division of Academic
Studies, Hellenic Navy Petty Officers Academy, Skaramagkas, Greece

Yao He
Institute of Functional Nano & Soft Materials (FUNSOM) and Jiangsu Key Laboratory of
Carbon-based Functional Materials & Devices Soochow University, Suzhou, Jiangsu, China

Young-Pil Kim
Dept. of Life Science, Hanyang University, Republic of Korea

Adriana Fontes
Departamento de Biofísica e Radiobiologia, Universidade Federal de Pernambuco, Recife,
Brazil

Beate. S. Santos and Claudilene R. Chaves
Departamento de Ciências Farmacêuticas, Universidade Federal de Pernambuco, Recife,
Brazil

Regina C. B. Q. Figueiredo
Centro de Pesquisa Aggeu Magalhães, Fundação Oswaldo Cruz, Recife, Brazil